Exploitation of Fungi

The fungi are a highly diverse kingdom of eukaryotic microbes. Recent advances in molecular genetics, together with the release of whole genome sequences of an increasing number of fungi, are facilitating their exploitation and commercialization. Fungi have the ability to secrete large quantities of proteins of commercial value, and their complex secondary metabolic pathways produce a diverse range of bioactive compounds that have had a major impact in the pharmaceuticals market. In addition, the fungi themselves are increasingly being developed as alternatives to conventional chemically based pest control strategies, and as bioremediation agents capable of transforming pollutants in the soil environment. With chapters written by international experts, this volume highlights current and future biological, biochemical and molecular exploitation of the fungi in biotechnology. It will have broad appeal, not only to mycologists and microbiologists, but also to biomedical scientists, biotechnologists, environmental and molecular scientists, plant pathologists and geneticists.

Geoff Robson is a Senior Lecturer in the Faculty of Life Sciences at the University of Manchester.

Pieter van West is a Senior Lecturer and a Royal Society University Research Fellow in the Institute of Medical Sciences at the University of Aberdeen.

Geoff Gadd is Professor of Microbiology, Head of the Division of Environmental and Applied Biology, and Deputy Research Director in the School of Life Sciences at the University of Dundee.

Exploitation of Fungi

Symposium of the British Mycological Society held at the University of Manchester September 2005

EDITED BY

G. D. ROBSON, P. VAN WEST
AND G. M. GADD

Published for the British Mycological Society

CAMBRIDGE
UNIVERSITY PRESS

CAMBRIDGE UNIVERSITY PRESS
Cambridge, New York, Melbourne, Madrid, Cape Town, Singapore, São Paulo

Cambridge University Press
The Edinburgh Building, Cambridge CB2 2RU, UK

Published in the United States of America by Cambridge University Press, New York

www.cambridge.org

Information on this title: www.cambridge.org/9780521859356

First published 2007

Printed in the United Kingdom at the University Press, Cambridge

A catalogue record for this publication is available from the British Library

ISBN 978-0-521-85935-6 hardback

Contents

Contributors

M. O. Ahonsi
River Basin Research Center,
Gifu University, Yanagido 1–1,
Gifu 501-1193,
Japan

P. Ambrosino
Dipartimento di Arboricoltura,
Botanica e Patologia Vegetale,
Sezione Patologia Vegetale,
Universita degli Studi di Napoli
Federico II, Via Universita,
100–80055 Portici (NA),
Italy

K. K. Anderson
Decision & Sensor Analytics,
Pacific Northwest National
Laboratory,
906 Battelle Blvd., Richland,
WA 99352,
USA

T. Anke
Institut für Biotechnologie und
Wirkstoff-Forschung, IBWF,
Kaiserslautern 67663,
Germany

D. B. Archer
School of Biology,
University of Nottingham,
University Park,
Nottingham NG7 2RD,
UK

S. V. Avery
School of Biology,
University of Nottingham,
University Park,
Nottingham NG7 2RD,
UK

S. E. Baker
Fungal Biotechnology Team MSIN:
K2–12,
Chemical and Biological Processes
Development Group,
Pacific Northwest National
Laboratory,
902 Battelle Blvd.,
Richland,
WA 99352,
USA

S. E. Barnes
School of Biology,
University of Nottingham,
University Park,
Nottingham NG7 2RD,
UK

B. W. Birren
The Broad Institute,
Massachusetts Institute of Technology,
77 Massachusetts Avenue,
Cambridge,
MA 02139–4307,
USA

D. Boss
Plant Pathology,
Institute of Integrative Biology,
ETH Zurich, LFW,
8092 Zurich,
Switzerland

D. Brown
North Carolina State University,
Department of Plant Pathology,
Campus Box 7251, Raleigh,
NC 27695–7251,
USA

K. S. Bruno
Fungal Biotechnology Team MSIN: K2–12,
Chemical and Biological Processes Development Group,
Pacific Northwest National Laboratory,
902 Battelle Blvd., Richland,
WA 99352,
USA

I. Carbone
North Carolina State University,
Department of Plant Pathology,
Campus Box 7244, Raleigh,
NC 27695–7244,
USA

R. Ciliento
Dipartimento di Arboricoltura,
Botanica e Patologia Vegetale,
Sezione Patologia Vegetale,
Universita degli Studi di Napoli Federico II,
Via Universita,
100–80055 Portici (NA),
Italy

S. Coughlan
Agilent Technologies Inc.,
Little Falls Site,
2850 Centerville Road,
Wilmington, DE 19808,
USA

R. J. Cox
School of Chemistry and Biological Sciences,
University of Bristol,
Bristol BS8 1TS
UK

Z. Dai
Fungal Biotechnology Team MSIN: K2–12,
Chemical and Biological Processes Development Group,
Pacific Northwest National Laboratory,
902 Battelle Blvd.,
Richland, WA 99352,
USA

D. S. Daly
Statistical Sciences,
Pacific Northwest National Laboratory,
3180 George Washington Way,
Richland, WA 99352,
USA

R. A. Dean
Center for Integrated Fungal Research,
Department of Plant Pathology,
1200 Partners Building II,
Box 7251,
North Carolina State University,
Raleigh,
NC 27695,
USA

G. Défago
Plant Pathology,
Institute of Integrative Biology,
ETH Zurich, LFW,
8092 Zurich,
Switzerland

J. Deng
North Carolina State University,
Department of Plant Pathology,
Campus Box 7251, Raleigh,
NC 27695–7251,
USA

S. Diener
North Carolina State University,
Department of Plant Pathology,
Campus Box 7253,
Raleigh, NC 27695–7253,
USA

N. Donofrio
North Carolina State University,
Department of Plant Pathology,
Campus Box 7251,
Raleigh, NC 27695–7251,
USA

D. J. Ebbole
Department of Plant Pathology and Microbiology,
Peterson Building, Rm 120,
MS# 2132 Texas A&M University,
College Station, TX 77843–2132,
USA

M. L. Farman
Department of Plant Pathology,
University of Kentucky,
1405 Veterans Drive,
Lexington, KY 40546–0312,
USA

S. Ferraioli
Dipartimento di Arboricoltura,
Botanica e Patologia Vegetale,
Sezione Patologia Vegetale,
Universita degli Studi di Napoli Federico II,
Via Universita,
100–80055 Portici (NA),
Italy

R. Fischer
Institute for Applied Life Sciences,
Applied Microbiology,
University of Karlsruhe,
Hertzstr. 16,
D-76187 Karlsruhe,
Germany

M. Fomina
Division of Environmental and Applied Biology, Biological Sciences Institute, School of Life Sciences, University of Dundee, Dundee DD1 4HN, Scotland, UK

G. M. Gadd
Division of Environmental and Applied Biology, Biological Sciences Institute, School of Life Sciences, University of Dundee, Dundee DD1 4HN, Scotland, UK

J. E. Galagan
The Broad Institute, Massachusetts Institute of Technology, 77 Massachusetts Avenue, Cambridge, MA 02139–4307, USA

S. Gigante
Dipartimento di Arboricoltura, Botanica e Patologia Vegetale, Sezione Patologia Vegetale, Universita degli Studi di Napoli Federico II, Via Universita, 100–80055 Portici (NA), Italy

T. Guillemette
Laboratoire de Microbiologie, UMR 77 Pathologie Végétale, Université d'Angers, 2 bd Lavoisier, 49045 Angers cedex, France

P. J. Harvey
University of Greenwich, Medway School of Science, Central Avenue, Chatham Maritime, Kent ME4 4TB, UK

S. H. M. Hutwimmer
Institute of Microbiology, University of Innsbruck, Technikerstrasse 25, A-6020 Innsbruck, Austria

H.-B. Jansson
Department of Marine Sciences and Applied Biology, University of Alicante, Apartado 99, E-03080 Alicante, Spain

R. Kulkarni
RTI, 3040 Cornwallis Road, Research Triangle Park, NC 27709, USA

S. Lanzuise
Dipartimento di Arboricoltura, Botanica e Patologia Vegetale, Sezione Patologia Vegetale, Universita degli Studi di Napoli Federico II, Via Universita, 100–80055 Portici (NA), Italy

L. L. Lasure
Fungal Biotechnology Team MSIN: K2–12,
Chemical and Biological Processes Development Group,
Pacific Northwest National Laboratory,
902 Battelle Blvd., Richland,
WA 99352,
USA

Y.-H. Lee
Seoul National University,
School of Agricultural Biotechnology,
Suwon, 441–744,
Korea

L. V. Lopez-Llorca
Department of Marine Sciences and Applied Biology,
University of Alicante,
Apartado 99, E-03080 Alicante,
Spain

M. Lorito
Dipartimento di Arboricoltura,
Botanica e Patologia Vegetale,
Sezione Patologia Vegetale,
Universita degli Studi di Napoli Federico II,
Via Universita,
100–80055 Portici (NA),
Italy

J. K. Magnuson
Fungal Biotechnology Team MSIN: K2–12,
Chemical and Biological Processes Development Group,
Pacific Northwest National Laboratory,
902 Battelle Blvd.,
Richland,
WA 99352,
USA

R. Marra
Dipartimento di Arboricoltura,
Botanica e Patologia Vegetale,
Sezione Patologia Vegetale,
Universita degli Studi di Napoli Federico II,
Via Universita,
100–80055 Portici (NA),
Italy

D. Martinez
Genome Annotation and Analysis,
Joint Genome Institute,
Los Alamos National Laboratory,
Los Alamos,
NM 87545,
USA

M. Maurhofer
Plant Pathology,
Institute of Integrative Biology,
ETH Zurich,
LFW,
8092 Zurich,
Switzerland

T. Mitchell
North Carolina State University,
Department of Plant Pathology,
Campus Box 7251,
Raleigh,
NC 27695–7251,
USA

M. E. Monroe
Biological Separations and Mass Spectrometry,
Pacific Northwest National Laboratory 3335 Q Avenue,
Richland,
WA 99352,
USA

Y. Y. Oh
North Carolina State University,
Department of Plant Pathology,
Campus Box 7251,
Raleigh,
NC 27695–7251,
USA

M. J. Orbach
Department of Plant Pathology,
University of Arizona,
Forbes Room 105,
PO Box 210036,
Tucson,
AZ 85721–0036,
USA

H. Pan
RTI,
3040 Cornwallis Road,
Research Triangle Park,
NC 27709,
USA

E. A. Panisko
Fungal Biotechnology Team MSIN: K2–12,
Chemical and Biological Processes Development Group,
Pacific Northwest National Laboratory,
902 Battelle Blvd.,
Richland,
WA 99352,
USA

E. J. Parker
Institute of Fundamental Sciences,
Massey University,
Private Bag 11 222,
Palmerston North,
New Zealand

A. Powell
North Carolina State University,
Department of Plant Pathology,
Campus Box 7251, Raleigh,
NC 27695–7251,
USA

M. Ruocco
Dipartimento di Arboricoltura, Botanica e Patologia Vegetale,
Sezione Patologia Vegetale,
Universita degli Studi di Napoli Federico II,
Via Universita,
100–80055 Portici (NA),
Italy

F. Scala
Dipartimento di Arboricoltura, Botanica e Patologia Vegetale,
Sezione Patologia Vegetale,
Universita degli Studi di Napoli Federico II,
Via Universita,
100–80055 Portici (NA),
Italy

C. E. Scheer
BEPHS Business Innovation,
University of Greenwich at Medway,
Central Avenue,
Chatham Maritime,
Kent ME4 4TB,
UK

M. Schröder
School of Biological and Biomedical Sciences,
Science Laboratories, South Road,
University of Durham,
Durham DH1 3LE,
UK

B. Scott
Centre for Functional Genomics,
Institute of Molecular Biosciences and Institute of Fundamental Sciences,
Massey University,
Private Bag 11 222,
Palmerston North,
New Zealand

C. Soderlund
Director of Bioinformatics,
Department of Plant Science,
303 Forbes Building,
Tucson, AZ 85721,
USA

I. Soriente
Dipartimento di Arboricoltura,
Botanica e Patologia Vegetale,
Sezione Patologia Vegetale,
Universita degli Studi di Napoli Federico II,
Via Universita,
100–80055 Portici (NA),
Italy

H. Strasser
Institute of Microbiology,
University of Innsbruck,
Technikerstrasse 25,
A-6020 Innsbruck,
Austria

N. J. Talbot
Department of Biological Sciences,
University of Exeter,
Hatherly Laboratories,
Prince of Wales Road,
Exeter EX4 4PS,
UK

A. Tanaka
Institute of Molecular Biosciences,
Massey University,
Private Bag 11 222,
Palmerston North,
New Zealand

E. Thines
Institut für Biotechnologie und Wirkstoff-Forschung,
Erwin-Schrödinger-Str. 56,
D-67663 Kaiserslautern,
Germany

M. Thon
Department of Computer Science,
320C Peterson Building MS# 2132,
Texas A&M University,
College Station,
TX 77843–2132,
USA

G. Turner
Department of Molecular Biology & Biotechnology,
University of Sheffield,
Sheffield SN10 2TN,
UK

D. Turrà
Dipartimento di Arboricoltura,
Botanica e Patologia Vegetale,

Sezione Patologia Vegetale,
Universita degli Studi di Napoli
Federico II,
Via Universita,
100–80055 Portici (NA),
Italy

D. Veith
Institute for Applied
Life Sciences,
Applied Microbiology,
University of Karlsruhe,
Hertzstr. 16,
D-76187 Karlsruhe,
Germany

F. Vinale
Dipartimento di Arboricoltura,
Botanica e Patologia Vegetale,
Sezione Patologia Vegetale,
Universita degli Studi di Napoli
Federico II,
Via Universita,
100–80055 Portici (NA),
Italy

H. J. Weitz
School of Biological Sciences,
Cruickshank Building,
University of Aberdeen,
Aberdeen AB24 3UU,
UK

C. F. Wend
Fungal Biotechnology Team MSIN:
K2–12,
Chemical and Biological Processes
Development Group,
Pacific Northwest National
Laboratory,
902 Battelle Blvd.,
Richland,
WA 99352,
USA

S. L. Woo
Dipartimento di Arboricoltura,
Botanica e Patologia Vegetale,
Sezione Patologia Vegetale,
Universita degli Studi di Napoli
Federico II,
Via Universita,
100–80055 Portici (NA),
Italy

J.-R. Xu
Department of Botany and Plant
Pathology,
915 West State Street,
Purdue University,
West Lafayette,
IN 47906,
USA

C. A. Young
Molecular Mycologist,
Forage Improvement Division,
The Samuel Roberts Noble
Foundation, Inc.,
2510 Sam Noble Pky,
Ardmore,
OK 73401,
USA

R. Zelger
The Research Centre for Agriculture
and Forestry Laimburg,
I-39051 Pfatten/Auer,
Italy

Preface

The fungi are a highly diverse kingdom of eukaryotic microbes that have been exploited commercially for decades due to their ability to secrete large quantities of proteins of commercial value, and because they possess complex secondary metabolic pathways producing a diverse range of bioactive compounds that have had a major impact in the pharmaceutical market. For example, penicillin and cephalosporin (antibiotics), cylosporin (immunosuppressant) and more recently the statins (cholesterol reducing agents) are estimated to be worth over $5 billion per annum, while enzymes and proteins produced commercially in fungal hosts are used in a diverse number of commercial markets including baking, brewing, detergent, textile and animal feed industries. In addition to the exploitation of fungal products, the fungi themselves are increasingly being developed as alternatives to conventional chemically based pest control strategies, as biocontrol agents active against commercially damaging insect pathogens and weeds, and as bioremediation agents capable of transforming organic and inorganic pollutants in the soil environment. Recent advances in the molecular genetics of the fungi together with the recent release of whole genome sequences of an increasing number of fungi will facilitate further the exploitation and commercialization of these important and ubiquitous eukaryotic microorganisms. The objective of this symposium volume is to highlight current and future biological, biochemical and molecular exploitation of the fungi in biotechnology, and act as an interface between current research and future commercialization.

The book is divided into five general parts, each dealing with an important area relevant to the overall topic, but with a considerable degree of complementarity between all the sections. The topics chosen provide a wide-ranging source of examples relating to the exploitation of fungi in a wide range of biomedical, industrial and environmental contexts. The first part, *Comparative and functional fungal genomics*, comprises chapters relating to genomic and proteomic analysis of industrial and plant pathogenic fungi, and how such information can be related to function. The second part on *Bioactive molecules* contains chapters that discuss the synthesis of some specific metabolites in fungi, e.g. pyridones and tetramic acids, as well as certain metabolites with important applications in agriculture. Molecular analysis of secondary metabolite production from endophytic fungi is also described together with a synthesis of modern genomic approaches that can elucidate secondary metabolism in *Aspergillus* species. The next part on *Protein folding and secretion* includes accounts of the roles of microtubules and motors in polarized fungal growth, cellular responses to protein-unfolding stress, as well as a genomic perspective on protein excretion and associated stresses in *Aspergillus*. The next part on *Fungal bioremediation* includes the Berkeley Award lecture on stress and the single cell, a topic relevant to many other research areas as well as fungal responses to toxicants. Other chapters provide accounts of novel bioluminescent fungal sensors for pollutants in the environment as well as the theory and practice of fungal bioremediation of xenobiotics and toxic metal pollutants. The final part on *Fungal biocontrol of pests* has accounts of the fungal control of subterranean pests, the development of mycoherbicides, interactions between *Trichoderma* spp. with plant hosts and fungal pathogens, and fungal parasites of invertebrates.

This book arises from the British Mycological Society Symposium on *Exploitation of Fungi* held at the University of Manchester in September 2005 and the editors would like to thank those authors who contributed enthusiastically to this book. Special thanks also go to Diane Purves in Dundee, who greatly assisted communication, collation, editing and formatting of chapters. The prime objective was to produce a wide-ranging volume that would highlight the importance of fungi in modern biotechnology as well as highlighting the modern approaches and tools that are now used to understand and manipulate fungal biology. We think that this has been achieved and hope the volume has broad appeal not only to mycologists of all persuasions but also to other microbiologists, biotechnologists and molecular biologists, who routinely rely on fungal model systems. It will be interesting to see how genomic revelations influence developments over the next few years.

Geoffrey D. Robson
Pieter van West
Geoffrey M. Gadd

I

Comparative and functional fungal genomics

1

Genome and proteome analysis of industrial fungi

S. E. BAKER, C. F. WEND, D. MARTINEZ, J. K. MAGNUSON, E. A. PANISKO, Z. DAI, K. S. BRUNO, K. K. ANDERSON, M. E. MONROE, D. S. DALY AND L. L. LASURE

Background

In order to decrease dependence on petroleum, the United States Department of Energy (USDOE) Office of the Biomass Program (OBP) is investing in research and development to enable its vision of the biorefinery. The biorefinery will decrease the use of petroleum through conversion of biomass such as crops or agricultural waste into fuels and products.

In 2004, the USDOE OBP asked researchers at the Pacific Northwest National Laboratory (PNNL) and the National Renewable Energy Laboratory (NREL) to prepare a list of the top ten building-block chemicals that can be derived from simple sugars by biological and/or chemical means. The resulting list of twelve building-block chemicals and the accompanying report (www.eere.energy.gov/biomass/pdfs/35523.pdf) form an informational foundation on which future DOE and industry bioproducts research will be built (Table 1.1).

How do fungi fit into the biorefinery? Analysis of the 'top ten' study indicates that nine of the top twelve chemical building blocks are currently produced, or may potentially be produced, by fungal fermentation processes. However, a significant barrier to the use of bio-based products is the economic feasibility – fuels and products must be price-competitive with those derived from petroleum. An obvious way to decrease the costs of biobased products from fungi is to make fermentation strains more productive and processes more efficient. Traditional strain improvement programmes typically span a timescale measured in decades and process development done through the use of batch cultures is extremely labour intensive.

How can current and future bioproducts mycology research in this area take advantage of the growing list of fungi whose genomes have been

Exploitation of Fungi, eds. G. D. Robson, P. van West and G. M. Gadd. Published by Cambridge University Press.

Table 1.1 *Top 12 sugar-derived chemical building blocks* ***(potentially derived from fungi in bold)***.

1, 4 succinic, fumaric and malic acids
2, 5 furan dicarboxylic acid
3 hydroxy propionic acid
aspartic acid
glucaric acid
glutamic acid
itaconic acid
levulinic acid
3-hydroxybutyrolactone
glycerol
sorbitol
xylitol/arabinitol

sequenced, as well as the cutting edge systems biology methods in order to bring the vision of the biorefinery to reality? With genome sequences available for several fungi of industrial importance and the ability to grow fungi at a steady state, the prospects for decreased time and cost investments for economical fungal bioprocesses that produce fuels and products have become considerably brighter.

What can fungal genomics offer industrial mycology?

At its most basic level, the genome sequence of an organism provides a high-level view of the biology of an organism. The presence or absence of genes or gene families can give glimpses into the lifestyle and ecological niches of a given organism. For example, the presence of genes that are predicted to be involved in sexual development in the 'asexual' opportunistic pathogen, *Aspergillus fumigatus*, has led researchers to hypothesize that this organism does have sex (Paoletti *et al.*, 2005). Since current or potential commodity chemicals found in the 'top ten' list are central metabolites, it is valuable to understand the central metabolic pathways present or deficient in a particular fungus of interest. Also, since the goal of the biorefinery is to use complex biomass sources of carbon (polysaccharides) it is important to understand the suite of plant polymer degrading enzymes present in the genome of an industrially relevant fungus in order to ascertain the substrate range of the organism.

The genomic sequence from an organism serves as the foundation for further study of the organism at the global scale. In particular, transcriptomics and proteomics methodologies are greatly enhanced by the presence of genomic sequence and predicted protein databases. Furthermore, encoded in the genome are all of the metabolic pathways that encode the ability of fungi to make an amazingly diverse array of compounds. The challenge that confronts the fungal research community at present is how to make use of all of the genomic information to further as many avenues of mycological research as possible.

The *Aspergillus niger* genome project

By the end of 2005, three different strains of *Aspergillus niger* will have been sequenced by three different organizations. Two strains have been sequenced by industry. A US company, Integrated Genomics, Inc., sequenced ATCC strain 9029, a wild-type strain that can produce citric acid. The sequence coverage for this strain is low (4–6X) and distributed across over 9000 contigs. The sequence was purchased by Pacific Northwest National Laboratory in 2004 and is currently available upon request. A derivative of the *A. niger* strain NRRL 3122 was sequenced by a Dutch company, DSM.

Most recently, the USDOE Joint Genome Institute (JGI) in collaboration with the PNNL Fungal Biotechnology Team began the first public *A. niger* sequencing project. ATCC strain 1015 was chosen as the target strain for sequencing. This choice was attractive for sequencing for a number of reasons. From a historical point of view, this strain was the first used for a patented citric acid process in 1917 (Currie, 1917). Furthermore, a derivative of the strain exists (ATCC strain 11414) that has been used for studies of citric acid production and fermentation morphology control (Perlman, Dorrell & Johnson, 1946; Shu & Johnson, 1947; Dai *et al.*, 2004). A genomic sequence and automated gene models are expected to be available through the JGI Genome Portal by early 2006.

The *Trichoderma reesei* genome project

Trichoderma reesei, the asexual anamorph of *Hypocrea jecorina* (Kuhls *et al.*, 1996), is one of the most prodigious generators of protein known. *T. reesei* was discovered in the south Pacific island of Samoa, degrading the canvas tents used by the US forces during World War II. The particular family of proteins for which *T. reesei* is most studied, cellulases (enzymes that break down cellulose), were the cause of the undesirable tent degrading capability of this ascomycete fungus. *T. reesei* has since become an important organism in biotechnology for the production of the now desirable cellulases and other valuable proteins.

The USDOE began working with two biotechnology companies, Genencor and Novozymes, to bring down the cost of cellulases so that they could be economically applied in the conversion of lignocellulosic material to simple sugars and then to ethanol or chemical products as alternatives to petroleum based fuels and products. To assist this effort, the JGI began sequencing the genome in 2002. The genome of *T. reesei* and the planned sequencing of two close relatives of *T. reesei*, *Trichoderma virens* and *Trichoderma atroviride*, both by the JGI, will allow researchers to use powerful comparative genomics techniques to help understand the complicated interplay of genes and regulatory elements involved in the production of important proteins.

As in all whole genome level projects, understanding the mass of data generated requires bioinformatics tools. To meet this need, the JGI has released version 1 of the genome to the public in an interactive portal available on the internet (http://genome.jgi-psf.org). Currently, the JGI is continuing to finish the genome at its Los Alamos National Laboratory site and will release version 2 to the public in early 2006. Version 2, containing updated annotations, will consist of fewer than 100 scaffolds, a ten-fold reduction over version 1.

Fungal proteomics

How can genome information be leveraged to gain high throughput global functional data? The rapidly developing field of proteomics – the study of the protein complement of an organism – has the potential to revolutionize how we study biological systems. Proteomic methodologies are being used in the study of the environment, as medical diagnostics, and in basic and industrial research settings (for example, Sadygov, Cociorva & Yates, 2004; Jacobs *et al.*, 2005; Ram *et al.*, 2005). One important issue for proteomic methodologies is the quality of the data, i.e. are we really seeing what we think we are seeing?

Current proteomics methodologies using both two-dimensional gel electrophoresis and liquid chromatography/mass spectrometry allow a snapshot of the proteins expressed by an organism at the particular point in time when the organism was processed for protein purification and analysis. The large amounts of data generated by 'shotgun' proteomic analysis often serve as the foundation for more detailed, hypothesis-driven research on individual proteins or groups of proteins. Increased quality of proteomics data will have dramatically positive effects on subsequent studies that use the data. Through the use of various statistical analysis techniques and internal protein/peptide standards we have been able to improve fundamentally the way that proteomic data is analyzed.

Current work at PNNL is focused on developing statistical analysis tools that allow for relative quantification of proteins between different samples. The statistical methodology casts a new physical description of measured peptide abundances as a mixed effects linear statistical model, and uses restricted maximum likelihood estimation (REML) to estimate model parameters and associated errors.

The case for filamentous fungal chemostat research in process development

In bioprocess research laboratories, development of a new process has traditionally employed iterative experiments consisting of culturing fungi in shake flasks, small fermentors and process prototype fermentors. One of the significant bottlenecks in this standard approach is the requirement for large numbers of repetitive experiments using batch fermentors. This is an inherently limiting step due to the amount of labour involved, equipment availability and that typically only a single variable can be examined per experiment. Large numbers of shake-flask experiments are often performed previous to or in parallel with the batch fermentor experiments to examine more nutritional and physical process variables. However, shake flasks are not amenable to process control and monitoring for dissolved oxygen, pH and other variables. An alternative to this standard approach is to use continuous culture technology, often referred to as chemostat or auxostat technology.

To date, relatively little chemostat research has been performed with filamentous fungi. Examples include fermentations to obtain various types of information, from examination of catabolic carbon flow, to stability determination of gene copies (Ng, Smith & McIntosh, 1973; Kristiansen & Sinclair, 1979; Wiebe & Trinci, 1991; Gallmetzer & Burgstaller, 2002). Some investigations have been performed in auxostats to evaluate maximum growth or product rates based upon a single limiting parameter (Simpson *et al.*, 1995; Swift *et al.*, 1998). Chemostat systems have been examined where clumps or flocs of filamentous morphology occur under certain agitation or flow rates (Amanullah, Tuttiett & Nienow, 1998).

For accelerating the rate of progress of investigation and optimization of novel bioprocesses continuous culture technology will be crucial. For example, growth rate and product formation rate can be measured in response to several carbon sources or different pH levels in a single continuous culture experiment. The medium composition can be optimized in an experiment by examining major (carbon source, nitrogen source) and minor (salts, micronutrients) medium components. The responses of the growth and product formation rate (efficiency) to critical parameters can be plotted and analyzed by statistical methods to arrive at the optimum. Translating this steady-state data back into the batch system where growth and production rate vary is a challenge to be addressed. This will be important for proving the utility of the method to industrial processes where the final process is nearly always operated in a batch mode on large scales.

An additional advantage of chemostat technology is the improvement of uniformity and reproducibility of biological samples. Proteomics research will benefit greatly from improvements in biological sample quality since this directly impacts the quality of the resulting proteomics data. A fungus grown in a chemostat can be examined under conditions of steady-state growth. Once steady-state growth is established, critical factors that may affect an organism's physiological state and therefore its proteome can be assessed by sampling, changing a parameter, re-establishing steady state and sampling again. Finally, chemostat technology will be useful in applying the principles of quantitative genetics to the improvement of desired characteristics in a target fungus through the use of selective pressure.

Conclusions

A systems biology approach is needed to utilize fully the tools available for the study of all fungi with potential roles to play in the biorefinery. While one challenge for industrial mycology lies in the development of the strains to supply the biorefinery with enzymes and products, another will be the integration of huge fungal organism datasets generated by genomics, proteomics and bioprocessing technology. Indeed, connecting metabolite and growth kinetic data from exploration of bioprocess parameter space with proteome and genome data should be a priority for this field.

A recent study funded by the USDOE OBP indicates that a number of building-block compounds derived from sugars that can feed into industrial chemical processes have the potential to be made through fungal fermentation processes. While central metabolism for the generic organism has been described, detailed descriptive and functional information for specific organisms is not well known. A combination of bioinformatics, bioprocess and proteomic experimentation and analysis for each fungus for which there is genomic information will be a powerful tool as we model how genetic alteration and growth parameter manipulation for improved fungal strains affects enzyme and organic acid production.

References

Amanullah, A., Tuttiett, B. & Nienow, A. W. (1998). Agitator speed and dissolved oxygen effects in xanthan fermentations. *Biotechnology and Bioengineering*, **57**, 198–210.

Currie, J. N. (1917). The citric acid fermentation of *Aspergillus niger*. *Journal of Biological Chemistry*, **31**, 15–37.

Dai, Z., Mao, X., Magnuson, J. K. & Lasure, L. L. (2004). Identification of genes associated with morphology in *Aspergillus niger* by using suppression subtractive hybridization. *Applied Environmental Microbiology*, **70**, 2474–85.

Gallmetzer, M. & Burgstaller, W. (2002). Efflux of organic acids in *Penicillium simplissimum* is an energy-spilling process, adjusting the catabolic carbon flow to the nutrient supply and the activity of catabolic pathways. *Microbiology*, **148**, 1143–9.

Jacobs, J. M., Adkins, J. N., Qian, W. J., Liu, T., Shen, Y., Camp, D. G. 2nd & Smith, R. D. (2005). Utilizing human blood plasma for proteomic biomarker discovery. *Journal of Proteome Research*, **4**, 1073–85.

Kristiansen, B. & Sinclair, C. G. (1979). Production of citric acid in continuous culture. *Biotechnology and Bioengineering*, **21**, 297–315.

Kuhls, K., Lieckfeldt, E., Samuels, G. J., Kovacs, W., Meyer, W., Petrini, O., Gams, W., Borner, T. & Kubicek, C. P. (1996). Molecular evidence that the asexual industrial fungus *Trichoderma reesei* is a clonal derivative of the ascomycete *Hypocrea jecorina*. *Proceedings of the National Academy of Sciences USA*, **93**, 7755–60.

Ng, A. M. L., Smith, J. E. & McIntosh, A. F. (1973). Conidiation of *Aspergillus niger* in continuous culture. *Archives of Microbiology*, **88**, 119–26.

Paoletti, M., Rydholm, C., Schwier, E. U., Anderson, M. J., Szakacs, G., Lutzoni, F., Debeaupuis, J. P., Latge, J. P., Denning, D. W. & Dyer, P. S. (2005). Evidence for sexuality in the opportunistic fungal pathogen *Aspergillus fumigatus*. *Current Biology*, **15**, 1242–8.

Perlman, D., Dorrell, W. W. & Johnson, M. J. (1946). Effect of metallic ions on the production of citric acid by *Aspergillus niger*. *Archives of Biochemistry*, **11**, 131–43.

Ram, R. J., Verberkmoes, N. C., Thelen, M. P., Tyson, G. W., Baker, B. J., Blake, R. C. 2nd, Shah, M., Hettich, R. L. & Banfield, J. F. (2005). Community proteomics of a natural microbial biofilm. *Science*, **308**, 1915–20.

Sadygov, R. G., Cociorva, D. & Yates, J. R. 3rd. (2004). Large-scale database searching using tandem mass spectra: looking up the answer in the back of the book. *Nature Methods*, **1**, 195–202.

Shu, P. & Johnson, M. J. (1947). Effect of the composition of the sporulation medium on citric acid production by *Aspergillus niger* in submerged culture. *Journal of Bacteriology*, **54**, 161–7.

Simpson, D. R., Wiebe M. G., Robson G. D. & Trinci A. P. J. (1995). Use of pH auxostats to grow filamentous fungi in continuous-flow culture at maximum specific growth-rate. *FEMS Microbiology Letters*, **126**, 151–7.

Swift, R. J., Wiebe, M. G., Robson, G. D. & Trinci A. P. J. (1998). Recombinant glucoamylase production by *Aspergillus niger* B1 in chemostat and pH auxostat cultures. *Fungal Genetics and Biology*, **25**, 100–9.

Wiebe, M. G. & Trinci A. P. J. (1991). Dilution rate as a determinant of mycelial morphology in continuous culture. *Biotechnology and Bioengineering*, **38**, 75–81.

2

The rice blast story: from genome sequence to function

R. A. DEAN, T. MITCHELL, R. KULKARNI, N. DONOFRIO, A. POWELL, Y. Y. OH, S. DIENER, H. PAN, D. BROWN, J. DENG, I. CARBONE, D. J. EBBOLE, M. THON, M. L. FARMAN, M. J. ORBACH, C. SODERLUND, J-R. XU, Y-H. LEE, N. J. TALBOT, S. COUGHLAN, J. E. GALAGAN AND B. W. BIRREN

Introduction

Rice blast disease, caused by the filamentous fungus *Magnaporthe grisea*, is a serious and recurrent problem in all rice-growing regions of the world (Talbot, 2003; Valent & Chumley, 1991). It is estimated that each year enough rice is destroyed by rice blast disease to feed 60 million people. Control of this disease is difficult; new host-specific forms develop quickly to overcome host resistance and chemical control is typically not cost effective (Ou, 1987). Infections occur when fungal spores land and attach themselves to leaves using a special adhesive released from the tip of each spore (Hamer *et al.*, 1988). The germinating spore develops an appressorium, a specialized infection cell, which generates enormous turgor pressure – up to 8 MPa – that ruptures the leaf cuticle allowing invasion of the underlying leaf tissue (de Jong *et al.*, 1997; Dean, 1997). Subsequent colonization of the leaf produces disease lesions from which the fungus sporulates and spreads to new plants. When rice blast infects young rice seedlings, whole plants often die, while spread of the disease to the stems, nodes or panicle of older plants results in nearly total loss of the rice grain. Recent reports have further shown that the fungus has the capacity to infect plant roots (Sesma & Osbourn, 2004). Different host-limited forms of *Magnaporthe* also infect a broad range of grass species including wheat, barley and millet.

The rice blast fungus has emerged as a seminal model to elucidate the molecular basis of infection-related development and host specificity (Valent & Chumley, 1991). The fungus is haploid, has a relatively small genome contained in seven chromosomes, can be induced to form

Exploitation of Fungi, eds. G. D. Robson, P. van West and G. M. Gadd. Published by Cambridge University Press.

appressoria *ex planta*, and is highly amenable to genetic and molecular genetic manipulation (Dean, 1997; Talbot, 2003).

Genome sequence

Efforts to sequence the genome began following the Second International Rice Blast Meeting held in Montpellier, France, in 1998. Here it was agreed that strain 70–15, a domesticated strain derived from Guy11, should be sequenced (Chao & Ellingboe, 1991). Initial efforts focused on creating deep BAC libraries, which were end sequenced, fingerprinted and assembled into a physical map and integrated with the genetic map (Martin *et al.*, 2002; Zhu *et al.*, 1999). Other work resulted in the generation of >25 000 ESTs derived from a number of different growth conditions and stages (Ebbole *et al.*, 2004). More recently, a whole genome shotgun (WGS) approach was used to create a draft sequence. In all, greater than seven-fold sequence coverage was produced. The draft genome sequence consists of 2273 sequence contigs longer than 2 kb, ordered and oriented within 159 scaffolds. The total length of all sequence contigs is 38.8 Mb, and the total length of the scaffolds, including estimated sizes for the gaps, is 40.3 Mb (Dean *et al.*, 2005). Current efforts are focused on an optical map as an aid to genome assembly and closing the gaps in the genome sequence.

Rapid expansion of repetitive elements

Novel pathogenic variants of *Magnaporthe* capable of infecting formerly resistant host plants arise with alarming frequency during rice cultivation. Such gains of virulence are often associated with transposon-mediated gene inactivation. Approximately 9.7% of the *Magnaporthe* genome assembly is comprised of repetitive DNA sequences. The majority of repetitive sequences in the assembly are retrotransposons comprising eight major families with at least 18 members. Five are retroelement families and three are DNA transposons. The repetitive elements are not uniformly distributed in the genome assembly, but form discrete clusters (Thon *et al.*, 2004).

These elements appear to have had a profound impact on genome evolution and also shed light on mechanisms of pathogenic variation. Investigation of the natural history of repetitive elements in *Magnaporthe* by the coalescent reconstructions revealed that most have undergone a rapid expansion in the genome at some point in their past. The ability to rapidly expand may be associated with the loss of repeat induced point (RIP) mutations, a genome defence mechanism that destroys duplicated sequences

(Selker, 2002). RIP is associated with meiosis, which occurs rarely in rice infecting strains under field conditions. Pot3 appears to be an exception in that it has undergone several rounds of expansion. This may reflect its more recent invasion of the *Magnaporthe* genome (Kang *et al.*, 2001).

Gene conservation, synteny and evolution

The *Magnaporthe* genome is estimated to contain 11 109 protein encoding genes encompassing 48% of the assembly (1 gene per 3.5 kb) (Dean *et al.*, 2005). This compares to 10 082 predicted in *Neurospora crassa* (Galagan *et al.*, 2003), a related pyrenomycete, and 9457 were predicted in the more distantly related plectomycete *Aspergillus nidulans* (www.broad.mit.edu/annotation/fungi/aspergillus/). Neither of these species causes plant disease. At the amino acid level, *Magnaporthe* orthologs in *N. crassa* and *A. nidulans* show an average identity of 47% and 46% respectively. Comparison of genes between *Magnaporthe* and *N. crassa*, as well as with predicted gene sets for which complete fungal genomes are available, suggest only ~65% conservation. However, conservation may be much higher. Only ~50% of the *Magnaporthe* predicted genes have EST support and ~1600 ESTs do not align with predicted ORFs, highlighting deficiencies in current gene prediction tools for filamentous fungi.

Colinearity of chromosome segments (synteny) is widely reported for plant and animal genomes, but is largely unknown in filamentous ascomycete fungi. Analysis of orthologous pairs of genes in *Magnaporthe* and *N. crassa* revealed that linkage group assignments were often conserved between the two species, notably chromosome 7 in *Magnaporthe* with linkage group I of *N. crassa*. Recently, a draft sequence of 38 BAC clones spanning chromosome 7 from *Magnaporthe* was completed, which when combined with the *Magnaporthe* WGS sequence, yielded a sequence assembly that was 4 Mb in length and contained only 50 gaps. This sequence served as the basis for a more detailed analysis of synteny. Using FISH software package, 21 syntenic blocks ($p < 0.001$) between chromosome 7 and the *N. crassa* genome that ranged in size from 5 to 16 orthologous gene pairs were identified (Calabrese, Chakravarty & Vision, 2003). Interestingly, all of the blocks were found on *N. crassa* chromosome 1, though the relative order of the syntenic blocks was not retained between the two chromosomes. Similar patterns of conserved synteny were found when the same analytical methods were used to compare chromosome 7 to two other filamentous fungi. Seventeen blocks were identified between chromosome 7 and the *Fusarium graminearum* genome, 14 of which were found on chromosome 2. The remaining three were found on chromosome 6, suggesting either a translocation of a large chromosomal segment in the *F. graminearum* lineage or an error in the genetic map. Only two syntenic blocks were identified in *A. nidulans*, reflecting its greater evolutionary distance to *Magnaporthe*.

On chromosome 7, transposable elements (TEs) are largely restricted to three clusters located in chromosomal segments that have a high recombination rate. These clusters are marked by more frequent gene duplications, and genes within the clusters have greater sequence diversity to orthologous genes from other fungi. In regions outside these clusters, significant levels of conserved synteny between chromosome 7 and other filamentous fungi were found, despite more than 200 million years of divergent evolution (Padovan *et al.*, 2005). Together these data suggest that TEs have a profound impact on the *Magnaporthe* genome by creating localized segments with increased rates of rearrangements, gene duplications and evolution.

Expansion of secreted proteins

During the evolution of *Magnaporthe* as a plant pathogen a number of gene families appear to have undergone expansion compared to non-pathogenic relatives. Notably, *Magnaporthe* contains nearly twice as many (1258 vs. 642 estimated using SignalP) potential secreted proteins compared with *N. crassa*. Several gene families expanded in *Magnaporthe* exhibited sequence similarity to proteins that are known to play a role in fungal pathogenicity, including several plant cell-wall degrading enzymes, as well as cutinases and proteinases. One particular family of expanded proteins in *Magnaporthe* contained a novel chitin-binding motif, CX7CCX5 C, found in 21 predicted proteins. These proteins may represent pathogen-associated molecular patterns, PAMPs, recognized by the host as pathogen effector proteins (Nimchuk *et al.*, 2003).

Novel signalling machinery

The signalling pathways cAMP and MAPK are well known to play a central role in infection related morphogenesis and pathogenesis (Dean, 1997; Xu, 2000). The core elements of these pathways are highly conserved, unlike upstream receptors and downstream transcription factors and their target genes. *Magnaporthe* contains a large class of G-protein-coupled receptors (GPCRs) with seven transmembrane helices (Kulkarni *et al.*, 2005). Within this class, a novel subclass of 12 receptors typified by PTH11, a cell surface integral membrane protein required for pathogenicity in *Magnaporthe* was identified. PTH11 has seven transmembrane spans and an N-terminal extracellular cysteine rich EGF-like domain (CFEM domain), a characteristic also seen in human GPCRs (Kulkarni, Kelkar & Dean, 2003). Only one member of this subclass was identified in *N. crassa*. In total, 61 GPCR-related proteins were identified in *Magnaporthe* that shared a common domain with homologues in *N. crassa* and other fungi belonging to one Ascomycota subphylum (Pezizomycotina). None were detected in other fungal groups (Basidiomycota or other Ascomycota subphyla including yeasts) or any other eukaryotic species.

Functional analysis

Inspection of the *Magnaporthe* genome sequence and comparisons to related organisms has provided valuable clues to the basis of fungal pathogenesis. Moreover, the genetic and experimental tractability of the pathosystem presents an opportunity to take a genome-wide approach to functionally dissect the underlying molecular mechanisms. Work is currently progressing on large-scale mutagenesis projects, analysis of secreted proteins and transcriptional profiling experiments. All data are stored in the Magnaporthe Grisea Oryza Sativa (MGOS) database available at www.mgosdb.org (Fig. 2.1).

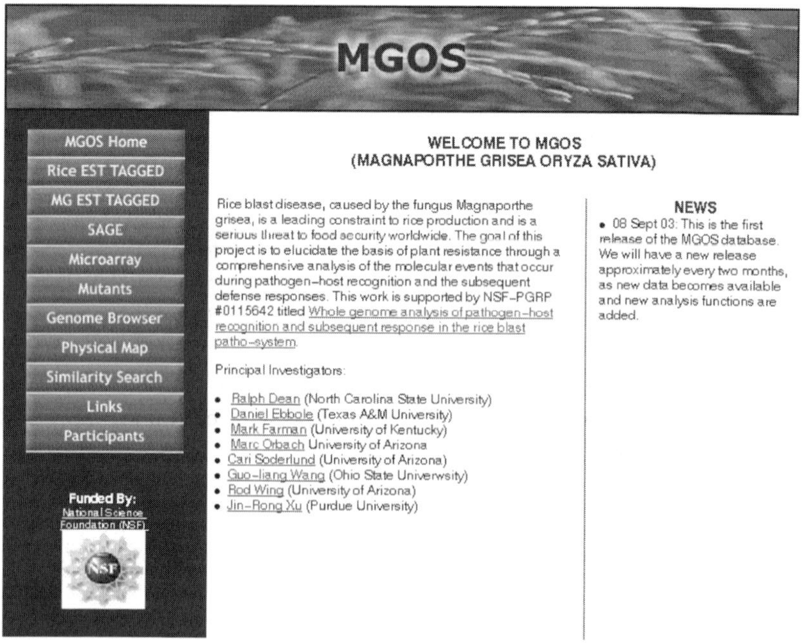

Figure 2.1 The Magnaporthe Grisea Oryza Sativa database. Screenshot of home page listing types of data available for querying.

Large-scale mutagenesis

More than 55 000 insertion mutants were created, the majority (>40 000) by *Agrobacterium tumefaciens* mediated transformation (ATMT) and the remainder by protoplast transformation. To minimize human error and facilitate flow of materials through a complex pipeline, PACLIMS was created (Donofrio *et al.*, 2005). PACLIMS is a web-based open-source software system that dynamically tracks the flow of materials, supports distributed operations, helps enforce standard operating procedures, records data, and provides a client/server for stand-alone operation and integration into existing laboratory information management systems (LIMS). All mutants are available from the Fungal Genetic Stock Center (www.fgsc.net). Southern analysis of >200 transformants indicated that the ATMT method is superior; most (>86%) have single DNA insertions in contrast to the apparent tandem insertions in the protoplast transformants. Flanking sequence analysis from over 200 ATMT strains revealed a clean insertion in the vast majority (>90%) of mutants. All mutants were analyzed for growth, pigmentation and sporulation phenotypes and more than half have been tested for auxotrophy – identified at a rate of 0.3% of strains tested. Around 1% of >14 000 ATMT mutants tested have defects in appressorium formation, and of these, half show reduced or non-pathogenic phenotypes on rice. DNA flanking insertion sites has been recovered from a number of mutants and includes mutations in a novel G-protein, a STAT-signal protein and a membrane protein.

Using a high throughput pathogenicity screen on young rice seedlings grown in glass tubes, 81 (0.3%) mutants out of ~29 000 transformants with attenuated ability to cause disease were identified with about one third causing no symptoms or small dark flecks. DNA flanking sequence has been isolated from ~30 mutants (Fig. 2.2). Insertions have occurred in several genes involved in regulating transcription and development, including kinesin, as well as genes encoding proteins containing SET and homeobox domains.

Secreted proteins

Proteins secreted or presented on the outside of the cell wall are crucial components for the ability of fungi to perceive and respond to the environment. Candidates were over expressed in *Magnaporthe* and purified by immuno precipitation. Several purified proteins, including cell wall degrading enzymes and proteinases, caused severe necrosis and hydrogen peroxide generation when placed on wound sites. Others induced hydrogen peroxide without causing a detectable necrosis response.

Genome-wide transcriptional profiling

A 22 000-element oligo-based microarray was designed in collaboration with Agilent Technologies that contains the Broad Institute predicted gene set, plus additional predicted features from other gene models (including GenScan) as well as ESTs for genes not predicted, yielding a total of 13 666 elements. The array also contains elements for ~7000 rice genes and is available from Agilent, stock item G4137 A (Fig. 2.3).

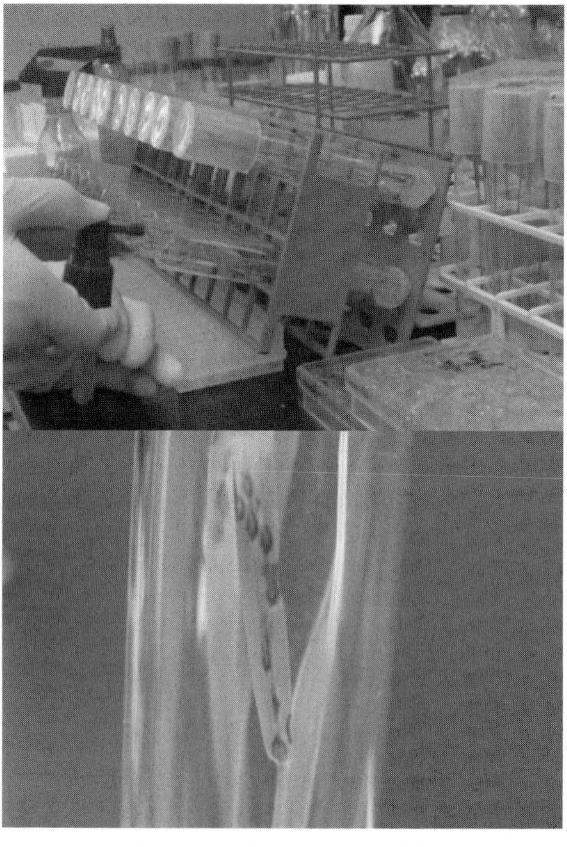

Figure 2.2 High throughput pathogenicity screen. Rice seedings are grown in glass tubes and spray inoculated with individual mutants (upper panel). Disease symptoms are recorded after seven days (lower panel).

Using this array, detailed transcription profiling experiments were conducted to identify genes expressed during infection, appressoria formation, under various nutritional stresses, and in mutated genome backgrounds. Protocols used and microarray data can be accessed through NCBI GEO (www.ncbi.nlm.gov/projects/geo) under accessions GSE2716 and GSE1945. Additional information can be found at MGOS (www.mgosdb.org). As anticipated, our results revealed that many hundreds of genes were differentially expressed. For example, 529–612 genes, representing ~4% of the total predicted gene set were up-regulated during spore germination. Similar numbers are also down-regulated (Table 2.1). Intriguingly, a number of TEs including MAGGY were found to be highly expressed in conidia. These findings were confirmed by RT-PCR. It is not known whether these elements are actively transposing, but transposon activity in germline cells is well documented (Prak *et al.*, 2003). In addition, ~2 and ~4% of genes were significantly differentially regulated in immature (7 h) and mature appressoria (12 h), respectively, compared to spores germinated on a non-inducing surface. In addition to genes expected such as those involved in melanin biosynthesis, genes encoding transporters and secondary metabolites were also found to be up-regulated. Similar numbers of genes were differentially regulated during growth in carbon and nitrogen limited environments. An NAD+ glutamate dehydrogenase gene was found to be up-regulated both during nitrogen starvation and

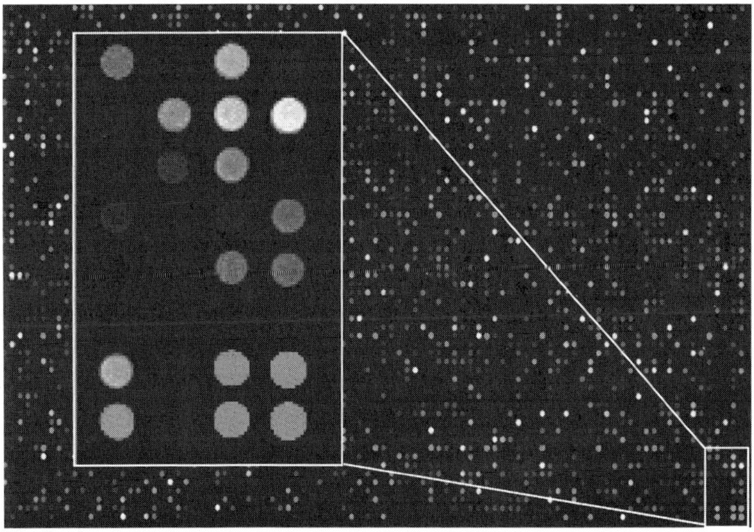

Figure 2.3 Agilent 22 000 element oligo array for *Magnaporthe*. Exploded view shows spots produce clear uniform signals.

Table 2.1 *Number of genes differentially expressed during germination and appressorium formation.*

Comparison[a]	2-fold up	2-fold down
Induced 7 h vs. spore[b]	588	652
Induced 12 h vs. spore	612	972
Non-induced 7 h vs. spore	603	493
Non-induced 12 h vs. spore	529	497
Induced 12 h vs. induced 7 h	321	187
Non-induced 12 h vs. non-induced 7 h	37	22
Induced 7 h vs. non-induced 7 h	194	85
Induced 12 h vs. non-induced 12 h	352	133

Notes:
[a] Spores were germinated on an appressorium inductive surface (induced) or non-inductive surface (non-induced) for 7 or 12 hours.
[b] First condition is numerator and second condition denominator.

appressorium formation. Targeted deletion of this gene revealed it is required for appressorium formation and pathogenicity. Whole genome expression analyses in mutant backgrounds, including *pmk1* and *mst12* have also been conducted (Bruno *et al.*, 2004; Park *et al.*, 2002). Under appressoria inducing conditions mutants exhibited differential expression of ~300 genes compared to the wild-type; however, the question still remains whether these genes are directly or indirectly regulated.

Other transcriptional profiling experiments have focused on *in planta* gene expression. At 48 h post inoculation (pi), the expression of 17 fungal genes could be detected in infected barley leaves, which rose to 348 at 96 h pi. Twelve genes were in common to all time points and included CYP1 (cyclophilin), a protein involved in signalling, which has been previously shown to be a pathogenicity determinant (Viaud, Balhadere & Talbot, 2002). Analysis across all microarray experiments to date revealed ~60 genes to be specifically expressed *in planta*. These include several transporters, and proteins involved in xylan and lipid metabolism.

Concluding remarks

Significant progress has been made over the past few years to elucidate fungal pathogenicity. The genome sequence of *Magnaporthe* and other filamentous fungi has provided unprecedented access to the elements required by a plant pathogenic fungus to attack and colonize its host plant. Analysis of the genome has revealed numerous secrets including a vast array of secreted proteins, novel surface receptors as well as the means of generating genetic variation in the absence of sexual reproduction. Functional studies, including DNA insertion mutants and genome-wide transcriptional profiling studies, are providing new clues to the genes required for infection-related development and pathogenicity. However, much remains in darkness. Gene prediction tools are inaccurate. Direct methods such as full length ESTs and genome tiling arrays are needed to define carefully the transcriptome. Moreover, for about half of the predicted genes there is no known function. Differential expression analysis may shed some light – such as provide information on sets of genes that exhibit coordinate expression. However, many genes are transcribed at very low levels or are regulated post-transcriptionally. Other functional approaches such as gene knockout, while often valuable to assess the role of a particular gene, more often than not reveal little because of functional redundancy. RNAi and other gene silencing techniques may offer some relief and have been shown to be effective in knocking down gene expression of some but not all genes examined in *Magnaporthe*. At the present time, little is known of the regulatory circuits governing gene expression including the identity of key transcription factors directing the infection process. In addition, our knowledge of the specific proteins from the fungus and host and how they interact to determine disease outcome remains largely a black box. Proteomics approaches, particularly exploiting protein arrays and chromatin-immunoprecipitation-chip (ChIP-chip) strategies will be key to further exploiting genome sequence information to advance knowledge of and model host–pathogen interactions (Conlon *et al.*, 2003; Zhu & Snyder, 2003).

Acknowledgements

The authors acknowledge the US Department of Agriculture, Cooperative State Research, Education and Extension Service and National Science Foundation for funding this work. We thank all members of the laboratories participating in this work and members of the Broad Institute sequencing, assembly and annotation teams. We also extend our thanks to the rice blast community at large.

References

Bruno, K. S., Tenjo, F., Li, L., Hamer, J. E. & Xu, J. R. (2004). Cellular localization and role of kinase activity of PMK1 in *Magnaporthe grisea*. *Eukaryotic Cell*, **3**, 1525–32.

Calabrese, P. P., Chakravarty, S. & Vision, T. J. (2003). Fast identification and statistical evaluation of segmental homologies in comparative maps. *Bioinformatics*, **19**, 74–80.

Chao, C. C. T. & Ellingboe, A. H. (1991). Selection for mating competence in *Magnaporthe grisea* pathogenic to rice. *Canadian Journal of Botany*, **69**, 2130–4.

Conlon, E. M., Liu, X. S., Lieb, J. D. & Liu, J. S. (2003). Integrating regulatory motif discovery and genome-wide expression analysis. *Proceedings of the National Academy of Sciences USA*, **100**, 3339–44.

de Jong, J. C., McCormack, B. J., Smirnoff, N. & Talbot, N. J. (1997). Glycerol generates turgor in rice blast. *Nature*, **389**, 244–5.

Dean, R. A. (1997). Signal pathways and appressorium morphogenesis. *Annual Review of Phytopathology*, **35**, 211–34.

Dean, R. A., Talbot, N. J., Ebbole, D. J., Farman, M. L., Mitchell, T. K., Orbach, M. J., Thon, M., Kulkarni, R., Xu, J. R., Pan, H., Read, N. D., Lee, Y. H., Carbone, I., Brown, D., Oh, Y. Y., Donofrio, N., Jeong, J. S., Soanes, D. M., Djonovic, S., Kolomiets, E., Rehmeyer, C., Li, W., Harding, M., Kim, S., Lebrun, M. H., Bohnert, H., Coughlan, S., Butler, J., Calvo, S., Ma, L. J., Nicol, R., Purcell, S., Nusbaum, C., Galagan, J. E. & Birren, B. W. (2005). The genome sequence of the rice blast fungus *Magnaporthe grisea*. *Nature*, **434**, 980–6.

Donofrio, N., Rajagopalon, R., Brown, D., Diener, S., Windham, D., Nolin, S., Floyd, A., Mitchell, T., Galadima, N., Tucker, S., Orbach, M. J., Patel, G., Farman, M., Pampanwar, V., Soderlund, C., Lee, Y. H. & Dean, R. A. (2005). 'PACLIMS' a component LIM system for high-throughput functional genomic analysis. *BMC Bioinformatics*, **6**, 94.

Ebbole, D. J., Jin, Y., Thon, M., Pan, H., Bhattarai, E., Thomas, T. & Dean, R. (2004). Gene discovery and gene expression in the rice blast fungus, analysis of expressed sequence tags. *Molecular Plant–Microbe Interactions*, **17**, 1337–47.

Galagan, J. E., Calvo, S. E., Borkovich, K. A., Selker, E. U., Read, N. D., Jaffe, D., FitzHugh, W., Ma, L. J., Smirnov, S., Purcell, S., Rehman, B., Elkins, T., Engels, R., Wang, S., Nielsen, C. B., Butler, J., Endrizzi, M., Qui, D., Ianakiev, P., Bell-Pedersen, D., Nelson, M. A., Werner-Washburne, M., Selitrennikoff, C. P., Kinsey, J. A., Braun, E. L., Zelter, A., Schulte, U., Kothe, G. O., Jedd, G., Mewes, W., Staben, C., Marcotte, E., Greenberg, D., Roy, A., Foley, K., Naylor, J., Stange-Thomann, N., Barrett, R., Gnerre, S., Kamal, M., Kamvysselis, M., Mauceli, E., Bielke, C., Rudd, S., Frishman, D., Krystofova, S., Rasmussen, C., Metzenberg, R. L., Perkins, D. D., Kroken, S., Cogoni, C., Macino, G., Catcheside, D., Li, W., Pratt, R. J., Osmani, S. A., DeSouza, C. P., Glass, L., Orbach, M. J., Berglund, J. A., Voelker, R., Yarden, O., Plamann, M., Seiler, S., Dunlap, J., Radford, A., Aramayo, R., Natvig, D. O., Alex, L. A., Mannhaupt, G., Ebbole, D. J., Freitag, M., Paulsen, I., Sachs, M. S., Lander, E. S., Nusbaum, C. & Birren, B. (2003). The genome sequence of the filamentous fungus *Neurospora crassa*. *Nature*, **422**, 859–68.

Hamer, J. E., Howard, R. J., Chumley, F. G. & Valent, B. (1988). A mechanism for surface attachment in spores of a plant pathogenic fungus. *Science*, **239**, 288–90.

Kang, S., Lebrun, M.-H., Farrall, L. & Valent, B. (2001). Gain of virulence caused by insertion of a Pot3 transposon in a *Magnaporthe grisea* avirulence gene. *Molecular Plant–Microbe Interactions*, **14**, 671–4.

Kulkarni, R. D., Kelkar, H. S. & Dean, R. A. (2003). An eight-cysteine-containing CFEM domain unique to a group of fungal membrane proteins. *Trends in Biochemical Sciences*, **28**, 118–21.

Kulkarni, R. D., Thon, M. R., Pan, H. & Dean, R. A. (2005). Novel G-protein-coupled receptor-like proteins in the plant pathogenic fungus *Magnaporthe grisea*. *Genome Biology*, **6**, R24.

Martin, S. L., Blackmon, B. P., Rajagopalan, R., Houfek, T. D., Sceeles, R. G., Denn, S. O., Mitchell, T. K., Brown, D. E., Wing, R. A. & Dean, R. A. (2002).

MagnaportheDB, a federated solution for integrating physical and genetic map data with BAC end derived sequences for the rice blast fungus *Magnaporthe grisea*. *Nucleic Acids Research*, **30**, 121–4.

Nimchuk, Z., Eulgem, T., Holt III, B. F. & Dangl, J. L. (2003). Recognition and response in the plant immune system. *Annual Review of Genetics*, **37**, 579–609.

Ou, S. H. (1987). In *Rice Diseases*. United Kingdom: Commonwealth of Agricultural Bureaux, pp. 109–200.

Padovan, A. C., Sanson, G. F., Brunstein, A. & Briones, M. R. (2005). Fungi evolution revisited, application of the penalized likelihood method to a bayesian fungal phylogeny provides a new perspective on phylogenetic relationships and divergence dates of ascomycota groups. *Journal of Molecular Evolution*, **60**, 726–35.

Park, G., Xue, C., Zheng, L., Lam, S. & Xu, J. R. (2002). MST12 regulates infectious growth but not appressorium formation in the rice blast fungus *Magnaporthe grisea*. *Molecular Plant–Microbe Interactions*, **15**, 183–92.

Prak, E. T., Dodson, A. W., Farkash, E. A. & Kazazian, H. H. Jr. (2003). Tracking an embryonic L1 retrotransposition event. *Proceedings of the National Academy of Sciences USA*, **100**, 1832–7.

Selker, E. U. (2002). Repeat-induced gene silencing in fungi. *Advances in Genetics*, **46**, 439–50.

Sesma, A. & Osbourn, A. E. (2004) The rice leaf blast pathogen undergoes developmental processes typical of root-infecting fungi. *Nature*, **431**, 582–6.

Talbot, N. J. (2003). On the trail of a cereal killer, exploring the biology of *Magnaporthe grisea*. *Annual Review of Microbiology*, **57**, 177–202.

Thon, M. R., Martin, S. L., Goff, S., Wing, R. A. & Dean, R. A. (2004). BAC end sequences and a physical map reveal transposable element content and clustering patterns in the genome of *Magnaporthe grisea*. *Fungal Genetics and Biology*, **41**, 657–66.

Valent, B. & Chumley, F. G. (1991). Molecular genetic analysis of the rice blast fungus *Magnaporthe grisea*. *Annual Review of Phytopathology*, **29**, 443–67.

Viaud, M. C., Balhadere, P. V. & Talbot, N. J. (2002). A cyclophilin acts as a virulence determinant during plant infection. *Plant Cell*, **14**, 917–30.

Xu, J. R. (2000). MAP kinases in fungal pathogens. *Fungal Genetics and Biology*, **31**, 137–52.

Zhu, H., Blackmon, B. P., Sasinowski, M. & Dean, R. A. (1999). Physical map and organization of chromosome 7 in the rice blast fungus, *Magnaporthe grisea*. *Genome Research*, **9**, 739–50.

Zhu, H. & Snyder, M. (2003). Protein chip technology. *Current Opinion in Chemical Biology*, **7**, 55–63.

II

Bioactive molecules

3

The biosynthesis of polyketides, acyl tetramic acids and pyridones by filamentous fungi

R. J. COX

Introduction

The filamentous fungi are proficient and copious producers of secondary metabolites. From the perspective of an organic chemist, the range and variety of chemical structures of these compounds is remarkable. Synthetic organic chemists have often used the very high structural complexity of fungal secondary metabolites to test their own abilities to mimic nature. From the perspective of a medicinal chemist, the diversity of compounds and structural types represents a pool of useful compounds often possessing unique biological properties. The range of structural types can, at first, appear baffling. However, most secondary metabolites produced by fungi fall into a relatively small number of classes: the alkaloids, derived from amines and amino acids; the terpenoids, derived from isopentenyl diphosphate; and the polyketides, generally derived from acetate. This system of classification is based on the biosynthetic origin of the compound in question, that is to say, a combination of the type of starting material and the type of chemical reactions used during biosynthesis. However, fungi also often combine different types of biosynthetic pathway during the manufacture of secondary metabolites. In Bristol, we have focused our efforts on understanding the biosynthesis of polyketides in fungi, but the inclusion of amino-acid derived moieties in the compounds we are interested in has also necessitated wider investigations.

Fungal polyketides

Fungi make some of the simplest and some of the most complex polyketides known (Simpson, 1995). It is useful to consider a heirarchy of complexity when considering the structures of fungal polyketides, because

Exploitation of Fungi, eds. G. D. Robson, P. van West and G. M. Gadd. Published by Cambridge University Press.

the complexity in structure is generated by enzymes and ultimately genes (Bingle *et al.*, 1999). The simplest structures are those such as orsellinic acid **1** (and its derivatives), a tetraketide, generated by many fungi in the genera *Aspergillus* and *Penicillium* (Fig. 3.1). Addition of an extra acetate gives a pentaketide such as tetrahydroxynaphthalene **2**, a compound widely distributed in fungi and which is involved in melanization and is a key component of apressorium formation and invasion of plant cells by organisms such as *Magnaporthe grisea*, *Colletotrichum lagenarium* and others. Similar aromatic compounds such as norsolorinic acid **3** are precursors for the mycotoxins sterigmatocyctin **4** and aflatoxin B1 **5** produced by species of *Aspergilli* and others (Minto & Townsend, 1997). Further complexity is represented by compounds such as 6-methyl salicylic acid **6**,

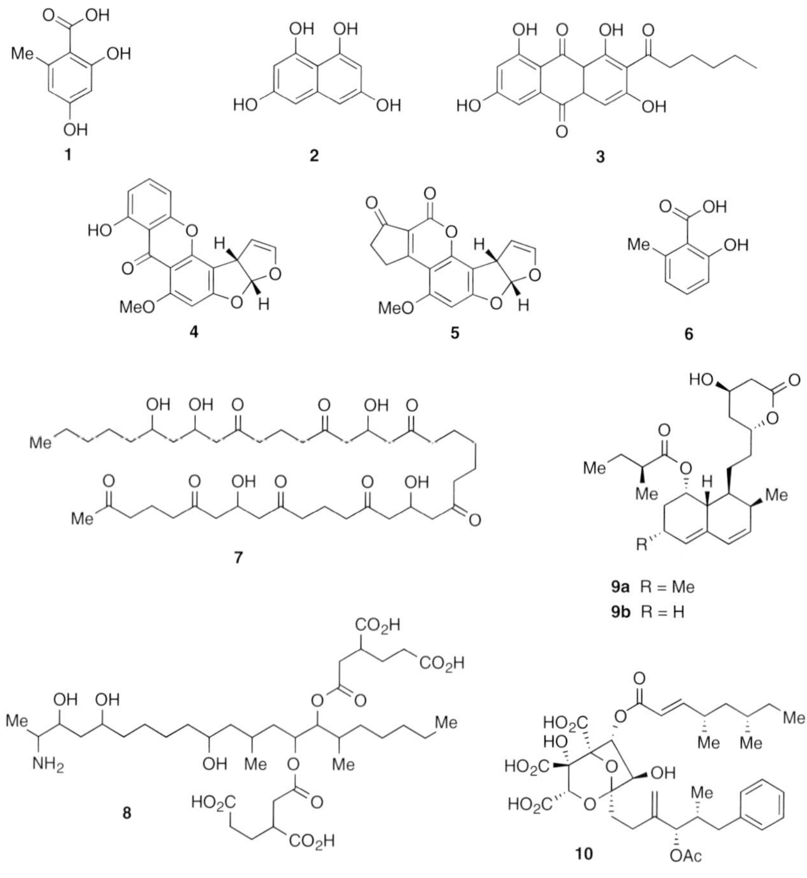

Figure 3.1 Polyketide structures.

where a reduction reaction occurs during biosynthesis. Further complexity is then observed in compounds such as T-toxin **7** produced by *Cochliobolus heterostrophus* (Yang *et al.*, 1996) and fumonisin **8** produced by *Gibberella fujikuroi* (Proctor *et al.*, 1999). In these compounds many more carbon atoms are used and many more reduction reactions occur. Finally, complex structures such as lovastatin **9** (Kennedy *et al.*, 1999) from *A. terreus* and squalestatin S1 **10** from *Phoma sp.* (Cox *et al.*, 2004) show important biological properties – both are inhibitors of mammalian sterol biosynthesis. Lovastatin **9a** is a comercial drug compound and potent inhibitor of HMG-CoA reductase, while squalestatin S1 **10** is an inhibitor of squalene synthase.

The chemical reactions of fungal polyketide biosynthesis

The chemical complexity of fungal polyketides arises as a result of the array of chemical reactions deployed during their biosynthesis. These can be understood to occur in two phases – those involved in the construction of the main carbon chain catalyzed by a multifunctional protein known as a polyketide synthase (PKS), and those reactions occurring after the PKS, catalyzed by numerous and diverse post-PKS processing enzymes (Fig. 3.2) (Staunton & Weissmann, 2001).

Fungal polyketide synthases

In fungi, these enzymes are large multifunctional proteins, possessing at least three recognizable catalytic domains, but often many more (Fig. 3.3). They have been studied most often at the genetic level where they are encoded by large single open reading frames (ORF) of 6–8 Kb. The core protein domains consist of an acyl transferase (AT), an acyl carrier protein (ACP) and a β-keto-acyl synthase (KS). The AT domain is responsible for loading the starter unit (usually acetate from acetyl CoA **11**) onto the active site thiol of the KS domain. The AT also loads the extender unit (malonate, derived from malonyl CoA **12**) onto the ACP – it is held as a thiolester bound to the terminal thiol of a prosthetic phosphopantetheine arm of the ACP. The KS then catalyzes the decarboxylative condensation of acetate and malonate to form the first polyketide intermediate – acetoacetate **13** as a thiolester bound to the ACP. These are the minimal reactions required for polyketide biosynthesis and are common with most PKS enzymes in plants and bacteria, but also common with fatty acid biosynthesis in all organisms. *C*-Methyl transferase domains (*C*MeT) can transfer a methyl group at this stage to give a methylated intermediate such as **14**. Reduction by a β-keto-reductase (KR) can then afford the β-hydroxy acyl species **15**, dehyration (DH) of which leads to the enoyl thiolester **16**. Finally, action of an enoyl reductase domain (ER) forms the fully saturated chain **17**. The cycle is then repeated to extend the chain (Simpson, 1995).

Programming of the PKS

All the chemical reactions of fungal PKS (apart from methyl transferase) are the same as those catalyzed by fatty acid synthases. However, a PKS has the ability to coordinate its activities – in some way the synthase is *programmed*. The simplest example of this is represented by chain-length selection. For example, the reactions carried out by orsellinic acid synthase (OAS) and by tetrahydroxynaphthalene synthase (THNS) are identical except that OAS catalyzes three extensions, giving the

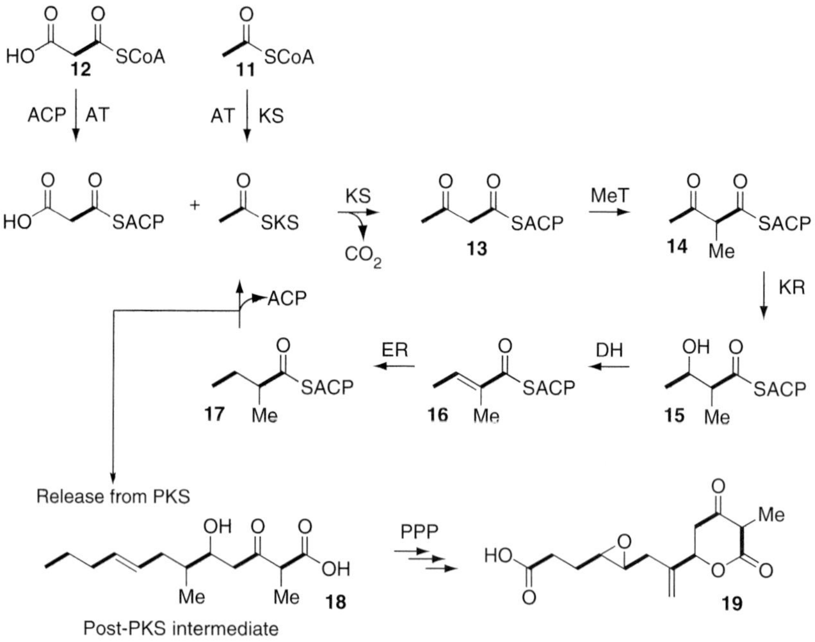

AT = Acyl Transferase	DH = β-Hydroxyacyl Dehydratase
ACP = Acyl Carrier Protein	ER = Enoyl Reductase
KS = β-Ketoacyl Synthase	MeT = Methyl Transferase
KR = β-Ketoacyl Reductase	PPP = Post PKS Processes

Figure 3.2 The chemical reactions catalyzed by fungal polyketide synthases. Post PKS processes often include cyclizations, oxidations and other modifications.

tetraketide **20**, while THNS catalyzes four extensions to give the pentaketide **21**. Both synthases contain single KS, AT and ACP domains which must act iteratively – the difference is in the number of iterations (Fig. 3.3). We refer to these, and similar synthases, as non-reducing synthases which produce *unreduced polyketides*.

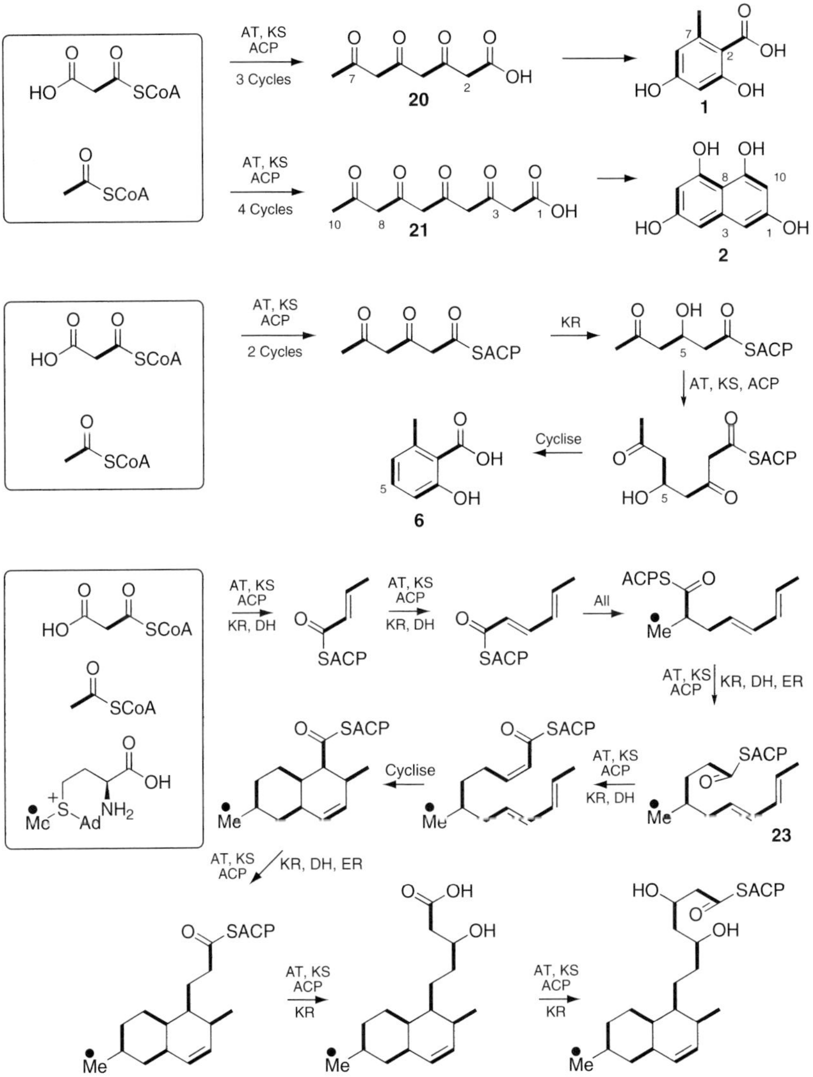

Figure 3.3 Programming in unreduced, partially reduced and highly reduced fungal polyketide biosynthesis. Catalytic domains as Fig. 3.2.

The next level of complexity in programming is exemplified by 6MSA synthase (6MSAS). Here, a KR domain is present in addition to KS, AT and ACP. However, the KR acts only once, while the KS domain acts three times. The KR is active only after the second extension to give the intermediate **22**. Further extension and cyclization then gives **6**. Such compounds are referred to as *partially reduced* polyketides.

In the case of lovastatin nonaketide biosynthesis, programming is much more complex because the KR, DH and ER domains must all be present and brought into play in different cycles during the eight rounds of chain extension. Methyl transferase domains are also subject to programmed control – for example during the biosynthesis of the lovastatin nonaketide, the methyl transferase acts only once, during the third round of chain extension to give enzyme-bound intermediate **23** (Fig. 3.3). These compounds belong to a large group of metabolites referred to as *highly reduced* polyketides. It is the highly reduced compounds which often show significant chemical complexity and corresponding biological activity – in particular these compounds are often likely drugs. Finally, PKS enzymes can also vary the starter unit, using other groups than acetate. For example, in the case of the squalestatin main chain, benzoic acid is the starter unit.

The programming of fungal PKS proteins is significantly more complex than that observed in bacterial PKS systems where each individual reaction is catalyzed by a different catalytic domain. This is exemplified in the case of 6-deoxy erythromycin B (6dEB) **24** biosynthesis in *Saccharopolyspora erythrae* (Wilkinson *et al.*, 2000). In this case the domains are arranged into six individual modules, each very similar to a single fungal PKS, to carry out the six rounds of chain extension and required reductive modifications (Fig. 3.4). A single module from these bacterial systems usually acts only once and must pass on its product to the next module. In contrast, a fungal iterative PKS must retain the growing substrate until it is fully formed. Fungal polyketide synthases could therefore be regarded as much more efficient than their modular bacterial analogues. However, the mechanism of programming in fungal polyketide synthases is not yet understood.

Post-PKS enzymes and reactions

Once the PKS has finished its work, the newly synthesized polyketide is generally released from the synthase. For example, in the case of aflatoxin B1 **5** biosynthesis, norsolorinic acid **3** is the first product of the pathway. The released compounds are then acted on by a series of other, usually discrete, enzymes (Minto & Townsend, 1997). In fungi the reactions are frequently oxidative and include epoxidation, hydroxylation and dehydrogenation, but other common reactions also include methylation of heteroatoms, cyclization and rearrangement reactions. In the case of aflatoxin biosynthesis, at least 11 individual post-PKS reactions occur (Fig. 3.5). The post-PKS enzymes of fungi are often more diverse and more frequently occurring than their bacterial counterparts – again leading to the observation of increased chemical diversity and complexity of the polyketides derived from fungi.

Genes and proteins

The genes encoding polyketide synthases and post-PKS processing enzymes are generally clustered, often with recognizable transport elements such as ABC-transporters and general efflux pumps. One of the earliest clusters investigated was that involved in lovastatin biosynthesis in *A. terreus* (Kennedy *et al.*, 1999). Lovastatin consists of two polyketide chains. Correspondingly, two polyketide synthases have been found within the lovastatin biosynthetic gene cluster, named lovB and lovF (Fig. 3.6). LovB encodes the lovastatin nonaketide synthase (LNKS), while lovF encodes the diketide synthase (LDKS). Each ORF encodes all the chemical reactivity required for biosynthesis of highly reduced polyketides, as well as a methyl-transferase domain required for methylation in each case. LovF is 7.6 Kb in

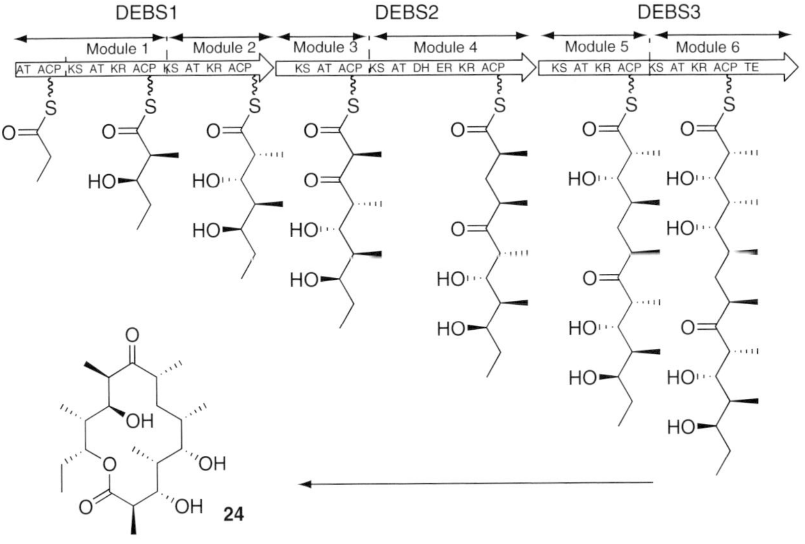

Figure 3.4 6dEB biosynthesis.

length, while lovB is longer at 9.1 Kb. Other PKS genes are similar in size and organization of the active site domains, which closely resemble the organization of mammalian fatty acid synthases (FAS). Post PKS processing enzymes are encoded by discrete ORFs, which are clustered

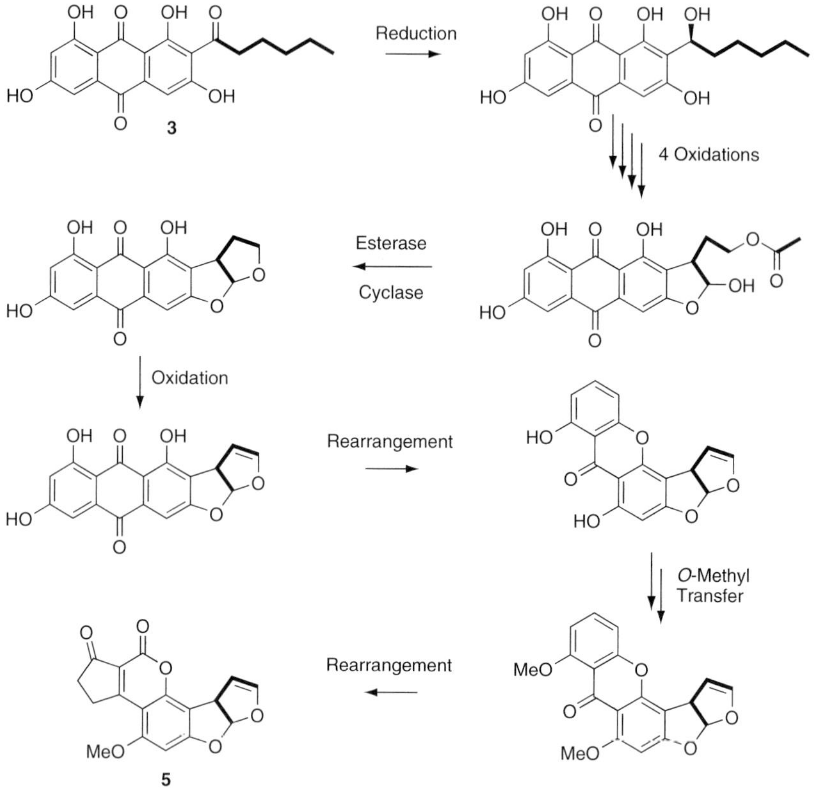

Figure 3.5 Aflatoxin post-PKS processing steps.

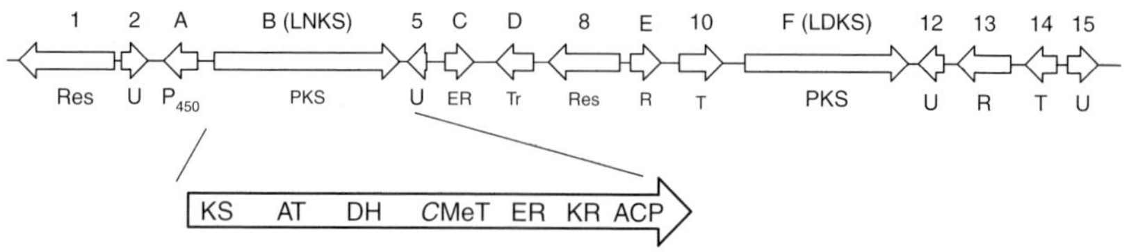

Figure 3.6 Part of the lovastatin gene cluster in *A. terreus*. ORFs are numbered, apart from those directly involved in biosynthetic steps which are indicated by letters. The ORFs encode the following proteins: Res, resistance; U, unknown; P_{450}, cytochrome P_{450} oxidase; PKS, polyketide synthase; ER, enoyl reductase; Tr, transferase; R, regulatory.

with the much larger PKS genes. In many cases the catalytic activity of the cognate proteins can be inferred by homology modelling – for example lovC appears to encode an ER protein, which makes chemical sense as the ER domain of LNKS appears to be unable to function correctly.

The approach to cloning and investigating fungal secondary metabolism gene clusters of interest

We began our genetic approach to understanding fungal polyketide biosynthesis in the late 1990s after a long period of more 'classical' biosynthetic studies involving the feeding of isotopically labelled precursor molecules to active fermentations of the fungal species under investigation. These types of experiment had shed light on the origin of the simple precursor molecules such as acetate, methionine and various starter units – and sophisticated versions could even track the origin of oxygen and nitrogen atoms. However, in many cases these experiments were unsuitable for determining later-stage intermediates. We wanted to be able to access the genes involved in fungal polyketide biosynthesis for two main reasons. Firstly, we wanted to extend our studies of biosynthetic processes, but secondly, we imagined that we might be able to engineer the biosynthetic genes to produce new compounds. In either case we needed to access fungal biosynthetic genes from a given organism which were involved in the biosynthesis of a particular compound of interest. When we began our studies very few fungal PKS genes were known – just examples of PKS genes involved in unreduced (tetrahydroxy naphthalene **2**) and partially reduced (6MSA **6**) compound biosynthesis were known. We wanted to develop a generic method able to rapidly clone PKS genes. We decided to exploit the observed homology in peptide sequence for particular catalytic domains belonging to different PKS proteins by developing degenerate PCR primers based on these regions of homology, which could be used to amplify fragments of the PKS genes of interest. Importantly, these PCR primers should be able to amplify fragments of PKS genes, but not fragments of the closely related FAS genes. We also wanted to determine if we could design PCR primers that were selective for PKS genes involved in the biosynthesis of different structural types of polyketide. For example, could we make pairs of primers selective for the amplification of the structurally interesting highly reduced polyketides, but which did not amplify either fragments of FAS genes or genes involved in the biosynthesis of unreduced or partially reduced polyketides?

The initial attempts of the Bristol group focused on the development of PCR primers selective for unreduced and partially reduced PKS genes

(Fig. 3.7) (Bingle, Simpson & Lazarus, 1999). Thus, sequences for the available PKS protein sequence KS domains were aligned and inspected for regions of high homology. In particular, we were interested in sequences which were conserved between PKS sequences for structurally similar compounds, but different from conserved sequences for FAS proteins or PKS proteins for structurally different compounds. Additionally, the primers should be located away from regions known to harbour introns and be far enough apart to yield PCR products that would be easily analyzed by gel electrophoresis (c. 700 bp in this case). Thus one primer pair (designated LC1/LC2c) was designed to be selective for unreduced PKS and another set (LC3/LC5c) was designed to be selective for partially reduced PKS. These primers incorporated degeneracy – i.e. included either mixed bases or inosines at defined positions in order to allow for codon variation in different template organisms.

In initial experiments, genomic DNA was prepared from a number of fungi known to be producers of polyketides. PCR reactions were then performed using the PCR primer pairs, and the products separated by gel electrophoresis. In these reactions, conditions usually had to be optimized for efficient product formation by varying the ratio of primer concentrations and by including varied amounts of Mg^{2+} ions and/or varying the

LC1 (5' KS NR primer)

```
Primer sequence      GAT CCI AGI TTT TTT AAT ATG
Degeneracy = 32        C    C      C   C   C
```

A. nidulans wA	**Asp**	**Pro**	**Arg**	**Phe**	**Phe**	**Asn**	**Met**
C. lagenarium PKS1	**Asp**	**Pro**	**Arg**	**Phe**	**Phe**	**Asn**	**Met**
P. patulum MSAS	**Asp**	Cys	Gln	**Phe**	**Phe**	Gly	Ile
A. parasiticus PKSL2	**Asp**	Ala	Ser	**Phe**	**Phe**	**Asn**	Ile

LC2c (3' KS NR primer)

```
Primer sequence      GT ICC IGT ICC GTG CAT TTC
Degeneracy = 4                      A       C
```

A. nidulans wA	Glu	**Met**	**His**	**Gly**	**Thr**	**Gly**	**Thr**
C. lagenarium PKS1	Glu	**Met**	**His**	**Gly**	**Thr**	**Gly**	**Thr**
P. patulum MSAS	Glu	Ala	**His**	Ala	**Thr**	Ser	**Thr**
A. parasiticus PKSL2	Glu	Ala	**His**	Ala	**Thr**	Ser	**Thr**

LC3 (5' KS PR primer)

```
Primer sequence      GCI GAA CAA ATG GAT CCI CA
Degeneracy = 8             G   G       C
```

P. patulum MSAS	**Ala**	**Glu**	**Gln**	**Met**	**Asp**	**Pro**	**Gln**
A. parasiticus PKSL2	**Ala**	**Glu**	**Gln**	**Met**	**Asp**	**Pro**	**Gln**
A. nidulans wA	**Ala**	Leu	**Gln**	Ala	**Asp**	**Pro**	Ala
C. lagenarium PKS1	**Ala**	Phe	**Gln**	Thr	**Asp**	**Pro**	Met

LC5c (3' KS PR primer)

```
Primer sequence      GT IGA IGT IGC GTG IGC TTC
Degeneracy = 4                      A       C
```

P. patulum MSAS	**Glu**	**Ala**	**His**	**Ala**	**Thr**	**Ser**	**Thr**
A. parasiticus PKSL2	**Glu**	**Ala**	**His**	**Ala**	**Thr**	**Ser**	**Thr**
A. nidulans wA	**Glu**	Met	**His**	Gly	**Thr**	Gly	**Thr**
C. lagenarium PKS1	**Glu**	Met	**His**	Gly	**Thr**	Gly	**Thr**

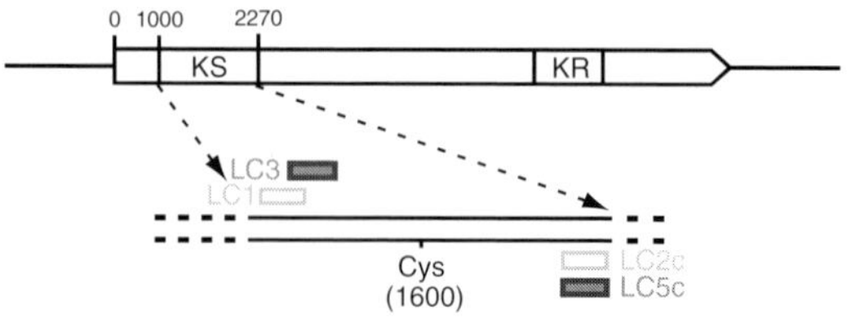

Approximate predicted PCR product size : LC1/2c - 700 bp
LC3/5c - 680 bp

Figure 3.7 PKS primers designed to selectively amplify regions of unreduced and partially reduced fungal PKS genes.

PCR annealing temperature. However, in all cases it was possible to amplify sequences of the expected length. The PCR products were then cloned and sequenced.

Sequence analysis of the PCR products showed that in all cases the cloned products were the expected 700 bp fragments of KS domains of PKS genes. We also conducted tree-plot analysis using both the known PKS sequences and the PCR products (Fig. 3.8). We found a clear relationship between the clades and the chemical structures of the PKS products (where known). Thus, the primers designed to amplify fragments of

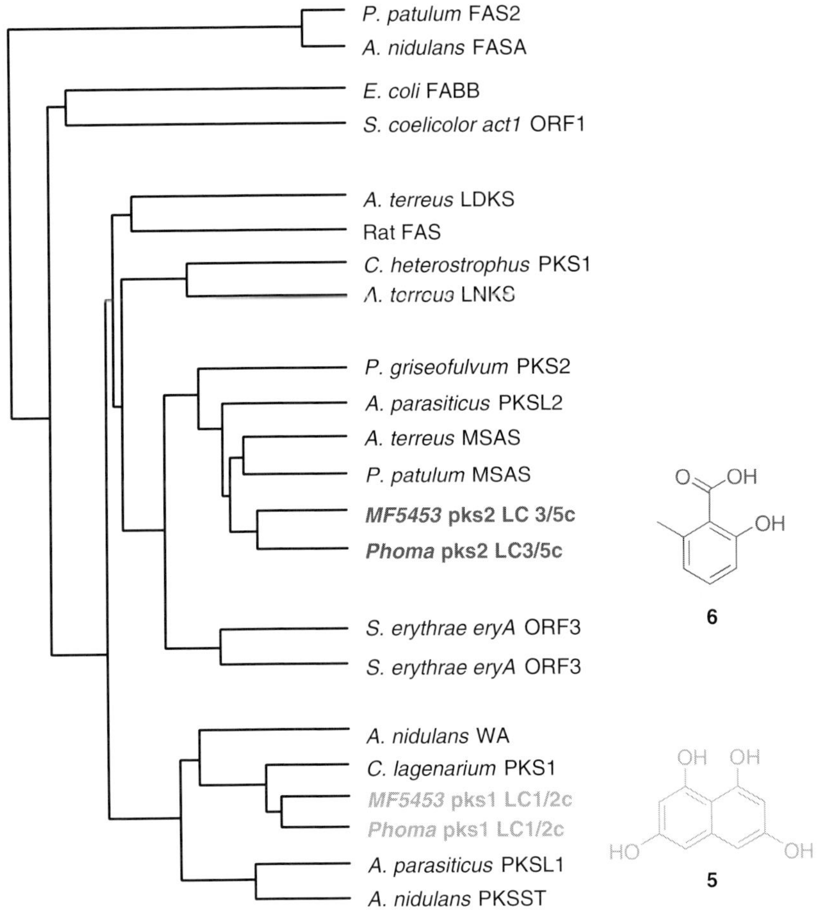

Figure 3.8 Tree-plot showing clustering of known PKS sequences into clades related to the chemical class of polyketide concerned. PCR products from LC1/2c and LC3/5c are highlighted.

unreduced PKS (LC1/LC2c) gave PCR products that closely grouped with the unreduced clade of known sequences. Likewise the LC3/LC5c primers amplified products which most closely grouped with the partially reduced PKS clade. This was an important result as it showed a clear link between sequence and chemical structure. Thus, while this methodology clearly worked, we were limited by the fact that we could not yet access PKS genes for the more structurally interesting highly reduced PKS because at the outset of our project no examples of highly reduced PKS genes were contained in the publicly available databases.

This situation was remedied in 1999 when Vederas and Hutchinson revealed the sequence of the lovastatin **9a** biosynthetic genes (Kennedy *et al.*, 1999). We then applied the same methodology to devise a pair of degenerate PCR primers selective for the KS domains of highly reduced polyketides (Nicholson *et al.*, 2001). In parallel, we also devised primers selective for *C*MeT domains using LNKS and LDKS *C*MeT domains as a guide. Here, we used sequences of *O*- and *N*-methyl transferases in our analysis in attempts to avoid amplifying these types of methyl transferases which are common in primary metabolism and DNA processing, for example. Once again trial PCR reactions with genomic DNA from diverse sources were followed by gel electrophoresis, cloning and sequencing. These PCR primers also appeared to be highly selective for their targets – the HRKS primers gave sequences corresponding to the LNKS/LDKS family in all cases. As previously observed, these sequences formed a distinct clade, separate from those of the partially reduced and unreduced clades. Interestingly, we also began to understand that single species of filamentous fungi could contain multiple PKS genes. Not only did we find that a single species could contain representatives of the non-reduced partially reduced and highly reduced PKS families, we often found more than one of each class.

The squalestatin synthase

We now had the ability to amplify fragments of PKS genes from any given filamentous fungus, and it appeared that the PCR primers could select subsets of PKS genes based on the chemical structure. We next set about using the fragments of PKS gene as probes to obtain full PKS genes. Squalestatin S1 **10** is a compound composed of two polyketide chains and is produced by a number of *Phoma* species. Each chain is methylated. It shows potent inhibition of mammalian squalene synthase and is a potential treatment for elevated cholesterol levels in humans. We reasoned that PCR products obtained from reactions containing *Phoma sp.* gDNA and primers selective for highly reduced PKS KS domains should yield products that could in turn be used to probe a genetic library of *Phoma* species.

We thus performed PCR reactions using *Phoma* genomic DNA and generated sequences corresponding to both KS and *C*MeT domains (Cox *et al.*, 2004). We used two species known to produce S1 as sources of template gDNA as a useful 'internal reference' – we expected to find very similar synthases in each organism. Sequencing of the PCR products from the two organisms confirmed that some of the PCR products were highly homologous. We next performed a classical secondary metabolism study – a time-course of S1 **10** production. We reasoned that when S1 **10** was being produced most rapidly, it was likely that mRNA encoding S1 biosynthetic proteins would be most abundant. Thus a cDNA library prepared from mRNA obtained in this period should be enriched in the required sequences. Thus cDNA and gDNA libraries prepared from the two organisms were probed with the *C*MeT sequences generated from the degenerate PCR reactions.

Sequence analysis of hybridized plaques revealed the presence of a 4.8 Kb fragment of a Type I PKS, containing the polyadenylated 3′ end of the sequence. This clone contained (5′→3′) a truncated fragment of a DH domain and complete *C*MeT, ER, KR and ACP domains. The 5′ portion of the PKS was then obtained using a process involving the rapid amplification of cDNA ends (RACE) to produce an overlapping clone, again of 4.8 Kb containing KS, AT, DH and *C*MeT domains. A contiguous, full-length clone was then constructed by digest and religation of the two cDNA fragments. The reconstructed 8142 bp PKS (Phoma PKS1) contained KS, AT, DH, *C*MeT, ER, KR and ACP domain encoding regions (Fig. 3.9). The translated gene, encoding a polypeptide of 2714 amino acids, shows high end-to-end amino acid homology with other known fungal PKS involved in the biosynthesis of highly reduced polyketides such as the lovastatin diketide synthase (LDKS, 59% identity), and the compactin diketide synthase (60% identity).

Close inspection of individual domains suggested they were catalytically competent. For example in the ER domain, the conserved NADPH binding motif (LxHx(G/A)xGGVG) was present (LIHAASGGVG). This contrasts with the situation in other fungal PKS genes, such as the LNKS, where the ER domain is probably inactive due to deleterious mutations in the NADPH binding region.

However, at this stage we had no proof that this PKS was involved in S1 **10** biosynthesis. Despite very many attempts, directed knockout of this

Phoma PKS were unsuccessful due to insurmountable difficulties in obtaining transformants. We thus turned to expression studies to investigate the role of the PKS. Thus a cDNA derived clone was transferred into pTAex-3, a vector previously used by the Ebizuka and Fujii groups for the expression of non-reduced fungal PKS in *Aspergillus oryzae*. Expression, induced by growth on starch, led to the production of a new compound in one *A.oryzae* transformant which was isolated by HPLC. Spectroscopic examination identified it as **25**, one of the two polyketides involved in S1 **10** biosynthesis.

The fusarin synthase

Fusarin C (**26**, Fig. 3.10) is a mycotoxin produced by *Fusarium moniliforme* (and other fusaria), a filamentous fungus involved in stem- and crown-rot diseases of cereals. The key chemical moiety is the core five-membered nitrogen-containing ring. Biosynthetic feeding studies suggest that it is derived from a polyketide, and probably an amino acid (aspartate or homoserine). The polyketide is multiply methylated, and thus the PKS responsible should have a *C*MeT domain. Interestingly, the polyunsaturated nature of the polyketide group suggests that KR and DH

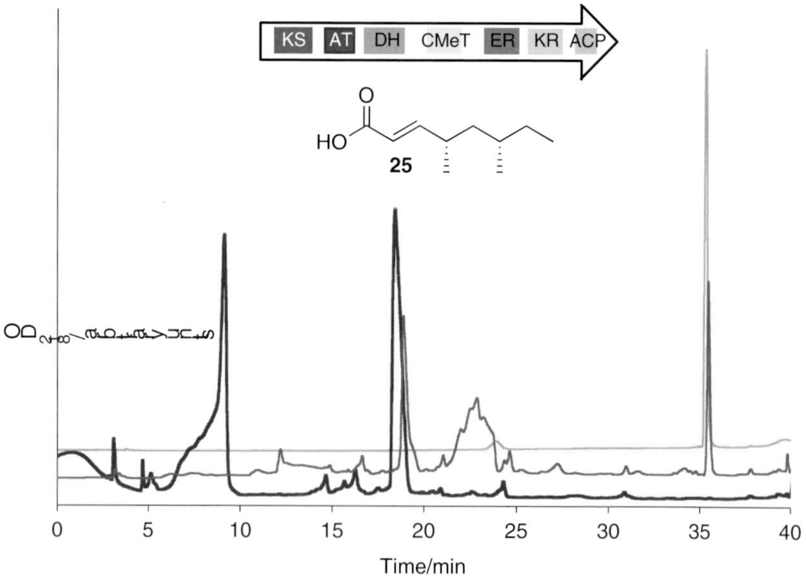

Figure 3.9 The squalestatin tetraketide synthase. HPLC trace showing expression experiments: WT *A. oryzae*; *A. oryzae* pTAex-3PhPKS; purified **25**.

domains must be present, but that the ER domain may be missing or defective.

We used the same strategy as used with the S1 **10** synthase: use of cDNA libraries generated during rapid fusarin C production and use of different producing strains – in this case *Fusarium moniliforme* and *Fusarium venenatum* – with the *C*MeT probe (Song *et al.*, 2004). This, once again, led to the rapid isolation of homologous clones from the genomic and cDNA libraries from both organisms. In this case, however, the gene isolated was significantly larger than the typical 8 Kb of most other fungal PKS genes. In this case a 12 Kb ORF was identified which contained all the typical PKS catalytic domains, but which also, at the 3′ end, contained a single non-ribosomal peptide synthase (NRPS) module consisting of condensation (C), adenylation (A), thiolation (T) and reduction (R) domains (Fig. 3.10).

In the PKS portion, there is an interesting story surrounding the ER activity. From the structure of fusarin C it is clear that ER must be inactive. However, in the gDNA sequence there appear to be two domains homologous to typical fungal PKS ER domains. However, examination of the cDNA sequence suggests that one of these domains is lost as part

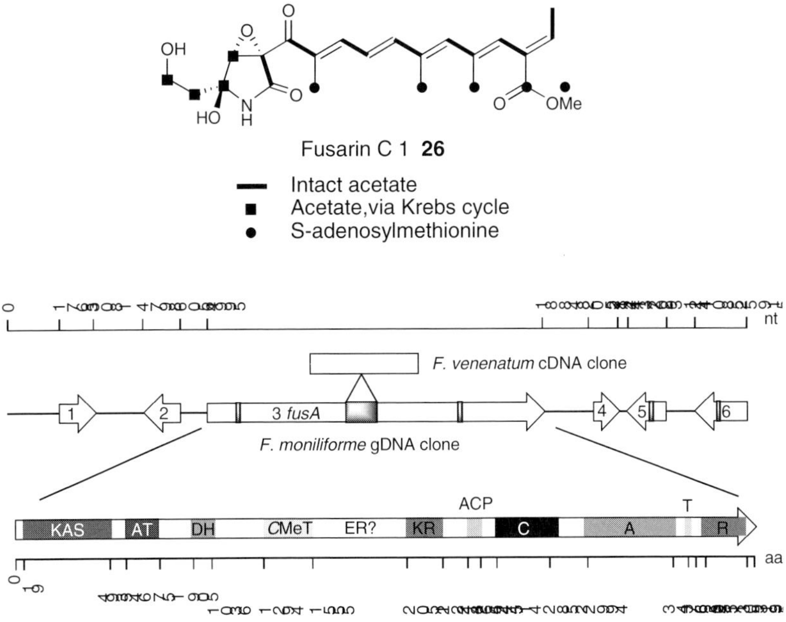

Figure 3.10 The fusarin biosynthetic gene cluster in *F. moniliforme*.

of a large (546 bp) intron, while the other remaining ER domain shows likely deleterious mutations. In this case we were successful in a directed knockout experiment and we showed that strains in which the PKS-NRPS was interrupted with a hygromycin cassette were incapable of making fusarin.

It appears likely that the iterative Type I PKS generates a tetramethylated polyunsaturated heptaketide **27** as its end product (Fig. 3.11). This is consistent with the active sites present in the PKS portion of FUSS. The likely inactivity of ER, supported by sequence analysis, corresponds with the polyene structure. We propose that the fully elaborated polyketide is then held by the ACP domain of the FUSS PKS as a thiolester.

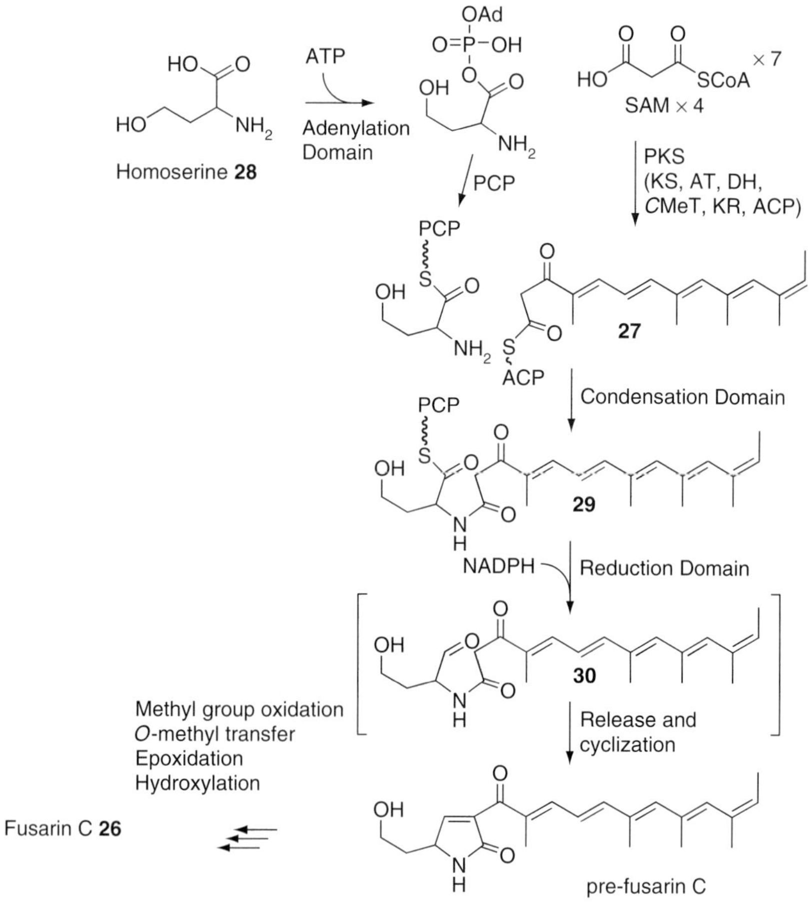

Figure 3.11 Reactions catalyzed by the fusarin PKS-NRPS.

Homoserine **28** is then adenylated and transferred to the thiolation domain (also known as peptidyl carrier protein, PCP) of the NRPS module where it is attached as another thiolester. At this stage the condensation domain catalyzes nucleophilic attack by the homoserine nitrogen on the ACP-bound polyketide and the polyketide is formally transferred to the NRPS portion of FUSS, forming an intermediate amide **29**.

The final biosynthetic step involves reduction of the PCP-bound thiolester by the NRPS reductive domain. In this case, the outcome of the reduction would be a free aldehyde **30** and such compounds are chemically susceptible to intramolecular aldol reactions. We have no evidence for a catalyzed cyclization, although this cannot be ruled out. This reductive mechanism obviates the requirement for a thiolesterase domain, which is present in other NRPS modules.

The tenellin synthase

Most recently we have examined the biosynthesis of the 2-pyridone tenellin **31** by *Beauveria bassiana*, an entopaphogenic filamentous fungus used in commercial insect pest control. Tenellin is a bright-yellow compound and is again derived from a mixed polyketide amino-acid pathway (Fig. 3.12). The core nitrogen-containing six-membered ring may be formed via a five-membered ring analogous to that of fusarin, formed from either phenylalanine or tyrosine. In this case we were more confident that our probing strategy using the *C*MeT probes would be successful and we used a single gDNA library formed from *B. bassiana* alone. PCR using the degenerate *C*MeT primers afforded a single product that was used as the probe. A single large PKS fragment (12 Kb) was isolated from the library containing the same domains as the fusarin synthase. Reconstruction *in silico* revealed a single large (c. 12 Kb) ORF showing high end-to-end homology to the fusarin PKS-NRPS. However, close sequence analysis, and examination of intron positions (by RT-PCR from mRNA preparations) showed some differences. Firstly, although a clear R domain appeared in the genomic sequence, preliminary sequence analysis of both gDNA and cDNA revealed an in-frame stop codon prior to this sequence, as well as a frame-shift. Secondly, homology between this ORF and the fusarin synthase appeared lower in the region of the ER domain – unsurprising since the fusarin synthase ER is inoperative, while the tenellin ER must be functional (Fig. 3.13).

Once again, the relationship between the domains present in the synthase and the chemistry that must occur is maintained. Here we propose that the tenellin synthase PKS component builds up a doubly methylated pentaketide, which is then condensed with phenylalanine (or tyrosine) by the NRPS C domain to give an intermediate amide **32**. This is not reduced, but cyclized and released to afford the acyl tetramic acid **33** as the first synthase-free intermediate. This must then undergo oxidative ring expansion to give **34**, and also *N*-hydroxylation and *para*-hydroxylation (if phenylalanine is the precursor) to finally form **31**. Two P_{450} genes are present close to the tenellin synthase and these could be involved in these processes. In the case of the tenellin synthase, experiments to achieve knockout or expression are ongoing, but the chemical evidence suggesting that this ORF is part of the tenellin synthase appears strong.

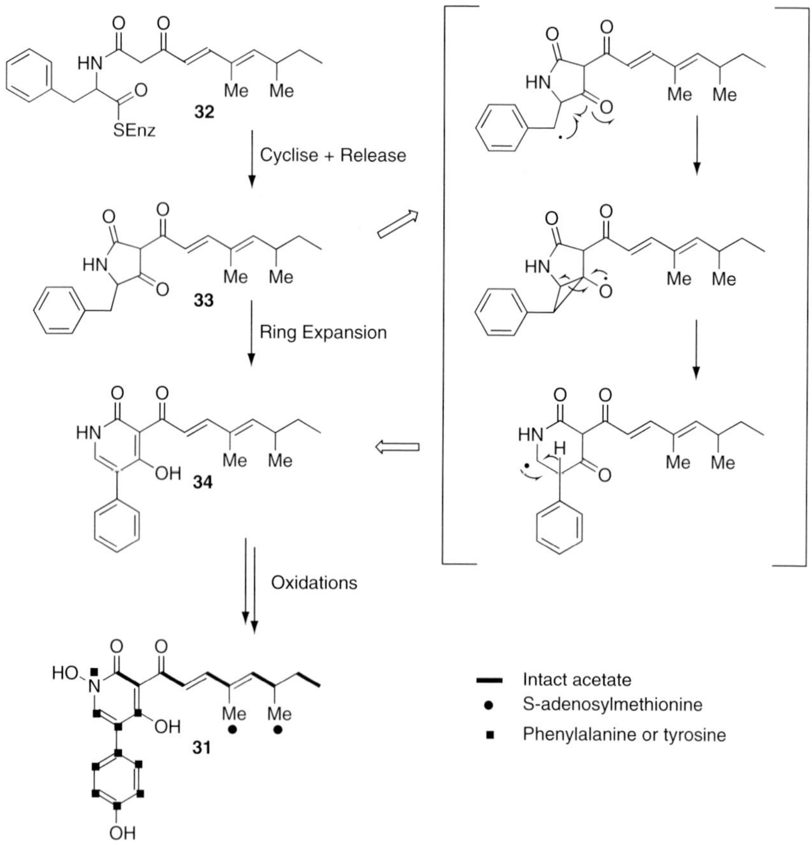

Figure 3.12 The predicted post synthase reactions leading to the tenellin skeleton. In square brackets is a possible radical mediated ring expansion which could be catalyzed by cytochrome P_{450} enzymes.

Future perspectives

Genomics approaches are now discovering polyketide and non-ribosomal peptide synthases much more rapidly than the gene probing approaches discussed above. However, this methodology is currently limited in scope due to cost and at present there is still a requirement for directed gene probing methods in particular species associated with particular compound production. The major challenge, which is brought into sharper relief by the numbers of synthases discovered through genomics, is that it is not possible to predict the structure of the compound produced by any given fungal biosynthetic gene cluster. While it is possible to predict whether a polyketide will belong to the unreduced, partially reduced or highly reduced families, it is not possible to predict chain length, methylation and reduction patterns or cyclizations. In the case of the post PKS genes, general classes can be predicted, such as enoyl reductase or P_{450}, but their exact specificity cannot be predicted. A case in point is the ace1 gene cluster of *Magnaporthe grisea* where an ORF very similar to the fusarin and tenellin synthases has been discovered. It is clear that the product of the ace1 protein is involved in avirulence signalling between rice plants and the fungal host, but the chemical structure cannot be predicted accurately. Thus, there is no strong link between the possession of genetic sequence and chemical structure. Expression and directed knockout strategies, as described above for SQTKS and the fusarin synthase will therefore remain of key importance.

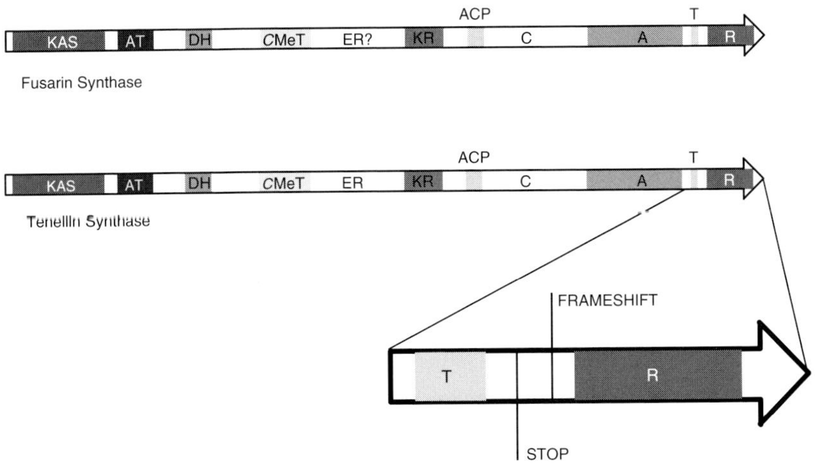

Figure 3.13 Comparison between the fusarin C and putative tenellin synthases.

Acknowledgements

The work described here has been funded, in part, by EPSRC, BBSRC, Glaxo and the EU. The work carried out in Bristol could not have been completed without the participation of the following coworkers and students: Lewis Bingle, Tom Nicholson, Frank Glod, Ying Zhang, Deirdre Hurley, Kate Harley, Dr Song Zhongshu, Dr Kirstin Eley, Dr Colin Lazarus, Dr Andy Bailey and Professor Tom Simpson FRS.

References

Bingle, L. E. H., Simpson, T. J. & Lazarus, C. M. (1999). Ketosynthase domain probes identify two subclasses of fungal polyketide synthase genes. *Fungal Genetics and Biology*, **26(3)**, 209–23.

Cox, R. J., Glod, F., Hurley, D., Lazarus, C. M., Nicholson, T. P., Rudd, B. A. M., Simpson, T. J., Wilkinson, B. & Zhang Y. (2004). Rapid cloning and expression of a fungal polyketide synthase gene involved in squalestatin biosynthesis. *Chemical Communications*, **20**, 2260–1.

Kennedy, J., Auclair, K., Kendrew, S. G., Park, C., Vederas, J. C. & Hutchinson, C. R. (1999). Modulation of polyketide synthase activity by accessory proteins during lovastatin biosynthesis. *Science*, **284**, 1368–673.

Minto R. E. & Townsend C. A. (1997). Enzymology and molecular biology of aflatoxin biosynthesis. *Chemical Reviews*, **97(7)**, 2537–55.

Nicholson, T. P., Rudd, B. A. M., Dawson, M., Lazarus, C. M., Simpson, T. J. & Cox R. J. (2001). Design and utility of oligonucleotide gene probes for fungal polyketide synthase. *Chemistry and Biology*, **8(2)**, 157–78.

Proctor, R. H., Desjardins, A. E., Plattner, R. D. & Hohn, T. M. (1999). A polyketide synthase gene required for biosynthesis of fumonisin mycotoxins in *Gibberella fujikuroi* mating population A. *Fungal Genetics and Biology*, **27(1)**, 100–12.

Simpson, T. J., (1995). Polyketide biosynthesis. *Chemistry and Industry*, **11**, 407–11.

Song, Z. S., Cox, R. J., Lazarus, C. M. & Simpson, T. J. (2004). Fusarin C biosynthesis in *Fusarium moniliforme* and *Fusarium venenatum*. *ChemBioChem*, **5(9)**, 1196 203.

Staunton, J. & Weissman, K. J. (2001). Polyketide biosynthesis: a millennium review. *Natural Product Reports*, **18(4)**, 380–416.

Wilkinson, B., Foster, G., Rudd, B. A. M., Taylor, N. L., Blackaby, A. P., Sidebottom, P. J., Cooper, D. J., Dawson, M. J., Buss, A. D., Gaisser, S., Bohm, I. U., Rowe, C. J., Cortes, J., Leadlay, P. F. & Staunton, J. (2000). Novel octaketide macrolides related to 6-deoxyerythronolide B provide evidence for iterative operation of the erythromycin polyketide synthase. *Chemistry and Biology*, **7(2)**, 111–17.

Yang, G., Rose, M. S., Turgeon, B. G. & Yoder, O. C. (1996). A polyketide synthase is required for fungal virulence and production of the polyketide T-toxin. *The Plant Cell*, **8(11)**, 2139–50.

4

Fungal metabolites as lead structures for agriculture

T. ANKE AND E. THINES

Introduction

In order to meet an increasing food and feed demand for a rising human population agricultural production must be increased, in which crop protection plays a key role. For many decades this field has been the domain of inorganic and synthetic organic chemistry. Many pests could be effectively controlled but high toxicity for humans and non-target organisms as well as in some cases a prolonged persistence in the environment, have made a search for less toxic and more environmentally safe compounds mandatory. The emergence of resistant pathogens, especially fungi, has added to the problem making an intensive search for agrochemicals with new target sites necessary. Today, the standards to be met by new agrochemicals are very high, especially with regard to efficiency and ecological safety. The price has to be competitive and, in general, much lower compared to pharmaceutical compounds. In addition, the active ingredients have to be produced in very large quantities. In 2004, 85 000 tons of agrochemicals were produced in Germany, among them 37 000 tons of fungicides (Hübenthal, 2005).

Natural products in plant protection

Natural products obtained from terrestrial and marine microorganisms, plants and animals are important sources of new chemical structures with biological activities useful for medicine or agriculture. In human medicine the natural products themselves, or semi-synthetic derivatives, play key roles in antibiotic therapy, organ transplantation, cancer therapy and other important fields. In the field of plant protection, their roles are much less conspicuous.

Exploitation of Fungi, eds. G. D. Robson, P. van West and G. M. Gadd. Published by Cambridge University Press.

While natural products, being part of the environment, are advantageous with regard to ecological safety, their major drawback is the production cost and the availability of the large amounts required. Thus, the number of plant protectants directly produced by extraction of plants or fermentation of microorganisms is scarce. Examples are the antibiotics blasticidin S, kasugamycin and validamycin A, all derived from streptomycetes, which have been used in the control of rice blast caused by *Pyricularia oryzae* (*Magnaporthe grisea*) and sheath blight caused by *Rhizoctonia solani*. Two more recent examples that have entered the market are the insecticides avermectin and spinosad, also produced by fermentations of actinomycetes. Both compounds exhibit low toxicities towards mammals and offer environmental benefits. Commercially successful fungicides have been developed based on the structure pyrrolnitrin, originally isolated from cultures of *Pseudomonas pyrrocinia* (Arima *et al*., 1965). Fenpiclonil is applied as seed dressing, while fludioxonil is applied both as a seed dressing and as a foliar spray against *Botrytis cinerea*. The mode of action of both compounds has been the subject of several investigations. It was suggested that both compounds have a new mode of action interfering with membrane transport (Stevens & Breckenridge, 2001).

Fungal metabolites as lead compounds for the chemical synthesis of agricultural fungicides

How do fungal metabolites fit into this scenario? Up to now there are no fungal metabolites being used as plant protectants that are produced by fermentation. However, as has been the case with other natural compounds, valuable leads have been obtained from the extremely rich diversity of their structures and biological activities. The most prominent examples are the strobilurins, which by now have become a major class of agricultural fungicides.

It all started when a search for new antibiotics from Basidiomycetes resulted in the discovery of two new antifungals, named strobilurins A (**1**) (Fig. 4.1) and B from cultures of *Strobilurus tenacellus* (Anke *et al*., 1977, for review see Anke & Steglich, 1999).

Soon after, in a search for the similar but optically active mucidin (Vondracek *et al*., 1970), the chiral oudemansin A (**3**) together with the achiral strobilurin A was detected in fermentations of *Oudemansiella mucida* (Anke *et al*., 1979). It was only much later that the identity of strobilurin A and mucidin, which is in fact achiral, was unequivocally proven (von Jagow, Gribble & Trumpower *et al*., 1986). Syntheses of strobilurin A as well as analogues and derivatives (e.g. **4** and **5**) (Fig. 4.2) started soon after in W. Steglich's group in Bonn and identified the *E*-ßmethoxyacrylate group as the toxophore with the possibility to exchange the enolether group to the corresponding oximether (Schramm, 1980; Anke *et al.*, 1988).

The strobilurins and oudemansins are potent inhibitors of fungal respiration. Their lack of toxicity towards mammals is surprising because the target within their mitochondrial respiratory chain, the ubihydroquinone oxidation (Q_p) centre of the bc_1 complex (Becker *et al.*, 1981), is present and sensitive. One reason might be a rapid excretion and metabolism to non-toxic products like the free acid as has been described for kresoxim methyl (**12**). This is also the case in strobilurin-resistant *Penicillium* spp. (Kettering, Sterner & Anke, 2004) which readily hydrolyse strobilurin A to the non-toxic acid (**6**) (Fig. 4.3). On the other hand, myxothiazol, an

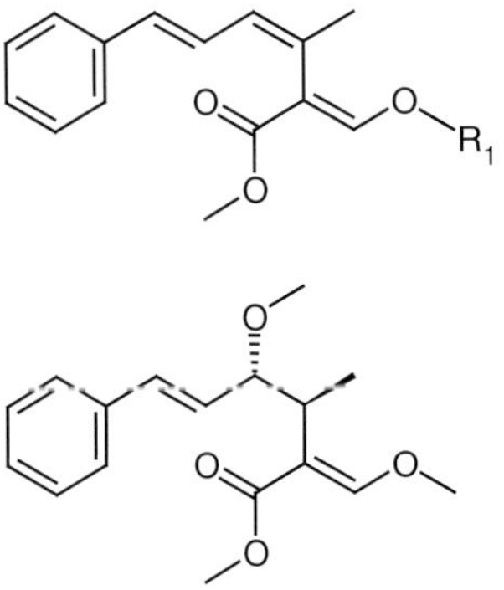

Strobilurin A **(1)** $R_1 = CH_3$

Strobilurin A acid **(2)** $R_1 = H$

Oudemansin A **(3)**

Figure 4.1 Structures of strobilurin A (**1**), strobilurin A acid (**2**) and oudemansin A (**3**).

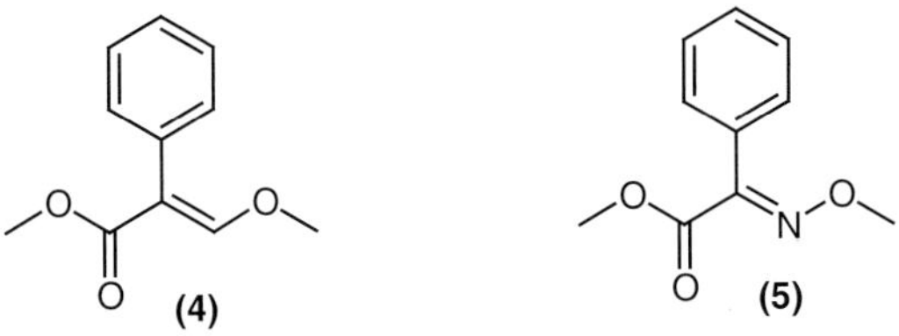

Figure 4.2 Synthetic analogues of strobilurin A.

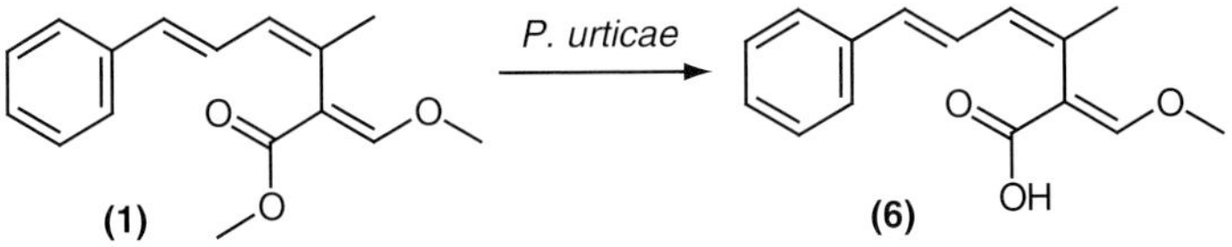

Figure 4.3 Hydrolysis of strobilurin A to give the non-toxic acid by *Penicillium urticae*.

antifungal antibiotic isolated from myxobacteria and sharing the same mode of action, has been found to be highly toxic to mammals (Gerth *et al.*, 1980).

S. tenacellus is resistant against its own product due to the exchange of three amino acid residues in positions 127, 254 and 261 within the strobilurin binding site of cytochrome b (Kraiczy *et al.*, 1996). The strobilurins are produced by a great number of Basidiomycetes growing in temperate and tropical climates all over the world. Their structures and antifungal activities, but not their mode of action, vary considerably from the simple strobilurin A to the elaborate strobilurins E, L, M and N (**7–11**) (Fig. 4.4)

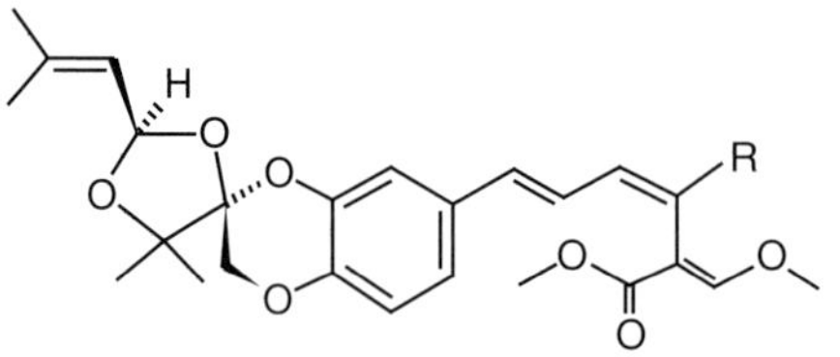

Strobilurin E (**7**) R = H

9-Methoxystrobilurin E (**8**) R = OCH_3

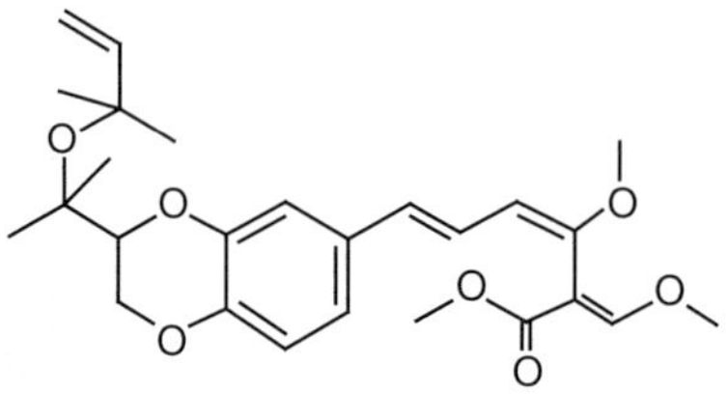

9-Methoxystrobilurin L (**9**)

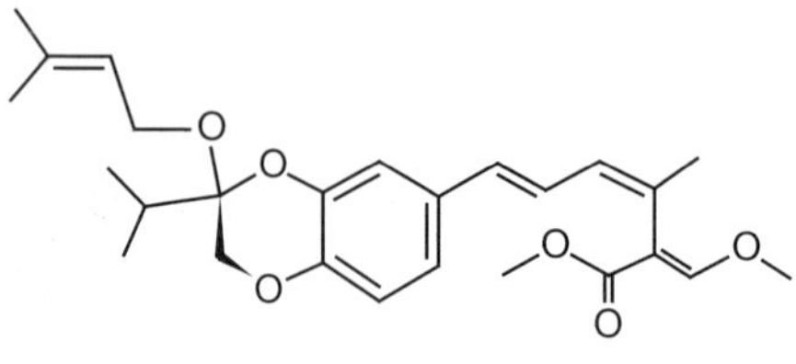

Strobilurin M (**10**)

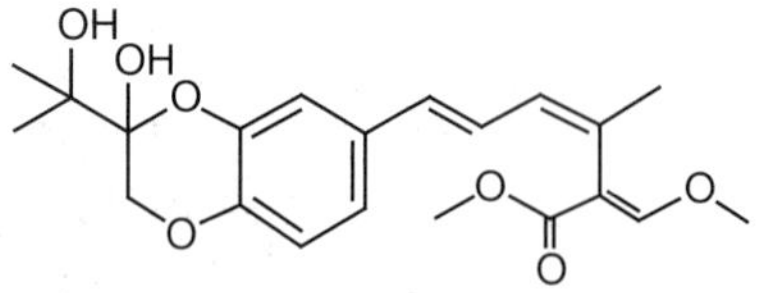

Strobilurin N (**11**)

Figure 4.4 Structures of the strobilurins E (**7**), M (**10**), N (**11**) the 9-methoxystrobilurins E (**8**) and L (**9**).

(Anke & Steglich, 1999). It is interesting that strobilurin production increases in the presence of competing fungi and that these can react with an enhanced production of another secondary metabolite (Kettering *et al.*, 2004). Since these findings were observed in cultures with only two partners, it is easily conceivable that, under natural conditions where a multitude of different organisms share the same substrate and habitat, a very complex chemical communication with primary and secondary metabolites takes place.

Synthetic strobilurin fungicides

The first patent for medical use of strobilurin A as antimycotic was filed by Hoechst in 1980. In 1981, first ICI and two years later BASF became interested in testing the compounds for potential uses as agricultural fungicides. While ICI were given a small sample of oudemansin, BASF started a close cooperation with our and Steglich's group beginning with a strobilurin A preparation and synthetic analogues. Both oudemansin and strobilurin A showed positive effects against a variety of phytopathogenic fungi; however, light instability turned out to be a problem especially for the strobilurins because of their triene side chain. This was soon solved by similar synthetic approaches both by BASF and ICI and followed a close competition for patents and structures (for review see Sauter, Steglich & Anke, 1999). The first products entering the market in 1996 were kresoxim methyl (BASF) (**12**) and azoxystrobin (Zeneca) (**13**) (Fig. 4.5) and the synthetic strobilurins have since developed into one of the major classes of fungicides, which only recently saved the South American soja crop from the Asian soja rust.

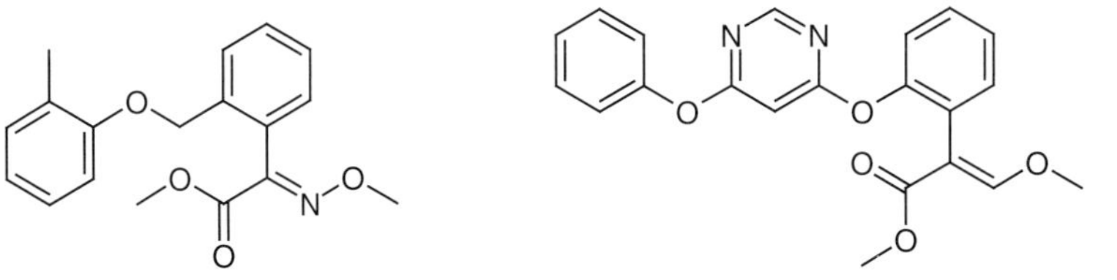

Kresoxim methyl (BASF AG) **(12)** Azoxystrobin=ICIA5504 (Zeneca) **(13)**

Figure 4.5 Kresoxim methyl **(12)** and azoxystrobin **(13)**, the first commercial strobilurin fungicides that entered the market.

Fungal antibiotics leading to new targets for fungicides

The existing or emerging resistance to commonly used agricultural fungicides offers one of the major challenges for modern fungicide research. One promising approach to overcome this problem is the identification of novel targets suitable for the development of selective plant protectants. These can be identified by a genome wide search for genes/proteins essential for the development or survival of a pathogen, the development of test systems and a subsequent screening for substances interfering with the newly found target. Inhibitors might be derived either by chemical synthesis or probing the pool of natural products. On the other hand, many new targets have been identified through mode of action studies of new antifungals. The latter will be discussed here, focusing on fungal secondary metabolites.

A search for new and less toxic antimycotics suitable for human or veterinary medicine resulted in two interesting leads.

The sordarins

Sordarin (**14**) (Fig. 4.6) was originally isolated from *Sordaria arenosa* and later together with derivatives from a number of other Ascomycetes including *Zopfiella marina*, *Hypoxylon croceum*, *Graphium putredinis*, *Xylaria longipes* and *Penicillium minioluteum* (for review see Anke & Erkel, 2002). The sordarins show remarkably high and selective activities against the human pathogens *Candida albicans, Cryptococcus neoformans, Histoplasma capsulatum* and *Coccidioides immitis* both in vitro and in vivo. The precise target of the sordarins within fungal protein biosynthesis has been investigated with sordarin derivatives (Capa *et al.*, 1998; Justice *et al.*, 1998). The sordarins and related compounds selectively interfere with fungal protein biosynthesis by interacting with elongation factor EF-2 and a site of the ribosome. Although sordarins like xylarin (**15**)

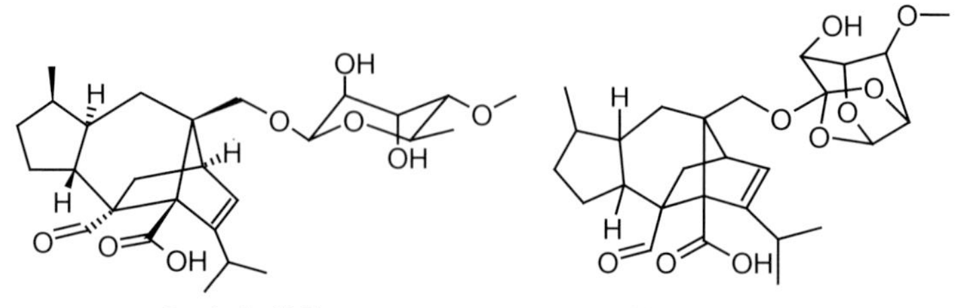

Sordarin (**14**) Xylarin (**15**)

Figure 4.6 Structures of sordarin (**14**) and xylarin (**15**), inhibitors of fungal protein biosynthesis.

act on plant pathogens such as *Ustilago* sp. (Schneider, Anke & Sterner, 1995), few leads for plant protectants have been found within this class, most likely because of their narrow spectrum of activity, which is largely restricted to yeasts or fungi with a yeast-like morphology, and due to their structural complexity, making synthetic approaches quite challenging.

Echinocandins and pneumocandins

The biosynthesis of the fungal cell wall and its components offers a variety of targets for highly selective antibiotics not toxic to plants and animals. Three enzyme activities are essential for fungal cell wall assembly: chitin synthases, (1,6)ß-glucan synthases and (1,3)ß-glucan synthases. While a variety of inhibitors of chitin synthases (polyoxins, nikkomycins) have been described from streptomycetes, the most effective inhibitors of (1,3)ß-glucan synthases have, to date, been derived from fungi (Debono & Gordee, 1994; Selitrennikoff, 1995; Current *et al.*, 1995). The producers are *Aspergillus* spp., *Zalerion arboricula* and *Coleophoma empetri*. The echinocandins and pneumocandins are composed of a cyclic peptide acylated with a long chain fatty acid. Of these, a semi-synthetic pneumocandin A_0 (**16**) derivative, caspofungin (Merck & Co, **17**), has been used in human therapy since 2001 while micafungin (Fujisawa) entered the market in 2002 (Fig. 4.7).

It is interesting that the acyl-moieties of the semi-synthetic echinocandins are decisive for their activities towards filamentous fungi. In theory, this could be exploited to optimize echinocandins for use against plant pathogens. However, an economically feasible synthesis would be a

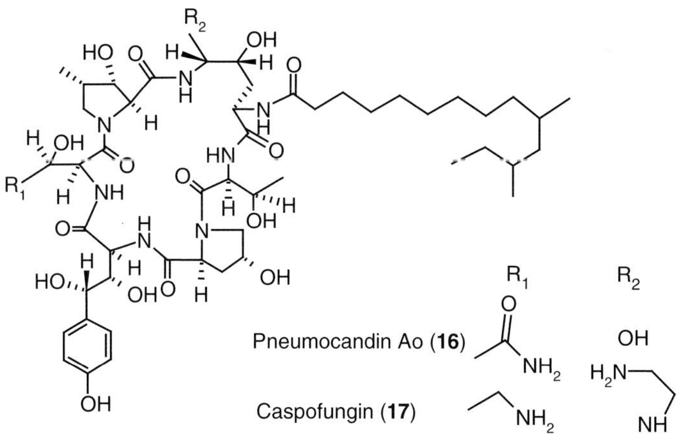

Figure 4.7 Structures of pneumocandin Ao (**16**) and caspofungin (**17**).

prerequisite although the target enzymes and test systems are available to screen libraries of synthetic chemicals or natural products for new leads.

Fungal antibiotics as inhibitors of infection-related processes in phytopathogenic fungi

Fungicidal compounds can be classified into those interfering with fungal vegetative growth, those preventing the pathogen from entering the host plant, and a third group that inhibits the final reproductive steps of development. In the past, natural products and chemicals have predominantly been screened in assays based on the detection of inhibition of fungal vegetative growth. However, screening approaches for compounds interfering exclusively with the pre-penetration phase of phytopathogenic fungi or their reproductive stages may in the future provide lead structures for novel plant protectants. Such compounds could interfere with conidial attachment, germination, germ tube formation, production and functioning of infection structures, or production of infective spores. Fungicides interfering exclusively with pre-penetration processes (appressorium function), notably inhibitors of dihydroxynaphthalene-melanin biosynthesis, have been used to protect rice plants against *Pyricularia oryzae* for approximately 30 years.

For foliar pathogens infecting plants from an air-borne spore inoculum, attachment to the leaves appears to be crucial for the successful colonization of the host. Therefore compounds interfering with adhesion should be useful as protective agents against fungal infection. To date, no fungal metabolites with such an activity have been obtained. However, the plant secondary metabolite zosteric acid has been identified as an inhibitor of attachment in *M. grisea* and *Colletotrichum lindemuthianum* (Stanley *et al.*, 2002), indicating that this early process in the invasive life-cycle of pathogenic fungi may be a target for plant protection.

In contrast, numerous inhibitors of conidial germination have been isolated from submerged cultures of fungi. Conidial germination is triggered in many plant pathogens by the physical or chemical constitution of the micro-environment and/or the loss of autoinhibitors. Several autoinhibitors from spores of phytopathogenic fungi are known to prevent conidial germination until they have been diluted or washed out of the spore. Some of these fungal metabolites are very effective, for example methyl-*cis*-3,4-dimethoxycinnamate, which prevents germination at concentrations in the pg ml^{-1} range (Staples, 2000). From *Colletotrichum* spp. mycosporine-alanine, gloesporone, (*Z*)- and (*E*)-ethyldiene-1,3-dihydroindole-2-one have been characterized as autoinhibitors, of which the latter two compounds prevent conidial germination also in other fungal species (Tsurushima *et al.*, 1995). Such compounds indicate that effective inhibitors can be found or developed that interfere with an essential developmental step in many fungal pathogens. Detailed knowledge of the physiological mechanisms underlying autoinhibition of germination may identify new targets for fungicides.

Several specific inhibitors of conidial germination not affecting vegetative mycelial growth in *M. grisea* have recently been identified. Flaviolin, a polyketide and a potential shunt product of the DHN-melanin biosynthesis, has been isolated from cultures of an unidentified ascomycete. Furthermore, tenuazoic acid from cultures of an endophytic *Cladosporium* sp. (Thines, Anke & Weber, 2004) was identified as an inhibitor of conidial germination of *M. grisea*. Interestingly, the same substance has been identified as a phytotoxin produced by *M. grisea* itself (Lebrun *et al.*, 1988), but has not yet been detected in ungerminated conidia of the fungus.

In a recent screening approach, novel inhibitors of conidial germination were identified from cultures of the Basidiomycete *Coprinus heptemerus*. Four diterpenes were found to interfere with conidial germination. Hepteromone G **(18)** (Fig. 4.8) was found to be the most active compound with protective activity in a rice leaf segment assay using *M. grisea* as pathogen (Kettering *et al.*, 2005).

In phytopathogenic fungi, the formation of infection structures is usually triggered after perception of environmental cues, such as the topography, surface hydrophobicity, or chemicals such as cutin monomers released from the surface of the plant. These signals are transduced in the cells by complex signalling cascades involving cAMP and mitogen-activated protein kinase pathways. The signal perception and the subsequent morphogenetic responses in appressorium differentiation provide targets in all fungi relying on appressoria for infection. One example of a

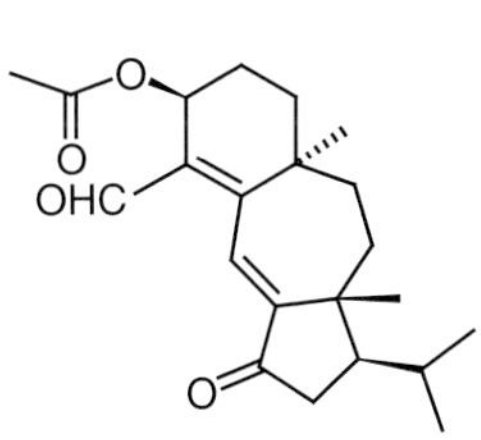

(18)

Figure 4.8 Structure of hepteromone G **(18)**, an inhibitor of conidial germination in *Magnaporthe grisea*.

successful compound interfering with appressorium formation in *Blumeria graminis* is quinoxyfen. This synthetic plant protectant, used against powdery mildew, inhibits the formation of the infection structure and host penetration without affecting mycelial growth (Wheeler *et al.*, 2000). Several fungal secondary metabolites have been identified as inhibitors of appressorium formation in *M. grisea* (Fig. 4.9). From cultures of *Gliocladium roseum*, four glisoprenins were found to inhibit the formation of the infection structures induced on the hydrophobic surface of GelBond sheets without affecting fungal vegetative growth (Thines *et al.*, 1997a; Sterner *et al.*, 1998). However, none of these compounds showed good protective activity in plant assays on rice or barley with *M. grisea* as pathogen. Additionally, neobulgarones, dimeric anthraquinone derivates from cultures of the Ascomycete *Neobulgaria pura*, have been identified as inhibitors of appressorium formation. Both glisoprenins and neobulgarones failed to inhibit appressorium formation on glass upon induction by cutin monomers or by cAMP, indicating that depending on the surface, the signalling pathways involved act partially independently of each other (Thines *et al.*, 1997b; Eilbert, Anke & Sterner, 2000a).

Several fatty acids have been identified from submerged cultures of fungi as inhibitors of appressorium formation in *M. grisea.* Monosaturated fatty

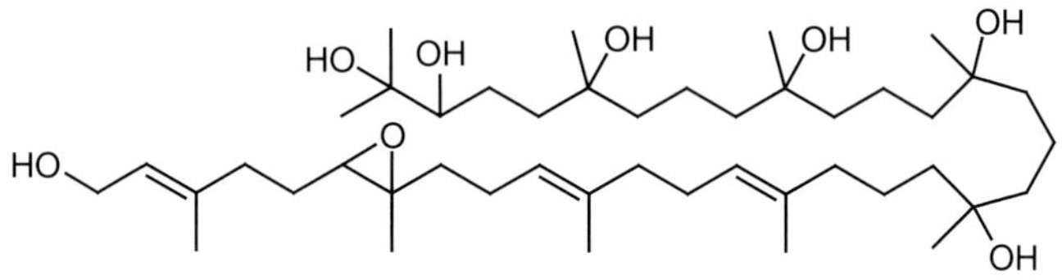

(19)

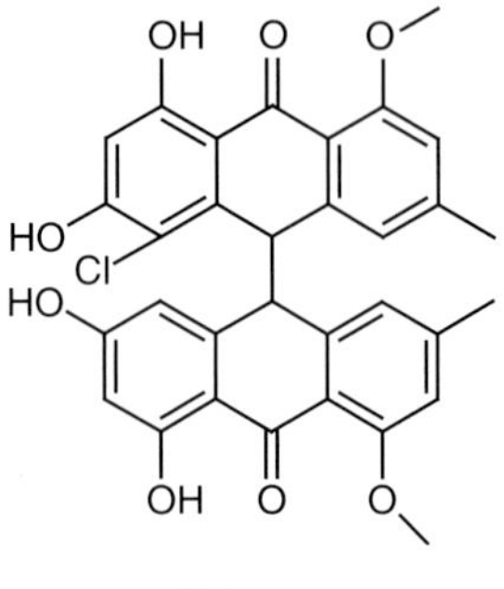

(20)

Figure 4.9 Glisoprenin C **(19)** and neobulgarone D **(20)**, inhibitors of appressorium formation in *Magnaporthe grisea* from natural sources.

acids with a chain length of 16, 18 or 20 carbon atoms showed good activity, amongst which fatty acids with a *cis*-configuration at their unsaturation, such as petroselinic acid and octadecenic acid, were the most effective inhibitors (Eilbert *et al.*, 1999). Germinating conidia treated with these fatty acids failed to form appressoria on hydrophobic surfaces or on glass upon induction with 1,16-hexadecanediol, while the formation of the infection structure on glass with addition of cAMP was not affected.

In contrast, two inhibitors of appressorium formation stimulated with cAMP or the membrane-permeable analogue chlorophenylthio-cAMP, have been identified in cultures of the Basidiomycete *Resupinatus leightonii*. Interestingly, both 4-germacradiene-2,6,12-triol and 1,6-farnesadiene-3,10,11-triol failed to inhibit appressorium formation on hydrophobic surfaces or upon induction with 1,16-hexadecanediol (Eilbert *et al.*, 2000b).

While these fungal metabolites have so far not encouraged the development of agricultural applications, they are valuable tools to study signalling pathways and to identify factors involved in the recognition of the host surface and subsequent differentiation events.

The maturation and functionality of the appressorium is essential for the penetration process in plant pathogenic fungi penetrating their hosts directly by means of a turgor generated within the infection structure. Turgor pressure is generated when a melanin layer is formed which retains the osmolyte inside the appressorium, while water is taken up. Mutants defective in melanin biosynthesis are unable to generate turgor pressure and therefore fail to penetrate the host cuticle. By using simple agar diffusion assays, several natural products have been identified as inhibitors of DHN-melanin biosynthesis (Thines *et al.*, 1995; Thines *et al.*, 1997c). Typical examples of inhibitors are scytalol D, a metabolite isolated from cultures of the anamorphic fungus *Scytalidium* sp. and cerulenin. Scytalol D may inhibit target enzymes of the biosynthetic pathway in a competitive manner because of its high structural similarity to melanin precursors (Thines *et al.*, 1997b), while cerulenin acts at an earlier stage of biosynthesis and is not a specific inhibitor of melanin biosynthesis.

To our knowledge, differentiation *in planta* and sporulation have not yet been examined in any large-scale screening approach. However, compounds selectively interfering with differentiation processes within infection-related morphogenesis have the advantage that they can prevent or cure fungal infections without toxicity against non-target organisms such as insects, animals, plants or benign or neutral fungal species. They would therefore fulfil the selectivity requirements for modern and environmentally friendly fungicides.

Outlook

The development of a new agricultural fungicide is a very costly and time consuming process. The requirements to be met are very high, exceeding even those of antimycotics used in medicine. As can be gleaned from the development of the strobilurins, it can be many years until the potential of a new group of antibiotics is realized. But the strobilurins are also a fine example of how natural products can play a very important role in the detection of new chemical entities and targets, which can then be exploited for the development of highly successful plant protectants. Approaches to improve the lead finding process could take advantage of the now available genomic sequences of plant pathogenic fungi, which allow the identification of genes/proteins essential for growth and development of pathogens but different in or absent from plants and animals. Once an essential target has been identified, it has to be shown that it is 'drugable', meaning amenable to inhibition by low molecular weight compounds. Among these, fungal products will occupy an important position since approximately 30% of all fungi produce antibiotics that have been optimized during evolution for the benefit of the producers in their microenvironment.

References

Anke, T. & Erkel, G. (2002). Non ß-lactam antibiotics. In *The Mycota X. Industrial Applications*, ed. H. D. Osiewacz. Berlin, Heidelberg: Springer Verlag, pp. 93–108.

Anke, T. & Steglich, W. (1999). Strobilurins and oudemansins. In *Drug Discovery from Nature*, eds. S. Grabley, & R. Thiericke. Berlin, Heidelberg: Springer Verlag, pp. 320–34.

Anke, T., Oberwinkler, F., Steglich, W. & Schramm, G. (1977). The strobilurins – new antifungal antibiotics from the basidiomycete *Strobilurus tenacellus* (Pers. ex Fr.) Sing. *Journal of Antibiotics*, **30**, 806–10.

Anke, T., Schramm, G., Steglich, W. & von Jagow, G. (1988). Structure-activity relationships of natural and synthetic *E*-ß-methoxyacrylates of the strobilurin and oudemansin series. In *The Roots of Modern Biochemistry – Energetics of the Cell, Fritz Lipmann's Squiggle and its Consequences*. Amsterdam: W. de Gruyter, pp. 657–62.

Anke, T., Hecht, H.-J., Schramm, G. & Steglich, W. (1979). Antibiotics from basidiomycetes. IX. Oudemansin, an antifungal antibiotic from *Oudemansiella mucida* (Schrader ex Fr.) Hoehnel (Agaricales). *Journal of Antibiotics*, **32**, 1112–17.

Arima, K., Imanaka, H., Kousaka, M., Fukuda, A. & Tamura, G. (1965). Studies on pyrrolnitrin, a new antibiotic. I. Isolation and properties of pyrrolnitrin. *Journal of Antibiotics*, **18**, 201–4.

Becker, W. F., von Jagow, G., Anke, T. & Steglich, W. (1981). Oudemansin, strobilurin A, strobilurin B and myxothiazol: new inhibitors of the bc_1 segment of the respiratory chain with a E-ß-methoxyacrylate system as common structural element. *FEBS Letters*, **132**, 329–33.

Capa, L., Mendoza, A., Lavandera, J. L., de las Heras, G. F. & Garcia-Bustos, J. F. (1998). Translation elongation factor 2 is part of the target for a new family of antifungals. *Antimicrobial Agents and Chemotherapy*, **42**, 2694–9.

Current, W. L., Tang, J., Boylan, C., Watson, P., Zeckner, D., Turner, W., Rodriguez, M., Dixon, C., Ma, D. & Radding J. A. (1995). Glucan biosynthesis as a target for antifungals: The echinocandin class of antifungals. In *Antifungal Agents, Discovery and Mode of Action*, eds. G. K. Dixon, L. G. Copping & D. W. Hollomon. Oxford: Bios Scientific Publishers, pp. 143–60.

Debono, M., Gordee, R. S. (1994). Antibiotics that inhibit fungal cell wall development. *Annual Review of Microbiology*, **48**, 471–97.

Eilbert, F., Thines, E., Anke, H. & Sterner, O. (1999). Fatty acids and their derivates as modulators of appressorium formation in *Magnaporthe grisea. Bioscience, Biotechnology and Biochemistry*, **63**, 879–83.

Eilbert, F., Anke, H. & Sterner, O. (2000a). Neobulgarones A-F from cultures of *Neobulgaria pura*, new inhibitors of appressorium formation in *Magnaporthe grisea. Journal of Antibiotics*, **53**, 1123–9.

Eilbert, F., Engler-Lohr, M., Anke, H. & Sterner, O. (2000b). Bioactive sesquiterpenes from the basidiomycete *Resupinatus leightonii. Journal of Natural Products*, **63**, 1286–7.

Gerth, K., Irschik, H., Reichenbach, H. & Trowitzsch, W. (1980). Myxothiazol, an antibiotic from *Myxococcus fulvus* (Myxobacterales). I. Cultivation, isolation, physico-chemical and biological properties. *Journal of Antibiotics*, **33**, 1474–9.

Hoechst-Patent: DE 3025268 (Priorität: 4. 7. 80) Verfahren zur Synthese von Strobilurin.

Hübenthal, A. (2005). Pflanzenschutz zwischen Nutzen und Risiko. *Nachrichten aus der Chemie*, **7/8**, 735–42.

Justice, C. J., Hsu, M., Tse, B., Ku, T., Baljovec, J., Schmatz, D. & Nielsen J. (1998). Elongation factor 2 as a novel target for selective inhibition of fungal protein synthesis. *Journal of Biological Chemistry*, **273**, 3148–51.

Kettering, M., Sterner, O. & Anke, T. (2004). Antibiotics in the chemical communication of fungi. *Zeitschrift für Naturforschung*, **59c**, 816–23.

Kettering, M., Valdivia, C., Sterner, O., Anke, H. & Thines, E. (2005). Heptemerones A~G, seven novel diterpenoids from *Coprinus heptemerus*: producing organism, fermentation, isolation and biological activities. *Journal of Antibiotics*, **58**, 390–6.

Kraiczy, P., Haase, U., Gencic, S., Flindt, S., Anke, T., Brandt, U. & Von Jagow, G. (1996). The molecular basis for the natural resistance of the cytochrome bc1 complex from strobilurin-producing basidiomycetes to center Qp inhibitors. *European Journal of Biochemistry*, **235**, 54–63.

Lebrun, M. H., Nicolas, L., Boutar, M., Gaudemer, F., Ranomenjanahary, S. & Gaudemer, A. (1988). Relationships between the structure and the phytotoxicity of the fungal toxin tenuazonic acid. *Phytochemistry*, **27**, 77–84.

Sauter, H., Steglich, W. & Anke, T. (1999) Strobilurine: Evolution einer neuen Wirkstoffklasse. *Angewandte Chemie*. **111**, 1416–38; Int. Ed. **39**, 1328–49 (1999).

Schneider, G., Anke, H. & Sterner, O. (1995). Xylarin, an antifungal *Xylaria* metabolite with an unusal tricyclic uronic acid moiety. *Natural Product Letters*, **7**, 309–16.

Schramm, G. (1980). Neue Antibiotika aus Höheren Pilzen. Dissertation, University of Bonn.

Selitrennikoff, C. P. (1995). *Antifungal Drugs: (1,3) ß-Glucan Synthase Inhibitors*. New York: Springer.

Stanley, M. S., Callow, M. E., Perry, R., Alberte, R. S., Smith, R. & Callow, J. A. (2002). Inhibition of fungal spore adhesion by zosteric acid as the basis for a novel, nontoxic crop protection technology. *Phytopathology*, **92**, 378–83.

Staples, R. C. (2000). Research on the rust fungi during the twentieth century. *Annual Review of Phytopathology*, **38**, 49–69.

Sterner, O., Thines, E., Eilbert, F. & Anke, H. (1998). Glisoprenins C, D and E, new inhibitors of appressorium formation in *Magnaporthe grisea* from cultures of *Gliocladium roseum*. *Journal of Antibiotics*, **51**, 228–31.

Stevens, J. T. & Breckenridge, C. B. (2001). Crop protection chemicals. In *Principles and Methods of Toxicology*, ed. W. Hayes. Philadelphia: Taylor & Francis, pp. 564–647.

Thines, E., Daussmann, T., Semar, M., Sterner, O. & Anke, H. (1995). Fungal melanin biosynthesis inhibitors: introduction of a test system based on the production of dihydroxynaphthalene (DHN) melanin in agar cultures. *Zeitschrift für Naturforschung*, **50c**, 813–19.

Thines, E., Eilbert, F., Sterner, O. & Anke, H. (1997a). Glisoprenin A, an inhibitor of the signal transduction pathway leading to appressorium formation in germinating conidia of *Magnaporthe grisea* on hydrophobic surfaces. *FEMS Microbiology Letters*, **151**, 219–24.

Thines, E., Eilbert, F., Sterner, O. & Anke H. (1997b). Signal transduction leading to appressorium formation in germinating conidia of *Magnaporthe grisea*: effects of second messengers diacylglycerols, ceramides and sphingomyelin. *FEMS Microbiology Letters*, **156**, 91–4.

Thines, E., Anke, H. & Sterner, O. (1997c). Scytalols A, B, C, and D and other modulators of melanin biosynthesis from *Scytalidium* sp. 36–93. *Journal of Antibiotics*, **51**, 387–93.

Thines, E., Anke, H. & Weber, R. W. (2004). Fungal secondary metabolites as inhibitors of infection-related morphogenesis in phytopathogenic fungi. *Mycological Research*, **108**, 14–25.

Tsurushima, T., Ueno, T., Fukam, H., Irie, H. & Inoue, M. (1995). Germination self-inhibitors from *Colletotrichum gloeosporioides* f. sp. *jussiaea*. *Molecular Plant-Microbe Interactions*, **8**, 652–7.

Von Jagow, G., Gribble, G. W. & Trumpower, B. L. (1986). Mucidin and strobilurin A are identical and inhibit electron transfer in the cytochrome bc1 complex of the mitochondrial respiratory chain at the same site as myxothiazol. *Biochemistry*, **25**, 775–80.

Vondracek, M., Capkova, J., Slechta, J., Benda, A., Musilek, V. & Cudlin J. Czech Pat 136495 (filed 26. 9. 1969/obtained 15. 5. 1970). Isolierung eines neuen antifungischen Antibiotikums. Bezug: Czech 136492.

Wheeler, I, Hollomon, D. W., Longhurst, C. & Green, E. (2000). Quinoxyfen signals a stop infection by powdery mildews. In *The BCPC Conference – Pests & Disease* 2000, Abstract 8A–4.

5

Molecular and genetic analysis of symbiosis expressed secondary metabolite genes from the mutualistic fungal endophytes *Neotyphodium lolii* and *Epichloë festucae*

B. SCOTT, C. A. YOUNG, A. TANAKA
AND E. J. PARKER

Introduction

Epichloë endophytes (*Epichloë*/*Neotyphodium* species) are an important group of clavicipitaceous fungi that form symbiotic associations (symbiota) with temperate grasses of the Pooideae subfamily (Scott, 2001a; Schardl, Leuchtmann & Spiering, 2004). These biotrophic fungi systemically colonize the intercellular spaces of leaf primordia, leaf sheaths and culms of vegetative tissue and the inflorescence of reproductive tissues. The asexual *Neotyphodium* species have no external growth stage and consequently form symptomless associations with their grass host where they are vertically transmitted through the seed following colonization of the developing ovule. Most of the sexual *Epichloë* species are also transmitted vertically but because of their ability to form external stromata on inflorescence tissue, can also be transmitted horizontally (Leuchtmann & Schardl, 1998; Schardl & Leuchtmann, 1999). The mating system is heterothallic (outcrossing) and in nature is mediated by anthomyiid flies (*Botanophila* spp.), which transfer spermatia between stromata (Bultman *et al.*, 1995; Bultman *et al.*, 1998). Formation of perithecia and release of ascospores into the environment provides a source of vegetative hyphae that give rise to conidia that subsequently infect new hosts, probably by way of colonization of the stigma and style of the host inflorescence (Chung & Schardl, 1997).

Taxonomy of epichloë endophytes

At least nine different *Epichloë* species are recognized including *E. typhina*, a broad host range species (Schardl & Wilkinson, 2000) and

Exploitation of Fungi, eds. G. D. Robson, P. van West and G. M. Gadd. Published by Cambridge University Press.

E. festucae, a natural symbiont of *Festuca* spp. (Leuchtmann, Schardl & Siegel, 1994). *E. festucae* is also capable of forming compatible associations with perennial ryegrass, *Lolium perenne* (Christensen *et al.*, 1997) and one isolate was recently identified as a natural symbiont of this host (Moon *et al.*, 2004). The asexual *Neotyphodium* species are predominantly inter-specific hybrids (heteroploids) that form mutualistic associations with their host (Schardl *et al.*, 1994; Tsai *et al.*, 1994; Moon *et al.*, 2000; Moon *et al.*, 2004). While estimates of genome size and analysis of gene copy number suggest that in some cases most of the ancestral genomes have been retained in the hybrid (e.g. LpTG2) in other cases (e.g. *N. coenophialum*) there would appear to have been substantial reorganization and loss of the ancestral genomes (Kuldau, Tsai & Schardl, 1999; Scott & Young, 2003). Molecular phylogenetic studies reveal that the asexual endophytes are derived from the sexual *Epichloë* species (Schardl *et al.*, 1994; Tsai *et al.*, 1994; Moon *et al.*, 2000; Moon *et al.*, 2004). *N. lolii*, an asexual symbiont of perennial ryegrass derived from *E. festucae*, is one of the few naturally occurring species that appears to be haploid (Christensen *et al.*, 1993; Schardl *et al.*, 1994). However, the very slow growth rate of *N. lolii* in axenic culture and difference in culture morphology to *E. festucae*, would suggest that *N. lolii* is a highly evolved derivative of *E. festucae*.

Benefits of the fungal–grass symbiosis

The major benefits for epichloë endophytes from mutualism are access to nutrients for growth from the intercellular space of the host and a means of dissemination through the seed. Other benefits of endophytism include protection from desiccation and protection from fungal-feeding insects, parasitism by other fungi, such as *Trichoderma* spp. and competition by other microbes. Benefits to the host include protection from animal and insect herbivory (Bacon *et al.*, 1977; Siegel *et al.*, 1990), resistance to nematodes (Kimmons, Gwinn & Bernard, 1990) and some fungal pathogens (Gwinn & Gavin, 1992), drought tolerance (Arachevaleta *et al.*, 1989) and greater field persistence (West *et al.*, 1988; Hill *et al.*, 1990).

Endophyte biological effects

While the molecular basis of many of the biological effects observed for epichloë endophytes are not well understood, the anti-herbivore effects are largely attributable to endophyte synthesis of a range of novel natural products. Metabolites identified to date include the anti-insect compounds, peramine (a pyrrolopyrazine) and loline alkaloids (saturated 1-aminopyrrolizidine alkaloids), and the anti-mammalian ergot alkaloids (e.g. ergovaline) and indole-diterpenes (e.g. lolitrem B) (Bush, Wilkinson & Schardl, 1997) (Fig. 5.1).

Peramine (Fig. 5.1), a potent insect feeding deterrent, is uniquely synthesized by the epichloë group of fungal endophytes in association with their host grasses. Taxonomically, it is the most widely distributed of the epichloë bioprotective metabolites. It has been detected in associations containing *E. typhina*, *E. festucae*, *E. amarillans*, *E. bromicola*, *E. elymi* and many of the asexual derivatives of these species (Lane, Christensen & Miles, 2000; Clay & Schardl, 2002).

Loline alkaloids (Fig. 5.1) have potent anti-insecticidal properties and can accumulate in grass tissue to levels as high as 2% of the plant dry mass (Craven *et al.*, 2001), far in excess of the fungal biomass, which is generally in the range of 0.2 to 1.0% of the plant biomass. With the exception of *N. uncinatum*, lolines have not been detected in axenic cultures of epichloë endophytes. Lolines have been detected in associations between *E. festucae* with *Festuca giganteum*, *N. coenophialum* with *Festuca arundinaceae*, *N. siegelii* and *N. uncinatum* with *Lolium pratense* (Blankenship *et al.*, 2001).

The ergot alkaloids are the most widely recognized group of metabolites, principally because of research carried out with *Claviceps purpurea*, the fungus responsible for ergotism from contaminated rye (Socic & Gaberc-Porekar, 1992). The most abundant metabolite of this class detected in epichloë–grass associations is the ergopeptine, ergovaline (Fig. 5.1).

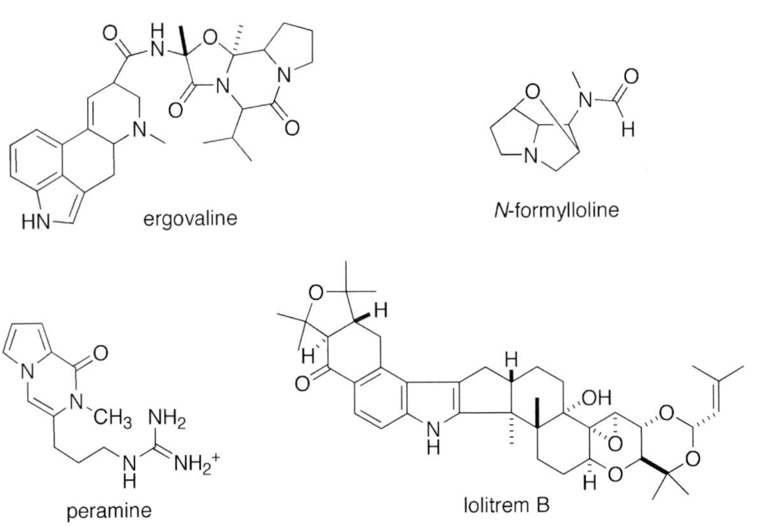

Figure 5.1 Chemical structures of ergovaline, *N*-formylloline, peramine and lolitrem B.

Ergovaline is similar in structure to ergotamine from *C. purpurea* except it contains valine instead of phenylalanine in the second position of the tripeptide. The ergot alkaloids are responsible for 'fescue toxicosis' in cattle (Bacon *et al.*, 1977).

The indole-diterpenes are a structurally diverse group of metabolites (Steyn & Vleggaar, 1985), many of which are potent mammalian tremorgens, because of their ability to interfere with neurotransmitter release (Selala *et al.*, 1991; Knaus *et al.*, 1994). These metabolites have a common structural backbone comprised of a cyclic diterpene derived from geranylgeranyl diphosphate (GGPP) and an indole moiety derived from either anthranilic acid or tryptophan (Acklin, Weibel & Arigoni, 1977; de Jesus *et al.*, 1983; Laws & Mantle, 1989; Byrne, Smith & Ondeyka, 2002). The structural diversity of these metabolites is achieved by additional prenylations, different patterns of ring substitutions and different ring stereochemistry. The most abundant group of these compounds found in seed and leaf tissue of grasses infected with epichloë endophytes are the lolitrems, principally lolitrem B (Fig. 5.1). The lolitrems are proposed to be responsible for 'ryegrass staggers' in sheep (Fletcher & Harvey, 1981).

This review provides an overview of recent developments in our understanding of the biosynthesis and genetics of two of these classes of bioprotective metabolites, peramine and the indole-diterpenes.

Non-ribosomal peptide synthetases in filamentous fungi

Non-ribosomal peptide synthetases (NRPS) are multi-functional enzymes that have a modular structure in which each module is a semi-autonomous unit involved in the recognition, activation and modification of a single amino acid residue of the final peptide (Marahiel, Stachelhaus & Mootz, 1997; von Döhren *et al.*, 1997). The basic structure of each repeating unit comprises an adenylation and thiolation domain for amino acid recognition and activation, respectively. A condensation domain located between two adjacent modules is responsible for catalyzing the formation of a peptide bond between the two adjacent amino acids. Further structural diversity of non-ribosomal peptides is generated by the presence of *N*-methylation, epimerization and reductase domains within modules.

Non-ribosomal peptides are a rich source of therapeutic agents for use in medicine because of their potent biological activity. Some of the best known metabolites are penicillin from *Penicillium chrysogenum* and cyclosporin from *Tolypocladium inflatum* (Tkacz, 2000). While little is known about their biological role in nature, several non-ribosomal peptides have been identified as fungal virulence factors. Sequence analysis of available fungal genomes reveals that filamentous fungi have up to 20 NRPSs (Yoder & Turgeon, 2001; Lee *et al.*, 2005). *Cochliobolus heterostrophus* has eight multimodular NRPS and three monomodular NRPS-like genes (Lee *et al.*, 2005). Deletion of each NRPS in this genome revealed that only one of these genes, NPS6, is required for normal virulence of *C. heterostrophus* on maize. Interestingly, this same gene is required for resistance to hydrogen peroxide, suggesting it may have a key role in nature for protecting the fungus from oxidative stress. Epichloë endophytes are also a rich source of non-ribosomal peptides (Panaccione, 1996; Panaccione *et al.*, 2001; Tanaka *et al.*, 2005). A recent study has shown that the gene for the synthesis of peramine, a potent insect feeding deterrent, has an important biological role in providing protection to the symbiotum from insect herbivory.

Molecular cloning of a peramine synthetase

The structure of peramine (Fig. 5.1) suggests that it is derived from proline and arginine via a diketopiperazine intermediate (Rowan & Gaynor, 1986; Rowan, 1993). This led to the hypothesis that the gene for peramine synthetase would encode a two module NRPS containing a methylation domain (Tanaka *et al.*, 2005). Using a degenerate PCR strategy Tanaka *et al.* (2005) isolated four unique adenylation domain products from *N. lolii* and *E. festucae*. One of these, clone ps9, fulfilled the criteria of a peramine synthetase as expression was upregulated *in planta* and sequences homologous to this clone were present in a selection of peramine producing endophytes. Using this clone as a probe *perA* was isolated from a cosmid library of *E. festucae*. As predicted, this 8.5 kb gene was found to encode a two module NRPS; module one contains an adenylation (A) and thiolation (T) domain and module two contains condensation (C), adenylation (A), methylation (M), thiolation (T) and reductase (R) domains (Fig. 5.2). Perennial ryegrass associations containing an

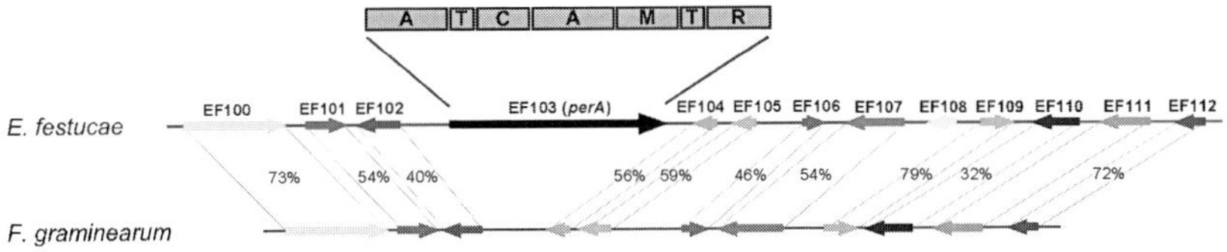

Figure 5.2 Organization of *Epichloë festucae* genes at the *perA* locus and conservation of synteny with *Fusarium graminearum*. The domain structure of *perA* is also shown. Adapted from Tanaka *et al.* (2005) with permission from Blackwell Publishing.

E. festucae mutant deleted for *perA* lacked detectable levels of peramine. Introduction of a wild-type copy of *perA* into the mutant restored the ability of the mutant to synthesize peramine, confirming that *perA* encodes a peramine synthetase. Furthermore, associations containing the *perA* mutant were susceptible to Argentine stem weevil (ASW) (*Listronotus bonariensis*) feeding damage, confirming that peramine is the metabolite responsible for ASW feeding deterrent activity.

No additional candidate genes for peramine biosynthesis were identified adjacent to the *E. festucae perA*. In fact, a comparison of the corresponding genome region in *Fusarium graminearum* demonstrated remarkable conservation of gene structure, order and orientation, with the exception of *perA*, which is absent from the genome of *F. graminearum* (Fig. 5.2). The conservation of synteny between *E. festucae* and *F. graminearum* suggests that the genome sequence of *F. graminearum* may be a particularly useful experimental system for predicting the genome structure of *E. festucae*. The presence of duplicate 12-bp repeats on either side of *perA* may be a 'footprint' of a former recombination event at this locus.

These results taken together led Tanaka *et al.* (2005) to propose that PerA alone is responsible for peramine biosynthesis (Fig. 5.3). To aid formation of the pyrrole ring they proposed that 1-pyrroline-5-carboxylate, rather than proline, is employed together with arginine as the initial substrates. The condensation domain of PerA is proposed to catalyze formation of the peptide bond between 1-pyrroline-5-carboxylate and arginine. The methylation domain of PerA is proposed to catalyze the *N*-methylation

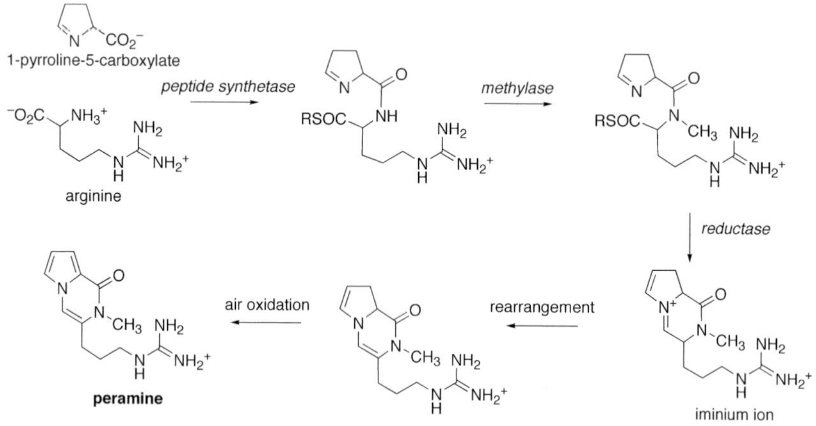

Figure 5.3 A proposed scheme for peramine biosynthesis. Adapted from Tanaka *et al.* (2005) with permission from Blackwell Publishing.

of the amino group of arginine. The reductase domain is proposed to reduce the thioester and cyclize the dipeptide to form an iminium ion that is concomitantly released. Deprotonation of this intermediate and oxidation of the pyrroline ring would give rise to peramine (Fig. 5.3).

The cloning of *perA* from grass endophytes now makes it possible to investigate how this gene is regulated in the plant and determine whether expression of *perA* changes in response to biotic stresses, such as insect herbivory. The availability of a *perA* isogenic mutant will also allow field studies to be carried out to test directly the ecological benefits of this secondary metabolite gene to the symbiotum. It will also be possible to investigate the taxonomic distribution of this gene and the evolution of peramine biosynthesis.

Indole-diterpene chemical diversity in endophytes

Lolitrems are an important subgroup of indole-diterpene metabolites that are relatively abundant in leaf sheath tissue and seeds of perennial ryegrass and some tall fescue (*Festuca arundinancea*) grasses containing *E. festucae*, *N. lolii* and some interspecific hybrids of these and other *Epichloë–Neotyphodium* species. The most abundant indole-diterpene found in the *N. lolii*–perennial ryegrass association is lolitrem B (Fig. 5.1). The link between lolitrem B and the mammalian mycotoxin disorder, 'ryegrass staggers', has led to a systematic analysis of the indole-diterpene profile in *N. lolii* infected perennial ryegrass seed, resulting in the isolation and chemical identification of many different indole-diterpene metabolites, including paspaline (Gatenby *et al.*, 1999), paspaline B (Munday-Finch, Wilkins & Miles, 1996a), 13-desoxy paxilline (Gatenby *et al.*, 1999), terpendole M (Gatenby *et al.*, 1999), lolicine A and B (Munday-Finch, Wilkins & Miles, 1998), lolilline (Munday-Finch *et al.*, 1997), lolitriol (Miles *et al.*, 1992; Munday-Finch *et al.*, 1998), lolitrems A, E, N and F (Gallagher, White & Mortimer, 1981; Miles *et al.*, 1994; Munday-Finch *et al.*, 1995; Munday-Finch *et al.*, 1996b; Munday-Finch *et al.*, 1998). The structural similarity between paxilline, an abundant metabolite of *Penicillium paxilli*, and these diverse indole-diterpenes suggests that the more complex indole-diterpenes found in grass endophytes are derived from either paxilline or proximate precursors of paxilline, such as paspaline.

Penicillium paxilli as a model experimental system

P. paxilli has proved to be a useful experimental system to dissect the biochemistry of indole-diterpene biosynthesis (Young *et al.*, 2001;

McMillan *et al.*, 2003). Using a combination of plasmid insertional mutagenesis combined with chromosome walking, a cluster of genes was isolated from this organism and shown by genetic analysis to be required for paxilline biosynthesis. Although the precise boundaries of the cluster have still to be defined, gene inactivation studies have shown that at least five gene products are required for paxilline biosynthesis (Fig. 5.4). These include *paxG* (encodes a GGPP synthase), *paxC* (prenyl transferase), *paxM* (FAD-dependent monoxygenase), and *paxP* and *paxQ* (cytochrome P450 monooxygenases). PaxG is proposed to catalyze the sequential condensation of dimethylallyl diphosphate (DMAPP) and isopentenyl diphosphate (IPP) to form geranyl diphosphate (GPP), GPP and IPP to form farnesyl diphosphate (FPP), and FPP and IPP to form GGPP (Fig. 5.4). Targeted deletion of *paxG* in *P. paxilli* gave rise to mutants with a paxilline negative phenotype (Young *et al.*, 2001). Interestingly, *P. paxilli* contains an additional GGPP synthase, *ggs1*, which is proposed to have a role in primary isoprenoid metabolism. The complete absence of any indole-diterpene in a *paxG* deletion background indicates that *ggs1* is unable, in vivo, to complement this defect, possibly due to a different cellular localization. The presence of two copies of GGPP synthase in both *Aspergillus flavus* (Zhang *et al.*, 2004) and *Fusarium fujikuroi* (Tudzynski &

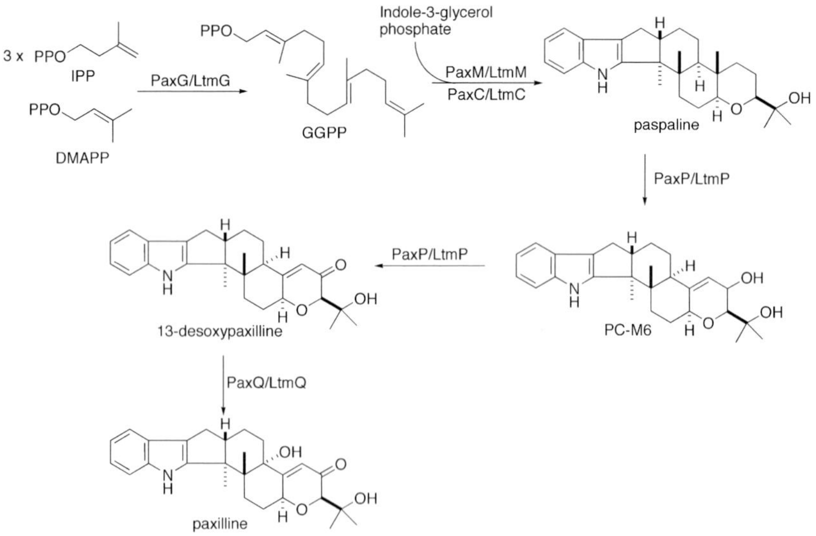

Figure 5.4 A proposed scheme for paspaline and paxilline biosynthesis in *Penicillium paxilli* and *Epichloë festucae*. The enzymes for biosynthesis of paxilline in *P. paxilli* are designated Pax (paxilline biosynthesis) and the orthologues in *E. festucae* Ltm (lolitrem biosynthesis).

Hölter, 1998), fungi that synthesize aflatrems and gibberellins, respectively, suggests that the presence of paralogous copies of GGPP synthase genes in filamentous fungi is a molecular signature for diterpene biosynthesis (Zhang *et al.*, 2004; Young *et al.*, 2005). PaxC and PaxM are proposed to catalyze the addition of indole-3-glycerol phosphate to GGPP and subsequent cyclization to form the first stable indole-diterpene, possibly paspaline (Fig. 5.4) (Parker & Scott, 2004). Although the precise catalytic roles of PaxC and PaxM in paxilline biosynthesis remain to be determined, chemical-feeding experiments with *Emericella* demonstrated that radiolabelled 3-geranylgeranylindole was incorporated into emindole DA and paxilline (Fueki *et al.*, 2004). These results suggest that 3-geranylgeranylindole may be an intermediate in the synthesis of paspaline.

At least two other genes are required for paxilline biosynthesis in *P. paxilli*. Deletion of *paxP* or *paxQ*, both coding for cytochrome P450 monooxygenase enzymes, gave rise to mutants which accumulated paspaline and 13-desoxypaxilline, respectively, confirming that these two genes are essential for paxilline biosynthesis (McMillan *et al.*, 2003). These results suggest that paspaline and 13-desoxypaxilline are the most likely substrates for the corresponding enzymes (Fig. 5.4). This hypothesis can be tested by chemical complementation experiments, where these and other potential intermediates such as paspaline B and PC-M6 (Munday-Finch *et al.*, 1996a), are fed to mutants such as CY2 and LM662 deleted for the entire cluster (Young *et al.*, 1998; Young *et al.*, 2001), but containing introduced copies of *paxP* and *paxQ*, respectively. Analysis of the genes and gene products for paxilline biosynthesis in *P. paxilli* made it possible to isolate *pax* gene orthologues from the grass endophytes, *N. lolii* and *E. festucae*.

Indole-diterpene gene clusters in epichloë endophytes

Using a degenerate PCR strategy, two unique GGPP synthase PCR products were amplified from the genomes of the lolitrem producing endophytes, *N. lolii* and *E. festucae*, whereas a single product was amplified from the lolitrem non-producing strain, *E. typhina* (Young *et al.*, 2005). Molecular cloning and characterization of the two copies of GGPP synthase in *N. lolii* and *E. festucae*, designated *ltmG* (GGPP synthase for lolitrem biosynthesis) and *ggsA* (GGPP synthase for primary metabolism), showed that they corresponded to orthologues of the *P. paxilli paxG* and *ggs1*, respectively. In both endophyte species, *ltmG* is linked to two other genes, *ltmM* (encodes an FAD-dependent monooxygenase) and *ltmK* (cytochrome P450 monooxygenase), to form a cluster of three biosynthetic genes (Fig. 5.5). *ltmM* was shown to be a functional orthologue of *paxM*, whereas *ltmK* appears to be a gene unique for lolitrem biosynthesis. All three *ltm* genes are weakly expressed in axenic culture but strongly expressed *in planta*, a result consistent with the preferential synthesis of indole-diterpene metabolites *in planta*, and indicative of a specific signalling mechanism between the grass host and endophyte.

Flanking this *ltm* gene cluster are large blocks of AT-rich retrotransposon-like DNA (Fig. 5.5). The strong GC-to-AT mutation bias associated with this DNA suggests that these sequences have undergone extensive repeat-induced point mutation (Selker *et al.*, 1987; Cambareri *et al.*, 1989). One of these elements, Rua, was shown to be an abundant and highly dispersed element within the genome of both *N. lolii* and *E. festucae*, but is absent from *E. typhina* (Scott & Young, 2003; Young *et al.*, 2005). The presence of these retrotransposon relics made it very difficult for Young *et al.* (2005) to chromosome walk beyond this region by screening genomic libraries or by inverse PCR. They found that clones representative of these regions were significantly under-represented in the available genomic libraries or when the clones were isolated they showed extensive chromosomal rearrangements.

Deletion of *ltmM* in *E. festucae* strain Fl1 confirmed that this gene is required for lolitrem biosynthesis (Young *et al.*, 2005). Targeted replacements of *ltmM* were obtained at a frequency of 6%, confirming that *E. festucae* is readily amenable to genetic analysis (Scott *et al.*, 2005).

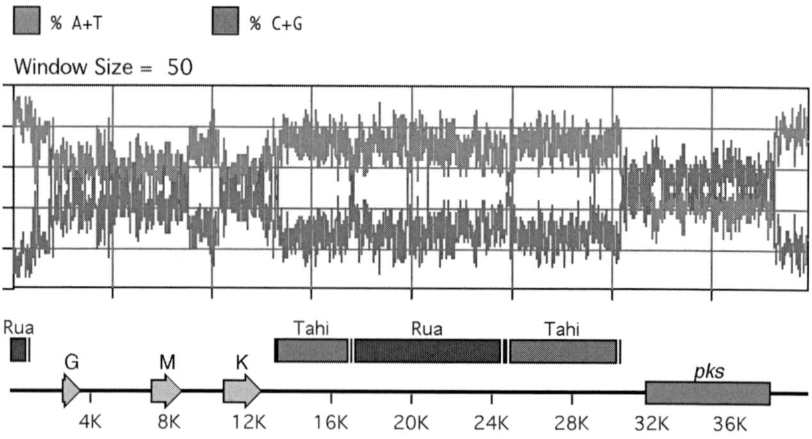

Figure 5.5 Physical organization of cluster one *ltm* genes and associated retrotransposon DNA in *Neotyphodium lolii*. The change in base composition across this *ltm* locus is shown above the map.

Mutants deleted for *ltmM* showed no differences from the wild-type strain in infection rates, hyphal morphology and growth, and relative fungal biomass when inoculated into perennial ryegrass seedlings, suggesting that *ltmM* has no direct role in the interaction between host and endophyte. Endophyte DNA biomass was shown to be approximately 1% of the plant DNA biomass as determined by real-time quantitative PCR (Fig. 5.6). Similar results were obtained for artificial associations between a deletion derivative of the LpTG-2 ergovaline synthetase gene, *lpsA*, and perennial ryegrass (Panaccione *et al.*, 2001). Analysis of plant extracts from associations containing endophyte strains deleted for *ltmM* lacked detectable levels of lolitrems B, A and E and several other UV fluorescent compounds that were present in extracts from wild-type associations (Young *et al.*, 2005). The absence of so many intermediates is consistent with the

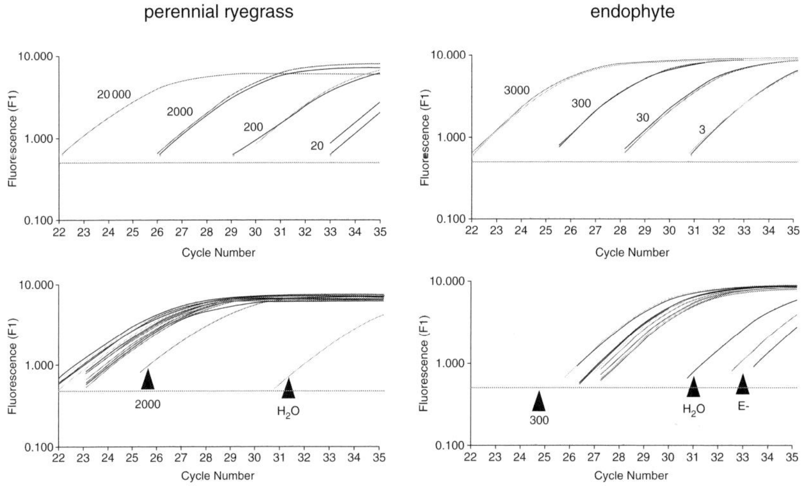

Figure 5.6 Real-time PCR analysis to determine the endophyte DNA biomass *in planta*. Top panel: standard curves for the estimation of endophyte DNA biomass. The standards are shown in pg. The perennial ryegrass standards were amplified with primers designed to the ryegrass β-tubulin gene. The endophyte standards were amplified with primers designed to the endophyte *ggsA* gene (Young *et al.*, 2005). Bottom panel: total DNA was extracted from pseudostems of *E. festucae* infected perennial ryegrass. The DNA biomass of the ryegrass and the endophyte was determined from a 10 ng total DNA sample. All samples were compared to the appropriate standard curves by calibration with an internal standard included in each run. The amplification products for the water (H_2O) and endophyte-free (E-) controls were non-specific as determined by analysis of the corresponding melting curve (data not shown).

proposed role of the *ltmM* gene product as an early step in the biosynthesis of the first stable indole-diterpene intermediate, paspaline (Fig. 5.4).

From a comparison of the structure of paxilline and lolitrem B, together with the knowledge that at least five gene products are required for paxilline biosynthesis in *P. paxilli*, we predict that at least ten gene products are required for the biosynthesis of lolitrem B (Young *et al.*, 2005). Given the difficulties of chromosome walking through AT-rich blocks of degenerate retrotransposon sequence, the alternative approach of DNA suppression subtractive hybridization was used to isolate candidate *ltm* genes on the basis of their homology with previously isolated and characterized paxilline biosynthesis genes from *P. paxilli* (Young *et al.*, unpublished results). Using this approach expressed sequence tags (ESTs) that shared high sequence identity to *paxC* and *paxP* were isolated and shown to map to three distinct contigs. Isolation and mapping of the corresponding genomic regions identified two additional *ltm* gene clusters. The first cluster contains five genes, including candidate orthologues of *paxP* (*ltmP*), *paxQ* (*ltmQ*), *paxD* (*ltmD*) and *paxC* (*ltmC*). The second cluster contains two novel genes that are proposed to be unique to lolitrem biosynthesis. These two new clusters are linked to one another, and to the previously isolated cluster (Young *et al.*, 2005), within a region of approximately 100 kb. Each cluster is separated from the other by a block of highly repetitive AT-rich retrotransposon sequence. These results show that in contrast to *P. paxilli*, the genes for indole-diterpene synthesis in epichloë endophtyes are organized in a very complex gene cluster. By analogy with the genome structure of *Plasmodium falciparum*, this cluster may be located in a rapidly evolving region of the genome such as a chromosomal sub-telomeric region (Gardner *et al.*, 2002).

Exploitation of endophyte indole-diterpene biosynthesis pathways

The molecular cloning of the genes for indole-diterpene biosynthesis in epichloë endophytes now allows elucidation of the likely biochemical function of the corresponding gene products using strategies such as targeted gene replacement in *E. festucae* and complementation of *P. paxilli pax* mutants, by expressing the *ltm* genes under the control of *pax* promoters. By analogy with what is known about paspaline and paxilline biosynthesis in *P. paxilli*, the gene products of *ltmG*, *ltmC* and *ltmM* are proposed to be required for early steps in endophyte indole-diterpene biosynthesis and *ltmP* and *ltmQ* for later steps (Fig. 5.4). However, the diversity of indole-diterpenes found in epichloë endophytes would suggest that the biosynthetic steps beyond paspaline are more likely to comprise a metabolic grid rather than a linear biosynthetic pathway (Gatenby *et al.*, 1999; Lane *et al.*, 2000) (Fig. 5.7). This hypothesis can now be tested by targeted deletion of the various steps in the pathway and analysis of the intermediates that accumulate in perennial ryegrass plants infected with these mutants. While this approach will identify likely substrates for the various gene products, confirmation that they are indeed the substrates will require chemical complementation experiments to be carried out. To date, growth conditions for reliable synthesis of indole-diterpenes in axenic cultures of epichloë endophytes have not been established, making it very difficult to carry out chemical feeding experiments. An alternative approach will be to transfer endophyte *ltm* genes into *P. paxilli* strains deleted for the entire *pax* cluster and use these derivatives for in vivo enzymatic assays to identify products synthesized following addition of various metabolic intermediates. This approach has been very successful for elucidating the steps involved in gibberellin biosynthesis in *F. fujikuroi* (Rojas *et al.*, 2001; Tudzynski *et al.*, 2001; Tudzynski *et al.*, 2002, 2003) and the later steps in paxilline biosynthesis in *P. paxilli* (Saikia, Parker and Scott, unpublished results).

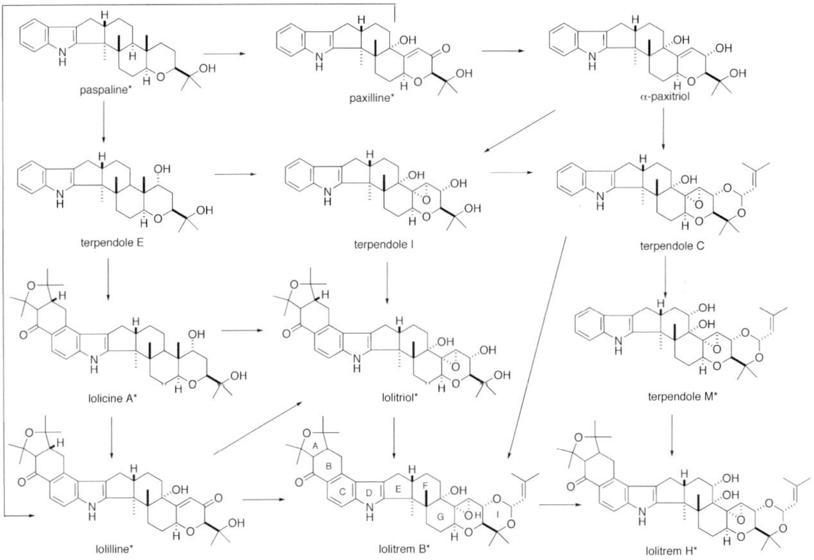

Figure 5.7 A proposed metabolic grid for the biosynthesis of indole-diterpenes in epichloë endophytes. Compounds marked with an asterisk have been isolated from endophyte-infected *Lolium perenne*. This figure is reproduced from Parker and Scott (2004) with permission from Marcel Dekker.

Currently, there is considerable interest in selecting natural endophyte isolates with desirable chemotypes for introduction into elite cultivars of either forage or turf grasses (Bouton & Easton, 2005). For forage grasses it is desirable to have endophytes that have either reduced synthesis or totally lack the ability to synthesize toxic mammalian alkaloids, yet retain the ability to synthesize other bioprotective metabolites, such as peramine (Tapper & Latch, 1999). For the turf grasses it is desirable to have endophytes that are highly toxic to grazing herbivores. The discovery and development of *N. lolii* strain AR1 as a non-toxic endophyte for use in forage grasses has established a benchmark for the development of novel endophytes in forage agriculture (Fletcher, 2005). The availability of cloned *ltm* genes will now allow field isolates to be screened for the presence, distribution and expression of these genes (Scott, 2001b; Scott, 2004). Molecular screening will provide the certainty required for predicting the toxin phenotype of naturally occurring endophytes, such as those that lack the ability to synthesize lolitrem B and/or ergovaline. This predictive chemistry approach will allow for very rapid screening and identification of isolates with chemotypes suitable for specific applications.

The ability to disrupt various steps in the lolitrem biosynthetic pathway opens the way for engineering 'designer' endophytes as well as enhancing our ability to better utilize the available endophyte germplasm. Disruption of an early step will completely eliminate indole-diterpene biosynthesis, thereby generating non-toxic endophytes. Disruption of steps later in the pathway will generate novel endophytes that have the capacity to accumulate pathway intermediates. Artificial plant associations containing these strains could be used for biopharming, as leaf and seed material would be enriched for these intermediates, making it easier to purify these compounds and to carry out the necessary biological testing required to identify desirable bioactivities.

Conclusions

The cloning of biosynthetic genes for the synthesis of the several different classes of alkaloids found in epichloë endophytes and elucidation of the corresponding biochemical pathways now opens up many new biotechnological opportunities for exploiting these economically important symbiotic fungi.

Acknowledgements

The research described in this review was supported by grants C10X0203 and MAU103 from the New Zealand Foundation for Research

(FRST), Science and Technology and the Royal Society of New Zealand Marsden Fund. Carolyn Young was a recipient of FRST Bright Futures Scholarship.

References

Acklin, W., Weibel, F. & Arigoni, D. (1977). Zur biosynthese von paspalin und verwandten metaboliten aus *Claviceps paspali*. *Chimia*, **31**, 63.

Arachevaleta, M., Bacon, C. W., Hoveland, C. S. & Radcliffe, D. E. (1989). Effect of the tall fescue endophyte on plant response to environmental stress. *Agronomy Journal*, **81**, 83–90.

Bacon, C. W., Porter, J. K., Robbins, J. D. & Luttrell, E. S. (1977). *Epichloë typhina* from toxic tall fescue grasses. *Applied and Environmental Microbiology*, **34**, 576–81.

Blankenship, J. D., Spiering, M. J., Wilkinson, H. H., Fannin, F. F., Bush, L. P. & Schardl, C. L. (2001). Production of loline alkaloids by the grass endophyte, *Neotyphodium uncinatum*, in defined media. *Phytochemistry*, **58**, 395–401.

Bouton, J. & Easton, S. (2005). In *Neotyphodium in Cool-Season Grasses*, eds. C. A. Roberts, C. P. West & D. E. Spiers. Oxford: Blackwell Publishing Ltd, pp. 327–40.

Bultman, T. L., White, J. F., Jr., Bowdish, T. I. & Welch, A. M. (1998). A new kind of mutualism between fungi and insects. *Mycological Research*, **102**, 235–8.

Bultman, T. L., White, J. F., Jr., Bowdish, T. I., Welch, A. M. & Johnston, J. (1995). Mutualistic transfer of *Epichloë* spermatia by *Phorbia* flies. *Mycologia*, **87**, 182–9.

Bush, L. P., Wilkinson, H. H. & Schardl, C. L. (1997). Bioprotective alkaloids of grass-fungal endophyte symbioses. *Plant Physiology*, **114**, 1–7.

Byrne, K. M., Smith, S. K. & Ondeyka, J. G. (2002). Biosynthesis of nodulisporic acid A: precursor studies. *Journal of the American Chemical Society*, **124**, 7055–60.

Cambareri, E. B., Jensen, B. C., Schabtach, E. & Selker, E. U. (1989). Repeat-induced G-C to A-T mutations in Neurospora. *Science*, **244**, 1571–5.

Christensen, M. J., Ball, O. J.-P., Bennett, R. J. & Schardl, C. L. (1997). Fungal and host genotype effects on compatibility and vascular colonization by *Epichloë festucae*. *Mycological Research*, **101**, 493–501.

Christensen, M. J., Leuchtmann, A., Rowan, D. D. & Tapper, B. A. (1993). Taxonomy of *Acremonium* endophytes of tall fescue (*Festuca arundinacea*), meadow fescue (*F. pratensis*) and perennial rye-grass (*Lolium perenne*). *Mycological Research*, **97**, 1083–92.

Chung, K.-R. & Schardl, C. L. (1997). Sexual cycle and horizontal transmission of the grass symbiont, *Epichloë typina*. *Mycological Research*, **101**, 295–301.

Clay, K. & Schardl, C. (2002). Evolutionary origins and ecological consequences of endophyte symbiosis with grasses. *The American Naturalist*, **160**, S99–S127.

Craven, K. D., Blankenship, J. D., Leuchtmann, A., Hignight, K. & Schardl, C. L. (2001). Hybrid fungal endophytes symbiotic with the grass *Lolium pratense*. *Sydowia*, **53**, 44–73.

de Jesus, A. E., Gorst-Allman, C. P., Steyn, P. S., van Heerden, F. R., Vleggar, R., Wessels, P. L. & Hull, W. E. (1983). Tremorgenic mycotoxins from *Penicillium crustosum*. Biosynthesis of Penitrem A. *Journal of the Chemical Society Perkin Transactions*, 1863–8.

Fletcher, L. (2005). In *Neotyphodium in Cool-Season Grasses*, eds. C. A. Roberts, C. P. West & D. E. Spiers. Oxford: Blackwell Publishing Ltd, pp. 229–41.

Fletcher, L. R. & Harvey, I. C. (1981). An association of a *Lolium* endophyte with ryegrass staggers. *New Zealand Veterinary Journal*, **29**, 185–6.

Fueki, S., Tokiwano, T., Toshima, H. & Oikawa, H. (2004). Biosynthesis of indole diterpenes, emindole, and paxilline: involvement of a common intermediate. *Organic Letters*, **6**, 2697–700.

Gallagher, R. T., White, E. P. & Mortimer, P. H. (1981). Ryegrass staggers: isolation of potent neurotoxins lolitrem A and lolitrem B from staggers-producing pastures. *New Zealand Veterinary Journal*, **29**, 189–90.

Gardner, M. J., Hall, N., Fung, E., White, O., Berriman, M., Hyman, R. W., Carlton, J. M., Pain, A., Nelson, K. E., Bowman, S., Paulsen, I. T., James, K., Eisen, J. A., Rutherford, K., Salzberg, S. L., Craig, A., Kyes, S., Chan, M. S., Nene, V. & Shallom, S. J. (2002). Genome sequence of the human malaria parasite *Plasmodium falciparum*. *Nature*, **419**, 498–511.

Gatenby, W. A., Munday-Finch, S. C., Wilkins, A. L. & Miles, C. O. (1999). Terpendole M, a novel indole-diterpenoid isolated from *Lolium perenne* infected with the endophytic fungus *Neotyphodium lolii*. *Journal of Agricultural and Food Chemistry*, **47**, 1092–7.

Gwinn, K. D. & Gavin, A. M. (1992). Relationship between endophyte infestation level of tall fescue seed lots and *Rhizoctonia zeae* seedling disease. *Plant Disease*, **76**, 911–14.

Hill, N. S., Stringer, W. C., Rottinghaus, G. E., Belesky, D. P., Parrott, W. A. & Pope, D. D. (1990). Growth, morphological, and chemical component responses of tall fescue to *Acremonium coenophialum*. *Crop Science*, **30**, 156–61.

Kimmons, C. A., Gwinn, K. D. & Bernard, E. C. (1990). Nematode reproduction on endophyte-infected and endophyte-free tall fescue. *Plant Disease*, **74**, 757–61.

Knaus, H.-G., McManus, O. B., Lee, S. H., Schmalhofer, W. A., Garcia-Calvo, M., Helms, L. M. H., Sanchez, M., Giangiacomo, K., Reuben, J. P., Smith, A. B., Kaczorowski, G. J. & Garcia, M. L. (1994). Tremorgenic indole alkaloids potently inhibit smooth muscle high-conductance calcium-activated channels. *Biochemistry*, **33**, 5819–28.

Kuldau, G. A., Tsai, H.-F. & Schardl, C. L. (1999). Genome sizes of *Epichloë* species and anamorphic hybrids. *Mycologia*, **91**, 776–82.

Lane, G. A., Christensen, M. J. & Miles, C. O. (2000) In *Microbial Endophytes*, eds. C. W. Bacon & J. F. J. White. New York: Marcel Dekker, pp. 341–88.

Laws, I. & Mantle, P. G. (1989). Experimental constraints in the study of the biosynthesis of indole alkaloids in fungi. *Journal of General Microbiology*, **135**, 2679–92.

Lee, B. N., Kroken, S., Chou, D. Y., Robbertse, B., Yoder, O. C. & Turgeon, B. G. (2005). Functional analysis of all nonribosomal peptide synthetases in *Cochliobolus heterostrophus* reveals a factor, NPS6, involved in virulence and resistance to oxidative stress. *Eukaryotic Cell*, **4**, 545–55.

Leuchtmann, A. & Schardl, C. L. (1998). Mating compatibility and phylogenetic relationships among two new species of *Epichloë* and other congeneric European species. *Mycological Research*, **102**, 1169–82.

Leuchtmann, A., Schardl, C. L. & Siegel, M. R. (1994). Sexual compatibility and taxonomy of a new species of *Epichloë* symbiotic with fine fescue grasses. *Mycologia*, **86**, 802–12.

Marahiel, M. A., Stachelhaus, T. & Mootz, H. D. (1997). Modular peptide synthetases involved in nonribosomal peptide synthesis. *Chemical Reviews*, **97**, 2651–73.

McMillan, L. K., Carr, R. L., Young, C. A., Astin, J. W., Lowe, R. G. T., Parker, E. J., Jameson, G. B., Finch, S. C., Miles, C. O., McManus, O. B., Schmalhofer, W. A., Garcia, M. L., Kaczorowski, G. J., Goetz, M. A., Tkacz, J. S. & Scott, B. (2003). Molecular analysis of two cytochrome P450 monooxygenase genes required for paxilline biosynthesis in *Penicillium paxilli* and effects of paxilline intermediates on mammalian maxi-K ion channels. *Molecular Genetics and Genomics*, **270**, 9–23.

Miles, C. O., Munday, S. C., Wilkins, A. L., Ede, R. M. & Towers, N. R. (1994). Large-scale isolation of lolitrem B and structure determination of lolitrem E. *Journal of Agriculture and Food Chemistry*, **42**, 1488–92.

Miles, C. O., Wilkins, A. L., Gallagher, R. T., Hawkes, A. D., Munday, S. C. & Towers, N. R. (1992). Synthesis and tremorgenicity of paxitrols and lolitriol: possible biosynthetic precursors of lolitrem B. *Journal of Agriculture and Food Chemistry*, **40**, 234–8.

Moon, C. D., Craven, K. D., Leuchtmann, A., Clement, S. L. & Schardl, C. L. (2004). Prevalence of interspecific hybrids amongst asexual fungal endophytes of grasses. *Molecular Ecology*, **13**, 1455–67.

Moon, C. D., Scott, B., Schardl, C. L. & Christensen, M. J. (2000). The evolutionary origins of *Epichloë* endophytes from annual ryegrasses. *Mycologia*, **92**, 1103–18.

Munday-Finch, S. C., Miles, C. O., Wilkins, A. L. & Hawkes, A. D. (1995). Isolation and structure elucidation of lolitrem A, a tremorgenic mycotoxin from perennial ryegrass infected with *Acremonium lolii*. *Journal of Agriculture and Food Chemistry*, **43**, 1283–8.

Munday-Finch, S. C., Wilkins, A. L. & Miles, C. O. (1996a). Isolation of paspaline B, an indole-diterpenoid from *Penicillium paxilli*. *Phytochemisty*, **41**, 327–32.

Munday-Finch, S. C., Wilkins, A. L. & Miles, C. O. (1998). Isolation of lolicine A, lolicine B, lolitriol, and lolitrem N from *Lolium perenne* infected with *Neotyphodium lolii* and evidence for the natural occurrence of 31-epilolitrem N and 31-epilolitrem F. *Journal of Agricultural and Food Chemistry*, **46**, 590–8.

Munday-Finch, S. C., Wilkins, A. L., Miles, C. O., Ede, R. M. & Thomson, R. A. (1996b). Structure elucidation of Lolitrem F, a naturally occurring stereoisomer of the tremorgenic mycotoxin Lolitrem B, isolated from *Lolium perenne* infected with *Acremonium lolii*. *Journal of Agricultural and Food Chemistry*, **44**, 2782–8.

Munday-Finch, S. C., Wilkins, A. L., Miles, C. O., Tomoda, H. & Omura, S. (1997). Isolation and structure elucidation of lolilline, a possible biosynthetic precursor of the lolitrem family of tremorgenic mycotoxins. *Journal of Agricultural and Food Chemistry*, **45**, 199–204.

Panaccione, D. G. (1996). Multiple families of peptide synthetase genes from ergopeptine-producing fungi. *Mycological Research*, **100**, 429–36.

Panaccione, D. G., Johnson, R. D., Wang, J., Young, C. A., Damrongkool, P., Scott, B. & Schardl, C. L. (2001). Elimination of ergovaline from a grass-*Neotyphodium* endophyte symbiosis by genetic modification of the endophyte. *Proceedings of the National Academy of Sciences USA*, **98**, 12820–5.

Parker, E. J. & Scott, D. B. (2004) In *Handbook of Industrial Mycology*, Chapter 14, ed. Z. An. New York: Marcel Dekker, pp. 405–26.

Rojas, M. C., Hedden, P., Gaskin, P. & Tudzynski, B. (2001). The *P450-1* gene of *Gibberella fujikuroi* encodes a multifunctional enzyme in gibberellin biosynthesis. *Proceedings of the National Academy of Sciences USA*, **98**, 5838–43.

Rowan, D. D. (1993) In *Agriculture, Ecosystems and Environment*, Vol. 44. Amsterdam: Elsevier Science Publishers B.V., pp. 103–22.

Rowan, D. D. & Gaynor, D. L. (1986). Isolation of feeding deterrents against stem weevil from ryegrass infected with the endophyte *Acremonium loliae*. *Journal of Chemical Ecology*, **12**, 647–58.

Schardl, C. L. & Leuchtmann, A. (1999). Three new species of *Epichloë* symbiotic with North American grasses. *Mycologia*, **91**, 95–107.

Schardl, C. L., Leuchtmann, A. & Spiering, M. J. (2004). Symbioses of grasses with seedborne fungal endophytes. *Annual Review of Plant Biology*, **55**, 315–40.

Schardl, C. L., Leuchtmann, A., Tsai, H.-F., Collett, M. A., Watt, D. M. & Scott, D. B. (1994). Origin of a fungal symbiont of perennial ryegrass by interspecific hybridization of a mutualist with the ryegrass choke pathogen, *Epichloë typhina*. *Genetics*, **136**, 1307–17.

Schardl, C. L. & Wilkinson, H. H. (2000). In *Microbial Endophytes*, eds. C. W. Bacon & J. F. White, Jr. New York: Marcel Dekker, Inc., pp. 63–83.

Scott, B. (2001a). *Epichloë* endophytes: symbionts of grasses. *Current Opinion in Microbiology*, **4**, 393–8.

Scott, B. (2001b). In *Molecular Breeding of Forage Crops*, ed. G. Spangenberg. Dordrecht: Kluwer Academic Publishers, pp. 261–74.

Scott, B. (2004). In *Molecular Breeding of Forage and Turf. Proceedings of the 3rd International Symposium on Molecular Breeding of Forage and Turf, Dallas, Texas, and Ardmore, Oaklahoma, USA.* May 18–22, 2003., Vol. 11, eds. A. Hopkins, Z.-Y. Wang, R. Mian, M. Sledge & R. E. Barker. Dordrecht: Kluwer Academic, pp. 133–44.

Scott, B. & Young, C. (2003). In *Clavicipitalean Fungi: Evolutionary Biology, Chemistry, Biocontrol and Cultural Impacts*, eds. J. F. White, Jnr., C. W. Bacon, N. L. Hywel-Jones & J. W. Spatafora. New York: Marcel Dekker Inc., pp. 425–43.

Scott, B., Young, C., Tanaka, A., Christensen, M., Tapper, B. & Bryan, G. (2005). In *Neotyphodium in Cool-Season Grasses*, eds. C. A. Roberts, C. P. West & D. E. Spiers. Oxford: Blackwell Publishing Ltd, pp. 93–101.

Selala, M. I., Laekeman, G. M., Loenders, B., Masuka, A., Herman, A. G. & Schepens, P. (1991). *In vitro* effects of tremorgenic mycotoxins. *Journal of Natural Products*, **54**, 207–12.

Selker, E. U., Cambareri, E. B., Jensen, B. C. & Haack, K. R. (1987). Rearrangement of duplicated DNA in specialized cells of Neurospora. *Cell*, **51**, 741–52.

Siegel, M. R., Latch, G. C. M., Bush, L. P., Fannin, F. F., Rowan, D. D., Tapper, B. A., Bacon, C. W. & Johnson, M. C. (1990). Fungal endophyte-infected grasses: alkaloid accumulation and aphid response. *Journal of Chemical Ecology*, **16**, 3301–15.

Socic, H. & Gaberc-Porekar, V. (1992). In *Fungal Biotechnology*, Vol. 4, eds. D. K. Arora, R. P. Elander & K. G. Mukerji. New York: Marcel Dekker, pp. 123–55.

Steyn, P. S. & Vleggaar, R. (1985). Tremorgenic mycotoxins. *Progress in the Chemistry of Organic Natural Products*, **48**, 1–80.

Tanaka, A., Tapper, B. A., Popay, A. J., Parker, E. J. & Scott, B. (2005). A symbiosis expressed non-ribosomal peptide synthetase from a mutualistic fungal endophyte of perennial ryegrass confers protection to the symbiotum from insect herbivory. *Molecular Microbiology*, **57**, 1036–50.

Tapper, B. A. & Latch, G. C. M. (1999). In *Ryegrass Endophyte: An Essential New Zealand Symbiosis. Grasslands Research and Practice Series No 7*, eds. D. R. Woodfield & C. Matthew. Napier: New Zealand Grassland Association, pp. 107–11.

Tkacz, J. S. (2000). In *Microbial Endophytes*, eds. C. W. Bacon & J. J. F. White. New York: Marcel Dekker, Inc.

Tsai, H.-F., Liu, J.-S., Staben, C., Christensen, M. J., Latch, G. C. M., Siegel, M. R. & Schardl, C. L. (1994). Evolutionary diversification of fungal endophytes of tall fescue grass by hybridization with *Epichloë* species. *Proceedings of the National Academy of Sciences USA*, **91**, 2542–6.

Tudzynski, B., Hedden, P., Carrera, E. & Gaskin, P. (2001). The P450-4 gene of *Gibberella fujikuroi* encodes ent-kaurene oxidase in the gibberellin biosynthesis pathway. *Applied and Environmental Microbiology*, **67**, 3514–22.

Tudzynski, B. & Hölter, K. (1998). Gibberellin biosynthetic pathway in *Gibberella fujikuroi*: evidence for a gene cluster. *Fungal Genetics and Biology*, **25**, 157–70.

Tudzynski, B., Mihlan, M., Rojas, M. C., Linnemannstoens, P., Gaskin, P. & Hedden, P. (2003). Characterization of the final two genes of the gibberellin biosynthesis gene cluster of *Gibberella fujikuroi: des* and *P450-3* encode GA4 desaturase and the 13-hydroxylase respectively. *Journal of Biological Chemistry*, **278**, 28635–43.

Tudzynski, B., Rojas, M. C., Gaskin, P. & Hedden, P. (2002). The gibberellin 20-oxidase of *Gibberella fujikuroi* is a multifunctional monooxygenase. *Journal of Biological Chemistry*, **277**, 21246–53.

von Döhren, H., Keller, U., Vater, J. & Zocher, R. (1997). Multifunctional peptide synthetases. *Chemical Reviews*, **97**, 2675–705.

West, C. P., Izekor, E., Oosterhuis, D. M. & Robbins, R. T. (1988). The effect of *Acremonium coenophialum* on the growth and nematode infestation of tall fescue. *Plant and Soil*, **112**, 3–6.

Yoder, O. C. & Turgeon, B. G. (2001). Fungal genomics and pathogenicity. *Current Opinion in Plant Biology*, **4**, 315–21.

Young, C., Itoh, Y., Johnson, R., Garthwaite, I., Miles, C. O., Munday-Finch, S. C. & Scott, B. (1998). Paxilline-negative mutants of *Penicillium paxilli* generated by heterologous and homologous plasmid integration. *Current Genetics*, **33**, 368–77.

Young, C. A., Bryant, M. K., Christensen, M. J., Tapper, B. A., Bryan, G. T. & Scott, B. (2005). Molecular cloning and genetic analysis of a symbiosis-expressed gene cluster for lolitrem biosynthesis from a mutualistic endophyte of perennial ryegrass. *Molecular Genetics and Genomics*, **274**, 13–29.

Young, C. A., McMillan, L., Telfer, E. & Scott, B. (2001). Molecular cloning and genetic analysis of an indole-diterpene gene cluster from *Penicillium paxilli. Molecular Microbiology*, **39**, 754–64.

Zhang, S., Monahan, B. J., Tkacz, J. S. & Scott, B. (2004). An indole-diterpene gene cluster from *Aspergillus flavus. Applied and Environmental Microbiology*, **70**, 6875–83.

6

What can genomics tell us about secondary metabolism in *Aspergillus*?

G. TURNER

Introduction

Genome resources for filamentous fungi have improved dramatically in the past few years. Since the publication of the *N. crassa* genome (Galagan *et al.*, 2003) the pace has accelerated, with many projects completed or nearing completion (Table 6.1), and more underway. The number of researchers investigating the molecular genetics of filamentous fungi is relatively small, and the number of species very large, such that our efforts are spread rather thinly compared to the *S. cerevisiae* community, for example. Nevertheless, there are indications that these new resources will be extremely beneficial for mycology, and will attract new researchers into the field. Unlike yeasts, filamentous fungi are known for their ability to produce a wide variety of secondary metabolites, which are often of importance to man as useful drugs or harmful toxins (Keller, Turner & Bennett, 2005). Studies over the past 20 years have led to the characterization of some of the biosynthetic pathways and the gene clusters which encode them, for example, the penicillin/cephalosporin and aflatoxin/sterigmatocystin pathways (Brakhage, 1998; Hicks, Shimuzu & Keller, 2002). Since these gene clusters often span substantial regions of the genome, their isolation and sequencing was a major undertaking (Keller & Hohn, 1997). The genome sequences now emerging provide us with easily recognizable gene clusters as a starting point for further investigations, show us the entire secondary metabolic capacity of any species, and pose new questions about how the secondary metabolic repertoire of genera and species has evolved. This review will focus on examples of the new data emerging from sequencing projects on the genus *Aspergillus*. Putative genes identified from bioinformatic analysis are named according to the current GenBank Accession numbers (e.g. Afu8g00170 for *A. fumigatus* and AN8412.2 for

Exploitation of Fungi, eds. G. D. Robson, P. van West and G. M. Gadd. Published by Cambridge University Press.

Table 6.1 *Fungal genome sequencing projects.*

Species	Reference
Aspergillus nidulans	1
Aspergillus fumigatus	2
Aspergillus oryzae	3
Aspergillus niger	4
Chaetomium globosum	1
Coccidioides immitis	1
Coprinus cinereus	1
Fusarium graminearum	1
Magnaporthe grisea	1
Neurospora crassa	1
Phanerochaete chrysosporium	5
Rhizopus oryzae	1
Stagonospora nodorum	1
Trichoderma reesei	6
Ustilago maydis	1

Notes:
1. Annotated and released sequences of filamentous fungi from the Fungal Genome Initiative at the Broad Institute www.broad.mit.edu/annotation/fungi/fgi/
2. TIGR www.tigr.org/tdb/e2k1/afu1/ and Sanger Centre www.sanger.ac.uk/Projects/A_fumigatus/
3. Japanese Consortium; available at website www.bio.nite.go.jp/dogan/MicroTop?GENOME_ID = ao
4. DSM proprietary sequence; available after publication early 2006
5. DOE genome.jgi-psf.org/whiterot1/whiterot1.home.html
6. DOE gsphere.lanl.gov/trire1/trire1.home.html

A. nidulans. Genome structure for *A. nidulans* can be viewed at www.broad.mit.edu/annotation/fungi/aspergillus/index.html and for *A. fumigatus* at www.cadre.man.ac.uk/Aspergillus_fumigatus/ and genes can be searched for using the GenBank Accession numbers.

Secondary metabolism in *Aspergillus* species

While it is difficult to define precisely what is meant by secondary metabolites (Chadwick & Whelan, 1992), also called natural products, the term is used to describe non-essential, low molecular weight metabolites, which often have biological activity against other organisms. Though the variety of such compounds is enormous (Turner, 1971; Turner & Aldridge, 1983; Frisvad & Samson, 1990) most of them can be grouped under a few headings that reflect their underlying structure and route of biosynthesis (Table 6.2).

Table 6.2 *Major groups of fungal secondary metabolites with examples.*

Non-ribosomal peptides (NRPS)	
β-lactams	antibiotics
cyclosporin	immunosuppressant
echinocandin	antifungal drug
destruxin	toxin
Polyketides (PKS)	
lovastatin	cholesterol lowering agent
aflatoxin	carcinogen
Indole alkaloids	
ergotamine	migraine treatment control of post-partum bleeding
Terpenes	
trichodiene	toxin
aristolochene	toxin

Secondary metabolism in the genetic model *A. nidulans* has been extensively studied for both β-lactam biosynthesis (non-ribosomal peptides) (Smith *et al.*, 1990; Brakhage, 1998; Penalva, Rowlands & Turner, 1998) and sterigmatocystin (a polyketide) (Keller & Hohn, 1997; Yu & Keller, 2005). These and other studies showed that fungal secondary metabolic pathway genes were typically found in clusters, and that the genes encoding the key enzymes of these pathways, namely the NRPS and PKS, could by recognized by their domain structure (Fig. 6.1). These features can be exploited in searching fungal genomes sequences for these kinds of secondary metabolic pathways.

Though alkaloid biosynthetic pathways have not been so well studied, the presence of dimethylallyl tryptophan synthase (DMATS) homologue(s) in a cluster is one indication of such a pathway (Grundmann & Li, 2005).

Terpene biosynthesis requires the action of a terpenoid cyclase on a polyisoprenoid compound such as farnesyl diphosphate. Terpene cyclases are found in both plants and fungi, but since they do not show strong conservation in primary sequence, it is likely that many are missed by simple bioinformatics approaches. Features of these key gene products are summarized below:

Non-ribosomal peptide synthetase

Non-ribosomal peptide synthetases (NRPS) are modular enzymes, where each module can recognize and adenylate an amino acid (Finking & Marahiel, 2004). The activated amino acid is then bound as a thioester to a

4′-phosphopantetheine cofactor, attached to a serine residue within a conserved motif found just after the adenylation domain. A peptide bond is formed between amino acids held on these regions, which are known as peptidyl carrier proteins (PCPs), since they hold amino acids or peptides attached to the enzyme (Fig. 6.1). This reaction involves a recognizable condensation domain, found after the PCP, though details of this reaction are unclear. The completed peptide is either released as a linear structure, as in the case of *A. nidulans* ACVS, which possess an integrated thioesterase domain near the C-terminus, or cyclized, as with cyclosporins of *Tolypocladium niveum*. There is usually a correspondence between the number of modules within an NRPS and the number of amino acids in the final peptide (for example, three modules for the three amino acids of ACV, 11 modules for the undecapeptide cyclosporin) though this is not always the case. Some peptide synthetase modules, exemplified by cyclosporin synthetase, also possess a methylation domain following the adenylation domain, leading to methylation of amino acid residues (for example, methylleucine and methylvaline are present in cyclosporins) (Weber *et al.*, 1994). *sidC* of *A. nidulans* encodes an NRPS which synthesizes the siderophore ferricrocin (Eisendle *et al.*, 2003). This NRPS contains three adenylation modules, while ferricrocin is a cyclohexapeptide

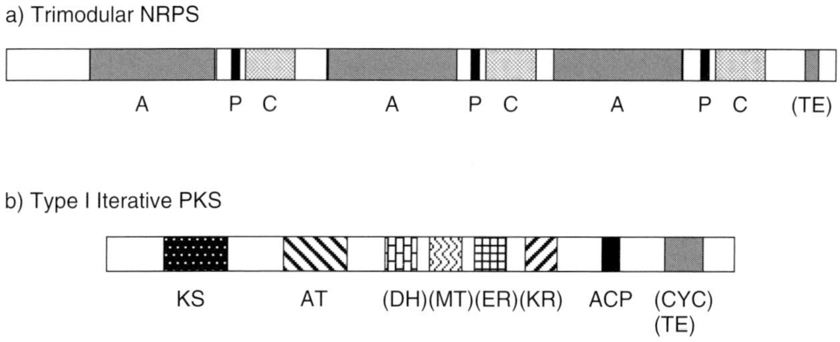

Figure 6.1 Domain structures typical of fungal non-ribosomal peptide synthetases and polyketide synthases. (a) NRPS domains: A, adenylation; P, peptidyl carrier; C, condensation; TE, thioesterase (present in ACV synthetase). Multimodular enzymes show repeating patterns of domains, where there is typically one module per amino acid of the non-ribosomal peptide. Some NRPS have a methyltransferase domain between the A and P domain in some modules, whereby the precursor amino acid is methylated. (b) PKS domains: KS β-ketoacyl synthase; AT, acyltransferase; DH dehydratase; MT, methyltransferase; ER, enoyl reductase; β-ketoreductase; ACP, acyl carrier protein; CYC, cyclase; TE, thioesterase. Domains in parentheses are often absent.

made with the composition Gly-Ser-Gly (acetylhydroxyornithine)$_3$. Unlike the typical domain pattern seen in ACVS (Fig. 6.1a) the domains of SidC are arranged in the pattern APCAPCAPCPCPC, which probably reflects iterative use of modules.

Polyketide synthase

Fungal polyketide synthases (PKS) are also multidomain (type I) enzymes which resemble eukaryotic fatty acid synthetases (Fig. 6.1b), and assemble acetyl CoA or malonyl CoA units to give an enormous variety of compounds (Cane & Walsh, 1999; see also R. Cox, this volume). Fungal PKS synthases seem to be limited to a single multidomain module (Kroken *et al.*, 2003), which can be used in an iterative manner to carry out multiple cycles of condensation, while bacterial type I (multidomain) PKS synthases have separate modules for each reaction cycle. The genes can be easily detected by the presence of the KS, AT and ACP domains, while some fungal PKS have additional domains (e.g. KR, DH, ER). Release of the completed polyketide can be by means of an integrated thioesterase domain, or by cyclization involving a C-terminal domain. A good example of the latter is the PKS encoded by the *wA* gene of *A. nidulans*. This synthesizes the aromatic compound naphthopyrone, a yellow spore pigment intermediate, which is then oxidized by a laccase encoded by *yA* to give the familiar green pigment of conidia (Fujii *et al.*, 2001).

Hybrid genes

As in prokaryotes, PKS-NRPS hybrid genes are also found. Many of the fungal genomes sequenced so far also reveal the presence of a type of hybrid gene which encodes a protein with PKS domains KS-AT followed by a single NRPS module with the structure C-A-P-TE (Kroken *et al.*, 2003). Though a gene of this type found in *Fusarium moniliforme* has been shown to produce fusarin C (Song *et al.*, 2004), the homologues found in other species, including *A. nidulans* (AN8412.2) and *A. fumigatus* (Afu8g00540) do not appear to be orthologues as judged by sequence identity, and their products have yet to be determined.

Dimethylallyltryptophan synthase

Dimethylallyltryptophan synthase (DMATS), encoded by a gene present in the ergot alkaloid cluster of *Claviceps purpurea* (Tudzynski *et al.*, 1999), and found in a number of related *Claviceps* species, is required for the prenylation of tryptophan to form indole alkaloids. Related prenyl transferase enzymes can also be detected in species that do not make clavine alkaloids, and seem to be involved in other prenylation reactions, where dimethylallyl diphosphate is a substrate for addition of isoprenyl groups to other aromatic compounds.

Terpene cyclases

These enzymes are involved in forming cyclic structures from the linear isoprenoid compounds such as geranyl diphosphate (monoterpenes) and farnesyl diphosphate (sesquiterpenes). For example, aristolochene cyclases of *Aspergillus terreus* and *Penicillium roquefortii* catalyze the cyclization of farnesyl pyrophosphate to form aristolochene, which is the precursor of a variety of sesquiterpenes (Caruthers *et al.*, 2000). Similarly, trichodiene synthase is required to form trichodiene from farnesyl diphosphate in *Fusarium sporotrichioides* (Rynkiewicz, Cane & Christianson, 2001). However, simple database searching using amino acid sequences alone is not as productive as with many other enzymes, since there is little conservation of primary sequence, and usually only tertiary structure comparisons reveal homology.

Secondary metabolic genes and clusters in *Aspergillus*

By using the primary sequence of the above enzymes and their conserved domains, it is relatively straightforward to search the genome databases for the major classes of secondary metabolic enzymes. NRPS and PKS genes are common in all *Aspergillus* species examined, though numbers vary between species. For example, *A. nidulans* has 27 PKS and 14 NRPS gene hits. These key genes are usually flanked by other putative secondary metabolic genes such as oxidoreductases, hydroxylases, methylases and drug transporters. While it is not possible to say precisely which of these genes are part of a putative biosynthetic cluster without further analysis, the end of the cluster can be estimated by the detection of clearly recognizable genes with known functions in, for example, primary metabolism or growth.

Since *Aspergillus* species are fairly closely related to each other, sharing a common ancestor less than 100 Myr ago (Kasuga, White & Taylor, 2002) and orthologues of most housekeeping genes are easy to find, it might be expected that similar orthologues would be detected for secondary metabolic genes and clusters. Interestingly, only a small proportion of the clusters appear to be orthologous between *Aspergillus* species, and even some of these are not always present in every *Aspergillus* species (Nierman *et al.*, 2005). For example, genes required for aflatoxin/sterigmatocystin biosynthesis are present in *A. nidulans* and *A. oryzae*, but absent from *A. fumigatus*. Similarly, the highly conserved penicillin biosynthetic cluster, found in *A. nidulans* and *P. chrysogenum*, is absent from *A. fumigatus*. While *A. fumigatus* has gene clusters for gliotoxin, clavines and prenylated indole derivatives, these seem to be absent from *A. nidulans*. This raises questions about how the secondary metabolic gene clusters have evolved within this genus. Whatever their ecological role, it seems that each species has developed/selected its own repertoire of secondary metabolites.

Gene clusters, biosynthetic pathways and their products

Before the availability of genome sequences, gene function for secondary metabolic clusters began with a known metabolite, and ended with gene sequence. Intensive research on the secondary metabolism in *A. nidulans* focused on the penicillin and sterigmatocystin pathways, and the structure and regulation of their biosynthetic gene clusters. In addition, the *wA* gene, long used as a marker in genetic crosses, was identified as encoding a PKS, and orthologues are present in other *Aspergillus* species. More recently, NRPS containing gene clusters which are involved in siderophore biosynthesis have been investigated, and are also highly conserved across different species.

When these clusters are accounted for, we are still left with the majority of the clusters unassigned. At present, our knowledge of amino acid recognition by fungal NRPS adenylation domains, and PKS gene structure, is not sufficient for us to predict accurately the products of these multienzyme complexes from primary amino acid sequence data. Nevertheless, we can sometimes make intelligent guesses about what kind of compound might be made by a gene cluster, and what secondary metabolites are known to be produced from a particular species. Together with some comparative genomics, predictions can be made as the first step in carrying out functional analysis. Confirmation of hypotheses requires deletion or overexpression of individual genes, followed by HPLC, mass spectrometry and NMR to analyze any changes in metabolic products.

This approach is already bearing fruit, good progress has been made since the release of the genome sequences, and is summarized below. Most of the post genome advances in discovering new gene/product relationships have been made with *A. fumigatus*. This reflects the more extensive literature on the chemistry of toxic secondary metabolites in this species compared to *A. nidulans* or *A. oryzae*, if the 'pre-genomic' penicillin and sterigmatocystin/aflatoxin pathways are discounted.

Gliotoxin

Gliotoxin is a cytotoxic compound with immunosuppressive activity produced by *A. fumigatus* as well as other fungal species (Sutton *et al.*, 1994). It is an epipolythidioxopiperazine (ETP), derived from phenylalanine and serine, with a disulphide bridge across a diketopiperazine ring. This suggested that it might be made via a dipeptide by means of a two-module NRPS. However, the genome of *A. fumigatus* has several NRPS that could be candidates, and primary sequence data could not predict which modules are able to recognize phenylalanine and serine. Fortunately, a gene cluster involved in sirodesmin biosynthesis was recently isolated from the plant pathogen *Leptosphaeria maculans* (Gardiner *et al.*, 2004). Sirodesmin is a phytotoxin that is also an ETP, though derived from tyrosine and serine, and comparison of the sirodesmin cluster with the *A. fumigatus* genome sequence has resulted in the identification of a very similar gene cluster. Each of these clusters includes a gene predicted to encode dimodular NRPS. Though full functional analysis of the putative gliotoxin cluster is still awaited at this time, this appears to be a very sound prediction (Gardiner & Howlett, 2005).

Fumitremorgins

A. fumigatus is known to produce a group of compounds that induce tremors in mammals. These include fumitremorgins A, B and C (Eickman *et al.*, 1975). The chemistry of these and related compounds has been studied in some detail, and labelling experiments have been used to clarify their biosynthetic route (Williams, Stocking & Sanz-Cervera, 2000; von Nussbaum, 2003). However, nothing was known about the genes and enzymes involved prior to the release of the genome sequence.

These earlier studies indicated that the pathway for biosynthesis of fumitremorgins might share steps with those for other related compounds in *A. fumigatus*, and for similar compounds in other species, in particular for brevianamides from *Penicillium brevicompactum*. All of these compounds appear to be derived from tryptophan and proline, again suggesting the involvement of a bimodular NRPS. Further, some prenylation steps are required, including prenylation of the tryptophan. The prediction would be that an enzyme similar to DMATS would be required. Prenylation is also needed for sirodesmin biosynthesis, but if the gliotoxin cluster is excluded, a good candidate would be any other cluster with both NRPS and DMATS encoding genes. One cluster which fulfils these criteria (adjacent to NRPS Afu8g00170) is present in the genome sequence, and also contains putative hydroxylase encoding genes and a putative methyltransferase (Fig. 6.2). Further, it has recently been demonstrated directly that the DMATS-like gene Afu8g00210 encodes an enzyme activity (FtmPT1) which can prenylate the *cyclo* L-tryptophan L-proline dipeptide intermediate brevianamide F (Grundmann & Li, 2005).

Recent data (Maiya, S., Grundmann, A., Li, S.-M. & Turner, G., unpublished) also suggests that insertion of multiple copies of the bimodular NRPS Af8g00170 of this cluster into *A. fumigatus* results in accumulation of a compound similar to brevianamide F.

Clavines

A gene cluster responsible for ergotamine biosynthesis was identified in *C. purpurea* (Tudzynski *et al.*, 1999; Haarmann *et al.*, 2005). While the biosynthetic route is quite well characterized, not all of the genes/enzymes for the individual steps have yet been established. The first committed step requires the prenylation of tryptophan by dimethylallyltryptophan synthase, and this enzyme provided the means of identifying the gene cluster. The pathway results in the formation of lysergic acid, which is futher modified by addition of a tripeptide to give ergotamine. This latter step requires the activity of a trimodular NRPS, also present in the cluster.

While ergopeptines such as ergotamine have not been reported in *A. fumigatus*, it is known to produce a number of clavines including fumigaclavine C. This would require steps in common with the ergotamine pathway of *Claviceps* species, though no NRPS would be needed. The best DMATS 'hit' in the genome is found in cluster Afu2g18040. This cluster also contains other sequences, notably a catalase Afu2g18030 and an

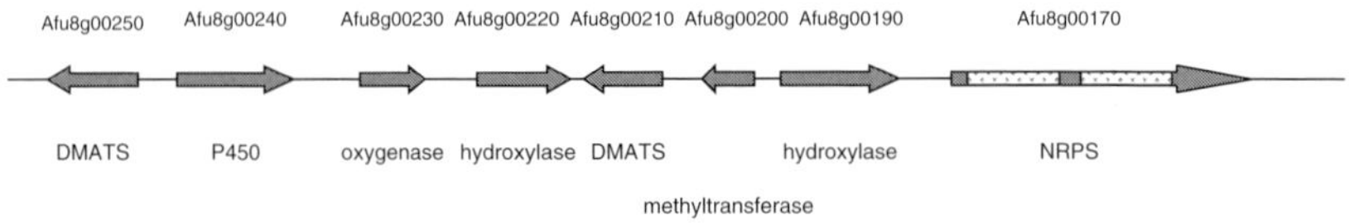

Figure 6.2 A putative prenylated indole pathway in *A. fumigatus*. This exemplifies a typical secondary metabolism cluster associated with one or more of the landmark genes, in this case NRPS and DMATS-related genes. The cluster was unknown until revealed by the *A. fumigatus* genome sequencing project (Nierman *et al.*, 2005). Enzymology following expression in *E. coli* has already shown that the DMATS-related gene Af8g00210 is able to prenylate the diketopiperazine *cyclo*-L-Trp L-Pro (brevianamide F) (Grundmann *et al.*, 2005). Af8g00170 is a bimodular NRPS probably responsible for synthesis of brevianamide F from tryptophan and proline. Other genes in the cluster are labelled according to best hits with known genes.

oxidoreductase Afu2g18000, which have high sequence identity to genes in the ergotamine cluster of *Claviceps*, making this pathway a good candidate for fumigaclavine biosynthesis. Recently, deletion of the DMATS-homologue Afu2g18040 (now named *dmaW*) was shown to lead to loss of fumigaclavine C from fungal extracts (Coyle & Panaccione, 2005). Fumigaclavine C biosynthesis also requires a further prenylation step, and genes encoding other DMATS-related enzymes, putative prenyl transferases, are also found in this cluster (Unsold & Li, 2005), and elsewhere in the genome (Nierman *et al.*, 2005).

Functional analysis of secondary metabolic clusters

The three examples above show how bioinformatics can contribute greatly to establishing gene/product relationships, particularly where there are clues about function from comparative genomics. In order to confirm predictions based on bioinformatics, and to elucidate the function of biosynthetic clusters where clues are few, it is necessary to carry out functional studies.

One of the first questions is whether a particular gene cluster is transcribed sufficiently to result in a detectable product. This problem is highlighted by interesting observations from *A. oryzae*. This species is used extensively in the food industry, for both enzyme production, and more directly for fermented products. It is a domesticated relative of *A. parasiticus*, a producer of the potent toxin and carcinogen aflatoxin. *A. oryzae* is known to possess the entire aflatoxin cluster, but, fortunately, this does not appear to be expressed at a detectable level. This loss of toxin, possibly resulting from differences in the AflR regulatory protein (Watson *et al.*, 1999), may be paralleled by loss of other unknown toxins during the long period of domestication and strain improvement. Indeed, *A. oryzae* possesses a greater number of predicted NRPS and PKS genes than *A. fumigatus* or *A. nidulans* (K. Kitamoto, personal communication; Nierman *et al.*, 2005).

Transcription of individual cluster genes can be conveniently tested using techniques such as RT-PCR, though microarrays are now becoming available for *Aspergillus* and will speed up expression testing of the large number of genes of interest under a variety of growth conditions. In addition to gene transcription, it is also important to find growth conditions under which the secondary metabolites of interest can be detected.

Methods for manipulation of *Aspergillus* genes within the host have been reviewed elsewhere (Turner, 1994; Riach & Kinghorn, 1996; Brookman & Denning, 2000; Archer & Turner, 2006) though the availability of the genome sequences has increased the need for more rapid and efficient methods and new tools, some of which are becoming available. For example, while gene disruption and promoter exchange requires homologous recombination to target vectors to a particular locus, ectopic integration resulting from non-homologous integration is common in fungi. This has been partly overcome by use of ku70/80 deletion strains, where key genes required for non-homologous recombination have been removed in *N. crassa* (Ninomiya *et al.*, 2004) and *A. nidulans* (B. Oakley, personal communication). Further, PCR based approaches for disruption and expression constructs are saving time in vector construction (Yang *et al.*, 2004; Yu *et al.*, 2004; Zarrin, Leeder & Turner, 2005).

Deletion of a gene followed by loss of a metabolite is possible if sufficient product is being made by the wild-type strain, and this may not always be the case. An alternative is therefore to overexpress a gene and look for enhanced product yield. This can be achieved by inserting additional copies of a gene, or exchanging the native promoter for a stronger promoter. For example, *gpdA* of *A. nidulans* is a strong promoter (Punt *et al.*, 1991), though it cannot be downregulated, and there is the risk that overproduction of an enzyme/product may be toxic for the host. An alternative is to use the *alcA* promoter of *A. nidulans* (Kennedy & Turner, 1996). This strong promoter can be induced by growth on ethanol, threonine or cyclopentanone, and repressed on rich media in the presence of high concentrations of glucose (Felenbok, Flipphi & Nikolaev, 2001). It is also possible to use this promoter in *A. fumigatus* (Romero *et al.*, 2003), though its use in *A. niger* is more problematic because it will not respond unless the regulatory gene *alcR* is also introduced (Nikolaev *et al.*, 2002).

After extraction of mycelium or supernatant with a solvent suitable for the compounds of interest, detection of secondary metabolites is most commonly done using HPLC. With this method alone, a standard is required so that retention times of standard and test can be compared, and standards are not always available for the wide range of compounds that might be found. Mass spectrometry can be used to check the mass of the test compound, though elucidation of structure may require further analysis by NMR.

Future prospects

Genes and gene clusters of unknown function

At this time, we still have to identify the function of the majority of putative secondary metabolic gene clusters revealed by the *A. nidulans* and *A. fumigatus* genome projects, and many more fungal genomes are now completed or nearing completion. Unlike the case for most of the fungal genes, which are common to many species, the wide diversity of secondary metabolic products further complicates this task. We still do not have a clear idea of the extent to which the unknown secondary metabolic gene clusters are expressed. While microarray approaches will be of great help for analyzing transcription on a genomic scale, analysis of the product spectrum and characterizing the as yet unknown metabolites is still quite challenging. However, while closely related species have different repertoires of products, sometimes more distantly related species share products. For example, penicillin is made by *A. nidulans*, *P. chrysogenum* and *Trichophyton* species, gliotoxin by *A. fumigatus* and *Trichoderma virens*, and diketopiperazines by *A. fumigatus* and *Penicillium brevicompactum*, so comparative genomics should help us to recognize common themes amongst the biosynthetic pathways, and give further insight into the distribution and evolution of secondary metabolic pathways.

Applications and exploitation

In addition to the relatively small number of fungal secondary metabolites that have already been exploited commercially, there has been continued interest in the future potential of a number of other pharmacologically active compounds, some of which are made by pathways whose details are not yet fully understood. For example, tryprostatins from *A. fumigatus* have activity against mammalian cells, blocking mitosis at G2/M (Cui *et al.*, 1996), and have been seen as lead compounds for development of anti-cancer drugs. Fumitremorgins, also produced by *A. fumigatus*, have been shown to reverse resistance to anti-cancer drugs by their interaction with multidrug resistance proteins (Rabindran *et al.*, 1998). Though unsuitable as therapeutic agents because of their toxic and tremorgenic activities, they may provide ideas for the development of new therapeutic agents. Establishing the gene/product relationships is only the first step in understanding the chemistry and enzymology of the biosynthetic pathways, but availability of the gene sequences for each step will greatly facilitate this, and provide tools for a wide range of biotransformations that may assist future drug design.

References

Archer, D. B. & Turner, G. (2006). Genomics of protein secretion and hyphal growth in *Aspergillus*. In *Fungal Genomics, The Mycota, Vol. 13*, ed. A. J. P. Brown. In press.

Brakhage, A. A. (1998). Molecular regulation of β-lactam biosynthesis in filamentous fungi. *Microbiology and Molecular Biology Review*, **62**, 547–85.

Brookman, J. L. & Denning, D. W. (2000). Molecular genetics in *Aspergillus fumigatus*. *Current Opinion in Microbiology*, **3**, 468–74.
Cane, D. E. & Walsh, C. T. (1999). The parallel and convergent universes of polyketide synthases and nonribosomal peptide synthetases. *Chemistry and Biology*, **6**, R319–R325.
Caruthers, J. M., Kang, I., Rynkiewicz, M. J., Cane, D. E. & Christianson, D. W. (2000). Crystal structure determination of aristolochene synthase from the blue cheese mold, *Penicillium roqueforti*. *Journal of Biological Chemistry*, **275**, 25533–9.
Chadwick, D. J. & Whelan, J. eds. (1992). *Secondary Metabolites: Their Function and Evolution*. CIBA Foundation Symposium 171. Chichester: John Wiley & Sons.
Coyle, C. M. & Panaccione, D. G. (2005). An ergot alkaloid biosynthesis gene and clustered hypothetical genes from *Aspergillus fumigatus*. *Applied and Environmental Microbiology*, **71**, 3112–18.
Cui, C. B., Kakeya, H., Okada, G., Onose, R. & Osada, H. (1996). Novel mammalian cell cycle inhibitors, tryprostatins A, B and other diketopiperazines produced by *Aspergillus fumigatus*. 1. Taxonomy, fermentation, isolation and biological properties. *Journal of Antibiotics (Tokyo)*, **49**, 527–33.
Eickman, N., Clardy, J., Cole, R. J. & Kirksey, J. W. (1975). The structure of fumitremorgin A. *Tetrahedron Letters*, **12**, 1051–4.
Eisendle, M., Oberegger, H., Zadra, I. & Haas, H. (2003). The siderophore system is essential for viability of *Aspergillus nidulans*: functional analysis of two genes encoding L-ornithine N^5-monooxygenase (*sidA*) and a non-ribosomal peptide synthetase (*sidC*). *Molecular Microbiology*, **49**, 359–75.
Felenbok, B., Flipphi, M. & Nikolaev, I. (2001). Ethanol catabolism in Aspergillus nidulans: a model system for studying gene regulation. *Progress in Nucleic Acid Research and Molecular Biology*, **69**, 149–204.
Finking, R. & Marahiel, M. (2004). Biosynthesis of nonribosomal peptides. *Annual Reviews of Microbioliology*, **58**, 453–88.
Frisvad, J. D. & Samson, R. A. (1990). Chemotaxonomy and morphology of *Aspergillus fumigatus* and related taxa. In *Modern Concepts in Penicillium and Aspergillus Classification*, ed. R. A. Samson & J. I. Pitt. New York: Plenum Press, pp. 201–8.
Fujii, I., Watanabe, A., Sankawa, U. & Ebizuka, Y. (2001). Identification of Claisen cyclase domain in fungal polyketide synthase WA, a naphthopyrone synthase of *Aspergillus nidulans*. *Chemistry and Biology*, **8**, 189–97.
Galagan, J. E., Calvo, S. E., Borkovich, K. A., Selker, E. U., Read, N. D., Jaffe, D., FitzHugh, W., Ma, L. J., Smirnov, S., Purcell, S. *et al.* (2003). The genome sequence of the filamentous fungus *Neurospora crassa*. *Nature*, **422**, 859–68.
Gardiner, D. M. & Howlett, B. J. (2005). Bioinformatic and expression analysis of the putative gliotoxin biosynthetic gene cluster of *Aspergillus fumigatus*. *FEMS Microbiology Letters*, **248**, 241–8.
Gardiner, D. M., Cozijnsen, A. J., Wilson, L. M., Pedras, M. S. & Howlett, B. J. (2004). The sirodesmin biosynthetic gene cluster of the plant pathogenic fungus *Leptosphaeria maculans*. *Molecular Microbiology*, **53**, 1307–18.
Grundmann, A. & Li, S.-M. (2005). Overproduction, purification and characterisation of FtmPT1, a brevianamide F prenyltransferase from *Aspergillus fumigatus*. *Microbiology*, **151**, 2199–207.
Haarmann, T., Machado, C., Lubbe, Y., Correia, T., Schardl, T. L, Panaccione, D. G. & Tudzynksi, P. (2005). The ergot alkaloid gene cluster in *Claviceps purpurea*: Extension of the cluster sequence and intra species evolution. *Phytochemistry*, **66**, 1312–20.
Hicks, J., Shimizu, K. & Keller, N. (2002). Genetics and biosynthesis of aflatoxins and sterigmatocystin. In *The Mycota. Vol. XI. Agricultural Applications*, ed. F. Kempken. Berlin: Springer, pp. 55–69.

Kasuga, T., White, T. J. & Taylor, J. W. (2002). Estimation of nucleotide substitution rates in eurotiomycete fungi. *Molecular Biology and Evolution*, **19**, 2318–24.

Keller, N. & Hohn, T. (1997). Metabolic pathway gene clusters in filamentous fungi. *Fungal Genetics and Biology*, **21**, 17–29.

Keller, N. P., Turner, G. & Bennett, J. W. (2005). Fungal secondary metabolism – from biochemistry to genomics. *Nature Reviews Microbiology*, **3**, 937–47.

Kennedy, J. & Turner, G. (1996). δ-(L-α-aminoadipyl)-L-cysteinyl-D-valine synthetase is a rate limiting enzyme for penicillin production in *Aspergillus nidulans*. *Molecular and General Genetics*, **253**, 189–97.

Kroken, S., Glass, N. L, Taylor, J. W., Yoder, O. C. & Turgeon, B. G. (2003). Phylogenomic analysis of type I polyketide synthase genes in pathogenic and saprobic ascomycetes. *Proceedings of the National Academy of Sciences USA*, **100**, 15670–5.

Nierman, W., Pain, A., Anderson, M. J. *et al.* (2005). Genomic sequence of the pathogenic and allergenic *Aspergillus fumigatus*. *Nature*, **438**, 1151–6.

Nikolaev, I., Mathieu, M., van de Vondervoort, P. J. I., Visser, J. & Felenbok, B. (2002). Heterologous expression of the *Aspergillus nidulans alcR–alcA* system in *Aspergillus niger*. *Fungal Genetics and Biology*, **37**, 89–97.

Ninomiya, Y., Suzuki, K., Ishii, C. & Inoue, H. (2004). Highly efficient gene replacements in *Neurospora* strains deficient for non-homologous end-joining. *Proceedings of the National Academy of Sciences USA*, **101**, 12248–53.

Penalva, M. A., Rowlands, R. T. & Turner, G. (1998). The optimization of penicillin biosynthesis in fungi. *Trends in Biotechnology*, **16**, 483–9.

Punt, P. J., Zegers, N. D., Busscher, M., Pouwels, P. H. & van den Hondel, C. A. (1991). Intracellular and extracellular production of proteins in *Aspergillus* under the control of expression signals of the highly expressed *Aspergillus nidulans gpdA* gene. *Journal of Biotechnology*, **17**, 19–33.

Rabindran, S. K., He, H., Singh, M., Brown, E., Collins, K. I., Annable, T. & Greenberger, L. M. (1998). Reversal of a novel multidrug resistance mechanism in human colon carcinoma cells by fumitremorgin C. *Cancer Research*, **58**, 5850–8.

Riach, M. B. R. & Kinghorn, J. R. (1996). Genetic transformation and vector developments in filamentous fungi. In *Fungal Genetics: Principles and Practice*, ed. C. J. Bos. New York: Marcel Dekker Inc., pp. 209–33.

Romero, B., Turner, G., Olivas, I., Laborda, F. & de Lucas, J. R. (2003). The *Aspergillus nidulans alcA* promoter drives tightly regulated conditional gene expression in *Aspergillus fumigatus* permitting validation of essential genes in this human pathogen. *Fungal Genetics and Biology*, **40**, 103–14.

Rynkiewicz, M. J., Cane, D. E. & Christianson, D. W. (2001). Structure of trichodiene synthase from *Fusarium sporotrichioides* provides mechanistic inferences on the terpene cyclization cascade. *Proceedings of the National Academy of Sciences USA*, **98**, 13479–81.

Smith, D. J., Burnham, M. K. R., Bull, J. H., Hodgson, J. E., Ward, J. M., Browne, P., Brown, J., Barton, B., Earl, A. J. & Turner, G. (1990). β-lactam antibiotic biosynthesis genes have been conserved in clusters in prokaryotes and eukaryotes. *EMBO Journal*, **9**, 741–7.

Song, Z., Cox, R. J., Lazarus, C. M. & Simpson, T. J. (2004). Fusarin C biosynthesis in *Fusarium moniliforme* and *Fusarium venenatum*. *Chemistry and Biochemistry*, **5**, 1196–203.

Sutton, P., Newcombe, N. R., Waring, P. & Müllbacher, A. (1994). In vivo immunosuppressive activity of gliotoxin, a metabolite produced by human pathogenic fungi. *Infection and Immunity*, **62**, 1192–8.

Tudzynski, P., Holter, K., Correia, T., Arntz, C., Grammel, N. & Keller, U. (1999). Evidence for an ergot alkaloid gene cluster in *Claviceps purpurea*. *Molecular and General Genetics*, **261**, 133–41.

Turner, G. (1994). Vectors for genetic manipulation. In *Aspergillus: 50 Years On*, ed. S. D. Martinelli & J. R. Kinghorn. Amsterdam: Elsevier, pp. 641–65.

Turner, W. B. (1971). *Fungal Metabolites*. London: Academic Press.

Turner, W. B. & Aldridge, D. C. (1983). *Fungal Metabolites II*. London: Academic Press.

Unsold, I. & Li, S.-M. (2005). Overproduction, purification and characterization of FgaPT2, a dimethylallyltryptophan synthase from *Aspergillus fumigatus*. *Microbiology*, **151**, 1499–505.

von Nussbaum, F. (2003). Stephacidin B – a new stage of complexity within prenylated indole alkaloids from fungi. *Angewandte Chemie International Edition*, **42**, 3068–71.

Watson, A. J., Fuller, L. J., Jeenes, D. J. & Archer, D. B. (1999). Homologs of aflatoxin biosynthesis genes and sequence of *aflR* in *Aspergillus oryzae* and *Aspergillus sojae*. *Applied and Environmental Microbiology*, **65**, 307–10.

Weber, G., Schorgendorfer, K., Schneider-Scherzer, E. & Leiner, E. (1994). The peptide synthetase catalyzing cyclosporine production in *Tolypocladium niveum* is encoded by a giant 45.8-kilobase open reading frame. *Current Genetics*, **26**, 120–5.

Williams, R. M., Stocking, E. M. & Sanz-Cervera, J. F. (2000). Biosynthesis of prenylated alkaloids derived from tryptophan. In *Biosynthesis: Aromatic Polyketides, Isoprenoids, Alkaloids (Topics in Current Chemistry 209)*, ed. F. J. Leeper & J. C. Vederas. Berlin/London: Springer, pp. 97–173.

Yang, L., Ukil, L., Osmani, A., Nahm, F., Davies, J., De Souza., C. P., Dou, X., Perez-Balaguer, A. & Osmani, S. A. (2004). Rapid production of gene replacement constructs and generation of a green fluorescent protein-tagged centromeric marker in *Aspergillus nidulans*. *Eukaryotic Cell*, **3**, 1359–62.

Yu J. H., Hamari, Z., Han, K. H., Seo, J. A., Reyes-Dominguez Y. & Scazzocchio C. (2004). Double-joint PCR: a PCR-based molecular tool for gene manipulations in filamentous fungi. *Fungal Genetics and Biology*, **41**, 973–81.

Yu, J. H. & Keller, N. (2005). Regulation of secondary metabolism in filamentous fungi. *Annual Reviews of Phytopathology*, **43**, 437–58.

Zarrin, M., Leeder, A. & Turner, G. (2005). A rapid method for promoter exchange in *Aspergillus nidulans* using recombinant PCR. *Fungal Genetics and Biology*, **42**, 1–8.

III

Protein folding and secretion

7

The role of microtubules and motors for polarized growth of filamentous fungi

R. FISCHER AND D. VEITH

Introduction

Polarized growth is the mechanism by which filamentous fungi extend their hyphae. Microtubules (MT) and filamentous actin (F-actin), in combination with their corresponding motor proteins, kinesins, dynein and myosins, play crucial roles in this process. The exact contribution of the MT cytoskeleton, however, is still under debate. In this review we will summarize recent advances in understanding the role of MTs and MT-dependent motor proteins in fungi with special emphasis on *Aspergillus nidulans*. Genetic, biochemical and cell biological approaches in *A. nidulans* and other fungi led to a modified view of many aspects within the past few years. There is increasing evidence that MT strings, which are visualized by immunostaining or GFP-tagging, consist of several MTs and their dynamics appears to be different in fast-growing hyphal tips as compared with young germlings. Whereas the spindle pole bodies were considered as the only or the main microtubule organizing centres (MTOCs) in filamentous fungi, it appears that several additional MTOCs are responsible for the generation of the MT array. In addition to new insights into the MT network and its dynamics, the roles of several kinesins have been elucidated recently and their interplay with dynein investigated. It became clear that MT functions are interwoven with those of the actin cytoskeleton and that three main structures are required for polarized growth, the Spitzenkörper (vesicle supply centre), the polarisome and probably cell end markers at the cortex. We propose a model for polarized growth, where the actin cytoskeleton and the polarisome are crucial for hyphal extension and the MT cytoskeleton continuously provides the building material within vesicles to the Spitzenkörper and determines growth directionality by delivery of cell end marker proteins.

Exploitation of Fungi, eds. G. D. Robson, P. van West and G. M. Gadd. Published by Cambridge University Press.

Polarized growth and the Spitzenkörper

One fascinating aspect of filamentous fungi is their continuous tip elongation. Whereas this phase of polarized growth only lasts a short time in the life-cycle of the budding yeast *Saccharomyces cerevisiae*, it is the main growth form of filamentous fungi. The cell wall network needs to be loosened and the membrane enlarged. These two processes are linked because enzymes, which are required outside the cell, are transported towards the tip within vesicles. These vesicles fuse with the cytoplasmic membrane and thus lead to the expansion of the membrane and the delivery of their contents (Bartnicki-Garcia *et al.*, 1995). However, the place where vesicles are generated may be far behind the growing tip. The involvement of vesicles for polarized growth has been proposed many decades ago, when Brunswick observed an accumulation of vesicles in the apical dome of fungal hyphae (Brunswick, 1924; Girbardt, 1957). This structure was named with the German word 'Spitzenkörper' (= apical body) or vesicle supply centre (VSC) (Bartnicki-Garcia *et al.*, 1995; Girbardt, 1957). The latter name refers to its proposed function as a transit station for vesicles from the hyphal body to the plasma membrane. The position of the organelle determines growth direction (Riquelme *et al.*, 1998). S. Bartnicki-Garcia (Riverside, CA, USA) and C. Bracker (West Lafayette, IN, USA) demonstrated fantastically with their groups the importance of this organelle for polarized growth.

Tip extension needs to be a well-controlled process, because secretion of cell-wall lytic enzymes, required for loosening the cell wall prior to extension, as well as the enzymes for cell wall synthesis, may be deleterious for the cell. In addition, new places of polarized growth need to be established for every branch formed (Riquelme & Bartnicki-Garcia, 2004). This shows that tip elongation is likely to require the full equipment of the cellular toolbox. Several organisms are studied extensively to understand the phenomenon. Most important are *S. cerevisiae* and the closely related but filamentously growing *Ashbya gossypii*, the fission yeast *Schizosaccharomyces pombe*, the basidiomycete and plant pathogen dimorphic *Ustilago maydis*, the human pathogenic *Candida albicans*, and several obligate filamentous fungi, such as *Neurospora crassa* and *Aspergillus nidulans* (Crampin *et al.*, 2005; Harris *et al.*, 2005; Martin & Chang, 2005; Philippsen *et al.*, 2005; Pruyne *et al.*, 2004; Steinberg *et al.*, 2001). Research on polarized growth currently has a very high impact, because it is considered as one major target for the development of antifungal drugs.

Morphology, organization and dynamics of the microtubule cytoskeleton

Microtubules are hollow tubes composed of 13 protofilaments, each of which is made up with the heterodimer αβ-tubulin as building block. MTs have an inherent instability and continuously elongate at their plus end, where αβ-tubulin dimers are added. One parameter, which determines the elongation rate, is the concentration of dimers in the cell. Both tubulin subunits contain a bound GTP. The nucleotide binding pocket is located at the interface between the α and β-tubulin subunits and thus this GTP is rather stable. On the other hand, GTP in the β-tubulin subunit is exposed and easily undergoes hydrolysis. Once β-tubulin contains GDP, the assembly is blocked and a catastrophic event occurs.

Microtubules are visible by electron microscopy (Jung *et al.*, 1998), but this method did not allow the study of MT behaviour in living cells. This became possible after the advent of the green fluorescent protein in *A. nidulans* and other fungi. X. Xiang's laboratory (Bethesda, USA) found that MTs extend with a speed of about 14 μm per minute, reach the cortex, pause for some time and undergo a catastrophic event. Subsequently, MTs shrink with a speed of about 30 μm per minute and MTs either depolymerize all the way to the MTOC or rescue occurs before and MTs elongate again (Han *et al.*, 2001). Slightly different values were recently obtained in the group of B. Heath (Sampson & Heath, 2005). They also observed that short MT fragments were able to slide towards the hyphal tip.

Microtubules are quite stiff structures and their orientation is basically dictated by the shape of the fungal cell. Hence, they are mostly aligned parallel to the growth axis (Fig. 7.1). Investigations of MT arrangement within the cell were carried out by immunofluorescence, and recently by using fluorescently labelled tubulin (Ding *et al.*, 1998; Fischer & Timberlake, 1995; Freitag *et al.*, 2004; Han *et al.*, 2001). In *A. nidulans* the number of MTs ranges from three to six. However, it appears that the structures, which can be seen after immunostaining or as GFP-labelled filaments, consist of several individual MTs. There is increasing evidence for this organization, especially coming from studies with *S. pombe*. Here, it was shown recently that the orientation of MTs can be opposite in one bundle and that a kinesin-like motor protein in combination with dynein is required for sliding of individual MTs within a bundle and maintenance of MT polarity (Carazo-Salas *et al.*, 2005). In *A. nidulans* Konzack *et al.* reported that fluorescence intensity varies along a MT and that the regions with low intensity do recover brightness after some time. Similarly, after bleaching of a given MT at one place, brightness returns quickly

(Veith *et al.*, 2005) (Fig. 7.1). In addition, we noticed with GFP-labelled MTs, that occasionally thin filaments detach from a MT for some time before they merge again to form a thick MT (Veith *et al.*, 2005) (Fig. 7.1). These observations are in agreement with a model that MT filaments consist of a bundle, and individual MTs within a bundle show certain individual behaviours and dynamics.

Origin of microtubules

Microtubules cannot efficiently assemble *de novo* under natural conditions, but require an initiation point, an MT organizing centre (MTOC). This point is characterized by a protein complex, whose characteristic component is γ-tubulin (Aldaz *et al.*, 2005; Job *et al.*, 2003; Doxsey *et al.*, 2005; Pereira & Schiebel, 1997). Originally discovered in *A. nidulans*, γ-tubulin was found in all eukaryotes studied and the concept of γ-tubulin mediated nucleation of MT polymerization is an accepted model (Oakley *et al.*, 1990; Oakley, 1995; Oakley, 2004; Oakley & Oakley, 1989). However, the exact mechanism is still under debate. It appears that γ-tubulin in higher eukaryotes forms a 2.2 MDa ring complex consisting of 12 to 14 (different

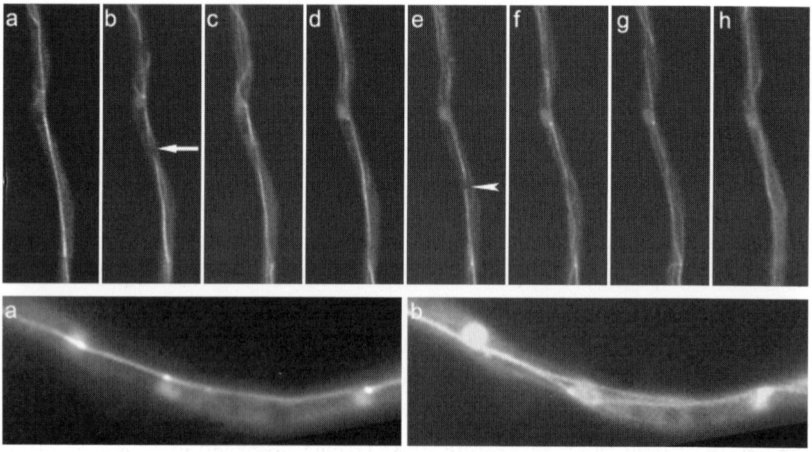

Figure 7.1 Visible MTs appear to be bundles of individual MTs. Upper panel (a–h): Time lapse of a bleaching experiment. GFP-labelled MTs were bleached with a laser beam in a confocal microscope at one point (arrow in b, arrow head in e) and quickly recovered (c, d, f–h). Lower panel (a, b): Time lapse of an MT, which splits into several strings. Three nuclei in a hyphal compartment are attached through their SPB to a MT. In (b) the single string resolves into several. Nuclei, the SPB and MTs were stained with GFP. Pictures in the lower panel were taken from Veith *et al.*, 2005.

numbers exist in the literature) γ-tubulin subunits associated with other proteins, the so called γTuRC (=γ-tubulin ring complex) (Aldaz *et al.*, 2005). The γ-TuRC acts as an initiator complex, where 13 tubulin protofilaments are emanating from. It has been known for a long time that fungal spindle pole bodies (SPB) are very active MTOCs (Jaspersen & Winey, 2004). The SPB is embedded into the nuclear envelope, divides prior to mitosis and, by definition, localizes at the poles of the mitotic spindle. SPBs consist of an inner and an outer plaque and they are able to polymerize MTs on both sides of the nuclear envelope. During mitosis the outer MTs are called astral microtubules, but also in interphase SPBs act as active MTOCs in *S. cerevisiae* as well as in filamentous fungi (Heath, 1981). It should be noted that the protein composition of the SPB in *S. cerevisiae* and filamentous fungi appears to be quite different, despite the apparent same function. We searched the *A. nidulans* genome for homologues of *S. cerevisiae* SPB-associated proteins, but could only identify very few (unpublished results). Further experiments are required to unravel whether this is due to low sequence conservation or to different protein compositions.

It seems that the SPBs are the only places from which the yeast *S. cerevisiae* polymerizes its MTs (see accompanying video in Hoepfner *et al.*, 2000). However, it has to be noted that cytoplasmic MTs appear not to play many important roles in *S. cerevisiae*, besides the positioning of the nucleus prior to mitosis (Maekawa & Schiebel, 2004). The cytoplasmic MT array is not very pronounced and usually limited to a few MTs growing out of the SPB into the cytoplasm. In contrast, filamentous fungi employ MTs for their fast, polarized growth during interphase (Horio & Oakley, 2005; Riquelme *et al.*, 2003) (Fig. 7.1). Nevertheless, it was assumed for a long time that SPB are the only place for MT initiation (Oakley, 2004; Sampson & Heath, 2005). This assumption was based on the finding that the intracellular αβ-tubulin pool is used for the assembly of spindle MTs as well as for cytoplasmic MTs. Indeed, cytoplasmic MTs are generally disassembled prior to mitosis and regenerate thereafter (Ovechkina *et al.*, 2003; Sampson & Heath, 2005). In order to determine the origin of new MTs, re-growth of MTs was observed in *S. pombe* after depolymerisation of MTs by drugs (Mata & Nurse, 1997). These studies revealed that in fission yeast MTs are generated from the SPB and other MTOCs around the nucleus and in the cytoplasm. During cell division an equatorial MTOC becomes very important (EMTOC) (Hagan, 1998; Sawin *et al.*, 2004; Venkatram *et al.*, 2005). The origin of MTs from the cell centre leads to an orientation with their plus ends towards the growing ends. Recently, another tool was used to determine the origin of MTs.

Using MT plus-end localizing proteins, such as homologues of the mammalian EB1, MT initiation was analyzed in the plant pathogenic basidiomycete *Ustilago maydis*. It was found that MT nucleation occurs at three places, at dispersed cytoplasmic sites, at a polar MTOC and at the SPB (Straube *et al.*, 2003).

In filamentous fungi, our knowledge of MT organization is restricted to a few species, such as the chytridiomycete *Allomyces macrogynus*, the basidiomycete *U. maydis* and the ascomycete *A. nidulans*, which is one of the best-studied examples. Whereas Sampson and Heath (2005) reported that MTs emanate only from SPBs, Konzack *et al.* (2005) demonstrated that MTOCs exist apart from the SPBs. This discrepancy may be due to different methods used. In the first study, the authors observed GFP labelled MTs and the location of nuclei was determined by the absence of cytoplasmic fluorescence. The authors of the second study used simultaneous labelling of nuclei with a red fluorescent protein and GFP labelled tubulin. In addition, a plus-end tracking protein, KipA, was used to determine the origin of MTs. MTOCs were found at the SPBs but also in the cytoplasm and at the septa of *A. nidulans* (Fig. 7.2). This organization recently received further evidence through the characterization of a novel MTOC-associated protein, ApsB (Veith *et al.*, 2005). Here, the authors demonstrated that MTOCs at septa are important for the production of the interphase cytoplasmic MT array (Fig. 7.2). These findings are in good agreement with the results obtained in *S. pombe* and *U. maydis*.

It is still an open question whether there are MTOCs at hyphal tips of filamentous fungi. Whereas γ-tubulin can be seen at the tips of *A. macrogynus* hyphae and thus MTs polymerize from the tip to the back (McDaniel & Roberson, 1998), γ-tubulin has not yet been detected at that place in e.g. *A. nidulans*. Nevertheless, using the kinesin motor KipA, Konzack *et al.* (2005) found that sometimes MTs do also polymerize from the tip. It has to be considered that a MT occasionally does not depolymerize upon contact with the cortex, but bends along the cortex towards the rear of the hypha. If this MT continued growth, it could explain the observed comets from the tip to the back of the hypha.

In *Neurospora crassa* MTs were seen recently as a fusion protein of β-tubulin with GFP (Freitag *et al.*, 2004; Mouriño-Pérez *et al.*, 2006). Here the number of MTs is much higher than in *A. nidulans* as is the number of nuclei, which was estimated to 200–400 in a fast growing hyphal tip. Consequently, investigations of MT dynamics and their origin

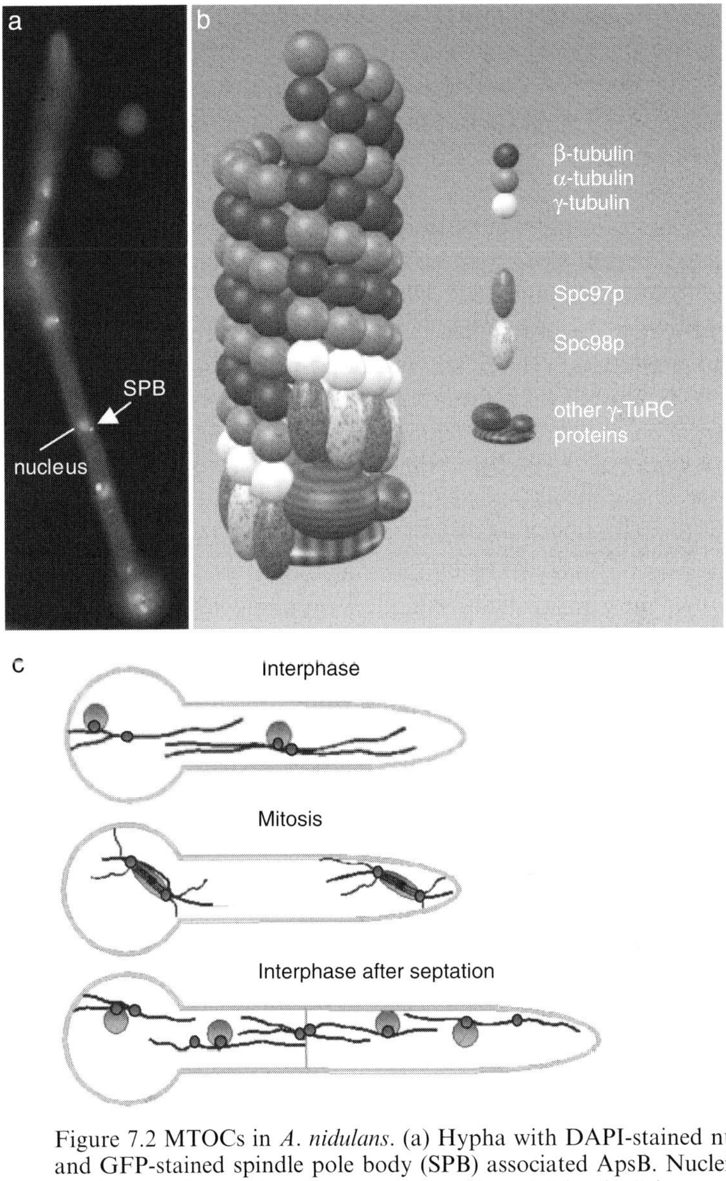

Figure 7.2 MTOCs in *A. nidulans*. (a) Hypha with DAPI-stained nuclei and GFP-stained spindle pole body (SPB) associated ApsB. Nuclei are evenly spaced and at each nucleus an SPB is visible. (b) Scheme of an MTOC with γ-tubulin and other proteins described in *S. cerevisiae*. Adapted from Oakley (2000) and Pereira & Schiebel (1997). (c) MTOCs are found at the nuclei, in the cytoplasm and at septa. Taken from Konzack *et al.* (2005).

are more complicated than in *A. nidulans*. However, interesting results can be expected from a comparison of *N. crassa* and *A. nidulans*, because mitoses are asynchronous in *N. crassa* but synchronous in *A. nidulans*. This suggests that regulation of MT dynamics may be carried out in different ways.

The microtubule plus end

Microtubules grow and shrink in a treadmilling manner if they are polymerized in vitro. In comparison, in vivo MTs are rather stable at the minus end and dynamics occur mainly at the MT plus end. Meanwhile, it is well accepted that this MT end consists of a large protein complex, which is involved in the regulation of MT dynamics as well as in the regulation of interactions with cortical actin, membrane proteins or proteins associated with the kinetochore of chromosomes (Akhmanova & Hoogenraad, 2005; Hestermann *et al.*, 2002; Schuyler & Pellman, 2001b). Given the diversity of interacting partners, it is obvious that the protein complex composition may vary depending on the function of the MT and is likely to be a highly controlled and organized structure. In fungi, the best-studied organisms with respect to the MT plus end are again *S. cerevisiae* and *S. pombe*. In *S. cerevisiae* MT–cortex interactions play important roles for the positioning of the mitotic spindle and nuclear migration (Schuyler & Pellman, 2001a). One of the most prominent examples for an MT plus-end associated protein is dynein (Fig. 7.3). It localizes to the MT tip and hitchhikes with the growing filament to the cell periphery. Once at the cortex, dynein is activated and pulls the attached MT towards the cortex. This leads to translocation of the nucleus (Maekawa *et al.*, 2003; Maekawa & Schiebel, 2004; Schuyler & Pellman, 2001a; Sheeman *et al.*, 2003). The kinesin motor protein Kip2 appears to be responsible for the plus end localization of several proteins, e.g. the CLIP170-like protein Bik1 (Carvalho *et al.*, 2004). Similar to the situation in *S. cerevisiae*, the CLIP170-like protein of *S. pombe*, Tip1 localized to MT plus ends. The responsible motor for this localization was Tea2 (Busch *et al.*, 2004). However, MTs do not play as important roles for polarized growth in yeasts as they do in filamentous fungi. Only some components have been found to localize at MT plus ends, among them are subunits of the dynein motor complex (Zhang *et al.*, 2002). Interestingly, conventional kinesin, KinA, is required for their MT tip localization (Zhang *et al.*, 2003) (Fig. 7.3). The CLIP170-like protein, ClipA, in *A. nidulans* does also accumulate at MT plus ends, and its localization is partly dependent on the Tea2/Kip2 homologue KipA (Efimov *et al.*, 2006).

The question is, which role plus-end localized proteins play for polarized growth? As mentioned above, MT–cortical interactions are necessary for dynein-dependent nuclear positioning prior to mitosis in *S. cerevisiae* (Carminati & Stearns, 1997). In *A. nidulans* dynein is also required for nuclear positioning and migration, and recently we showed that the

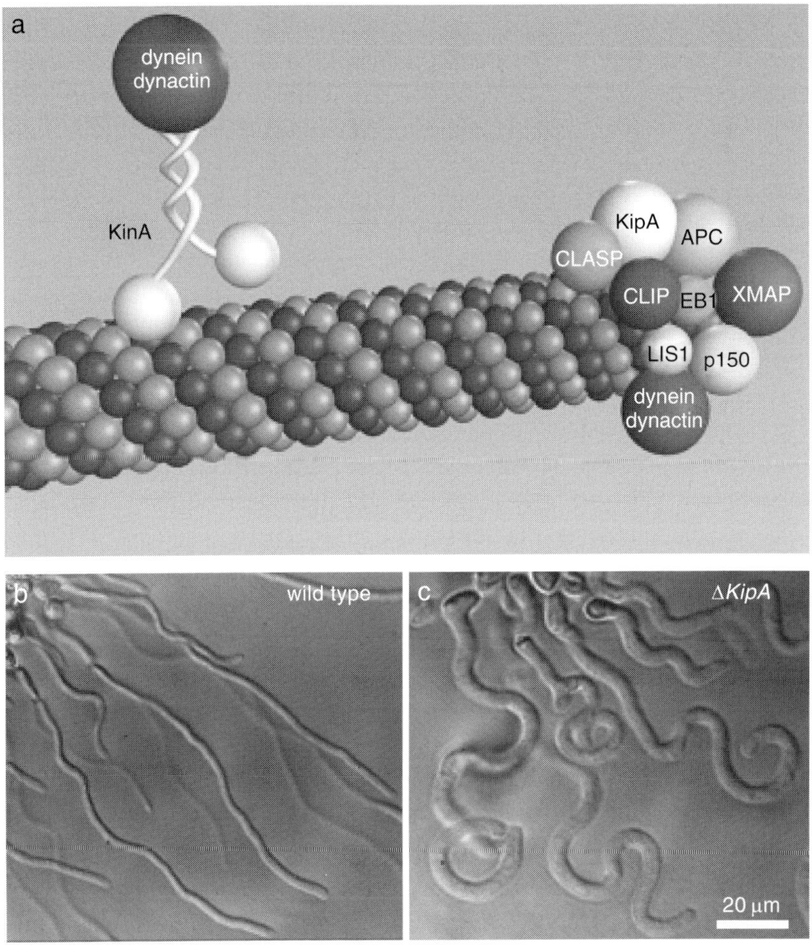

Figure 7.3 The microtubule plus end protein complex. (a) Motor proteins such as KinA or KipA deliver their cargo (here dynein) at the tip of the MT (Zhang *et al.*, 2003). A direct interaction between KinA and dynein or dynactin has not yet been verified. Modified after Hestermann *et al.* (2002). (b, c) When KipA, which was suggested to be involved in the delivery of cell end markers, is missing, hyphae loose directionality. Wild-type hyphae (b) grow straight, whereas hyphae of the *KipA* deletion strain (c) grow in curves. Pictures (b, c) were taken from Konzack *et al.* (2005).

interaction of MT-plus ends with the cortex contribute to the dynamics of mitotic spindles (Veith *et al.*, 2005; Xiang *et al.*, 1994; Xiang & Fischer, 2004). Whether interphase nuclei are also pulled through MT–cortex interactions is not clear yet. Whereas the role of the MT-plus end protein complex is quite obvious for force generation to translocate organelles, a role for polarized growth is less obvious. Some new ideas came from observations in growing tips of *A. nidulans* and visualization of MTs. Konzack *et al.* (2005) described how MTs merge at one point in the apex. Given that vesicles constantly travel towards the vesicle supply centre, the position of MT ends determines the vesicle supply centre location. In the *kipA* (*tea2*/*kip2*) mutant, MTs did not merge in one point and hyphae grew in meandering curves rather than straight. This was explained by the lack of proteins, normally transported by KipA, at the plus end, which mediate cortical contact. There is good evidence for such a situation in *S. pombe*. It was shown that the cortex protein Tea1 is transported by Tea2 (Browning *et al.*, 2003; Martin & Chang, 2003; Sawin & Snaith, 2004). If either of the two genes is deleted, *S. pombe* cells appear curved or T-shaped (Browning *et al.*, 2000; Snell & Nurse, 1994). Hence Tea1 and other proteins were named cell polarity determinants or cell end marker proteins. However, to prove such a model in *A. nidulans*, cargoes of KipA have to be identified and characterized. Another crucial piece in the puzzle is the identification of cortex proteins. Whereas cortical contacts of MTs involved in nuclear migration require the cortical protein ApsA in *A. nidulans* (Num1 in *S. cerevisiae*) (Veith *et al.*, 2005), this interaction appears not to be necessary for polarized growth (unpublished results).

The microtubule lattice

Microtubule function and dynamics are not only determined by the plus and minus end, but also by the filament lattice, which in higher eukaryotes can be decorated with a number of different microtubule-associated proteins (MAPs), which in turn may control the activity of associated motor proteins (Baas *et al.*, 1994; Baas & Qiang, 2005; Cassimeris & Spittle, 2001) (Fig. 7.4). Despite the abundance of those proteins in higher eukaryotes, it is not clear yet whether proteins like Tau exist in filamentous fungi. Proteins such as katanin and spastin could be conserved proteins, because sequences with high similarity can be found in the *A. nidulans* and in the *A. fumigatus* genomes (Konzack, unpublished results). Experimental data for the role of this class of MT-associated proteins are not available for filamentous fungi yet.

Microtubule-dependent motor proteins

Microtubules and their dynamics are in principle sufficient to create force and transport cargoes (attached to the growing end) in a cell. However, two classes of motor proteins have evolved, which guarantee fast MT-dependent movement in the cell, the minus-end directed dynein and the plus-end directed kinesins (Fig. 7.4). Some kinesins also move in the opposite direction. Both motor classes are characterized by a motor domain, in which ATP is hydrolyzed (Hirokawa, 1998). Within the protein, the location of the motor domain can be the N- or C-terminal as well as the middle region. How chemical energy is converted into conformational changes and how force is generated is best understood for conventional kinesin. Interested readers should refer to several recent reviews (Schliwa & Woehlke, 2003; Woehlke & Schliwa, 2000; Yildiz & Selvin, 2005).

Whereas all fungi employ a single dynein for their transport processes, their genomes usually contain several kinesin-encoding genes. For instance, *A. nidulans* harbours 11 and *N. crassa* 10 different kinesins (Fuchs & Westermann, 2005; Rischitor *et al.*, 2004). BimC was the first kinesin discovered in *A. nidulans* and defines the entire class of BimC-like kinesins (Enos & Morris, 1990). The gene was discovered in a screen for temperature-sensitive *A. nidulans* mutants with defects in mitosis (bim = block in mitosis). BimC is a C-terminal motor, which forms a tetramer with two motor domains opposite to each other. Because every head domain binds to an MT, this arrangement allows cross-linking of adjacent MTs. This feature is very important during mitosis, where mitotic spindle MTs slide along each other to distribute chromosomes (Kapitein *et al.*, 2005). Whereas BimC was discovered in a genetic screen (Morris, 1976), four

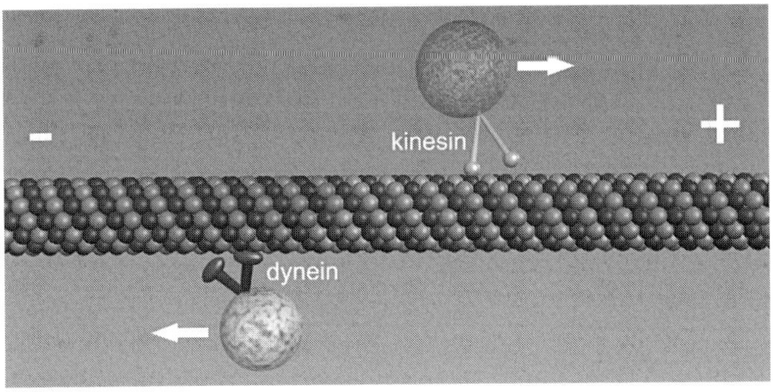

Figure 7.4 Kinesin and dynein transport cargoes along MTs.

other kinesins were isolated in reverse genetic approaches and will be discussed in the following chapters.

A second motor with functions in mitosis is the C-terminal kinesin-like protein KlpA, similar to *S. cerevisiae* Kar3 (Prigozhina *et al.*, 2001). The gene was isolated through a PCR approach and characterized subsequently. Deletion of *klpA* alone did not produce any severe phenotype but suppressed a *bimC* mutation (O'Connell *et al.*, 1993).

Another kinesin with a function in mitosis is the Kip3 family member KipB, where its motor domain is localized closer to the N-terminus. Gene deletion did not cause any defect in hyphal extension or organelle movement, but in chromosome segregation (Rischitor *et al.*, 2004). This was surprising, because a similar motor in *S. cerevisiae*, Kip3, is involved in nuclear migration (Miller *et al.*, 1998). However, *A. nidulans* KipB results are in good agreement with results for the homologous proteins in *S. pombe*, Klp5 and Klp6 (West *et al.*, 2002).

Two motors with N-terminal motor domains and pronounced roles in polarized growth are the conventional kinesin, KinA, and the CENP-E family kinesin, KipA. Deletion of *kinA* resulted in slower hyphal growth, which is similar to effects in other fungi (Lehmler *et al.*, 1997; Requena *et al.*, 2001; Seiler *et al.*, 1997; Wu *et al.*, 1998). It is generally accepted that this motor transports vesicles towards the extending tip and provides cell wall components (Seiler *et al.*, 1999). In addition, it appears to be involved in other cellular processes related to polarized growth, namely mitochondrial and nuclear distribution. Whereas nuclear distribution was affected in *N. crassa* and *A. nidulans*, mitochondrial distribution was changed in *N. haematococca* (Wu *et al.*, 1998). This may be due to the fact that mitochondrial movement depends on the actin cytoskeleton in *A. nidulans* (Suelmann & Fischer, 2000) and on the MT cytoskeleton in *N. crassa* (Fuchs *et al.*, 2002; Fuchs & Westermann, 2005). Whether mitochondrial distribution is also altered in *N. crassa* conventional kinesin mutants, has not been studied yet. The mechanism of how conventional kinesin may contribute to mitochondrial or nuclear distribution is not yet clear, but it could be that the effects are indirect. It was shown in *A. nidulans* that KinA is required for transportation of dynein subunits to the plus end of MTs (Zhang *et al.*, 2003). Dynein is a crucial motor for nuclear migration and prevention of dynein from the MT plus ends could cause the observed nuclear clustering (Xiang *et al.*, 1994). In addition, it has to be considered that conventional kinesin may well be involved in delivering other components of the MT plus end complex. Lack of conventional kinesin could thus influence the dynamics of MTs as well as their cortical interaction.

KipA of *A. nidulans* is similar to Tea2 in *S. pombe* and is characterized by an N-terminal motor domain (Konzack *et al.*, 2005). It accumulated at MT plus ends and appeared to reach this place by an intrinsic motor activity. Mutant proteins, in which a crucial residue for ATP hydrolysis was replaced, lost the ability to accumulate at MT tips but decorated them evenly. These findings were in agreement with studies of Tea2 in *S. pombe* (Browning *et al.*, 2003). Gene deletion caused a surprising phenotype in *A. nidulans*. Delta *kipA* strains grew as well as wild type strains, but hyphal morphology was changed. In contrast to MTs in wild type, MTs in the *kipA* deletion strain did not merge in one place at the apex. This was interpreted as a reason why hyphae would meander. If MTs do meet in one point, they would deliver the vesicles, which are transported along them at one place, the Spitzenkörper. Hence, hyphae would grow straightly. If MTs do not merge in one point, vesicles could be delivered at different places, and arbitrarily a majority could be deposited on the left, in the middle or on the right side of the hypha. If deposited at the left, the hypha would grow to the left. In *S. pombe* it was shown, that Tea2 transports e.g. Tea1 and Tip1, which are called 'cell end markers' and which are required for polarized growth (Browning *et al.*, 2003; Busch *et al.*, 2004). KipA could transport similar cargoes. However, no such protein has been identified yet.

Deletion of any kinesin motor (besides *bimC*) did not cause severe phenotypes. Actually, even a strain in which KinA, KipA and KipB were deleted was still viable, although hyphal growth and development were quite severely affected (Konzack *et al.*, 2005). This shows that kinesins can substitute for each other to some extent, which was recently shown nicely in the case of the Unc-104 homologues, Nkin-2 and Nkin-3 from *N. crassa*. Whereas Nkin-2 associates with mitochondria and connects mitochondria with MTs, Nkin-3 was found in the cytoplasm. Surprisingly, after depletion of Nkin-2, Nkin-3 was upregulated and also bound to mitochondria and MTs (Fuchs & Westermann, 2005). Homologues of these two motors do also exist in *A. nidulans* and are currently being investigated in our laboratory.

As mentioned above, fungi usually contain only a single dynein protein, although in some basidiomycetes the heavy chain is encoded by two genes (Eshel *et al.*, 1993; Martin *et al.*, 2004; Straube *et al.*, 2001; Xiang *et al.*, 1994; Yamamoto & Hiraoka, 2003). Dynein has a crucial role in nuclear migration, but is also implicated in vesicle transport (Seiler *et al.*, 1999). Because dynein moves towards the MT minus end, it is difficult to imagine that it is directly involved in polarized growth, given that

MTs are mainly oriented with their plus-ends to the membrane. Indeed, deletion of dynein does not cause an immediate block of hyphal extension and the impact on colony growth could partly be due to the lack of nuclei and other organelles, which are translocated with the help of dynein (Xiang *et al.*, 1994).

Besides the concerted action of the cytoskeleton and associated motor proteins to translocate organelles, cytoplasmic streaming has to be considered as another mechanism to push forward the cytoplasm and organelles. Mouriño *et al.* (2006) showed recently in *N. crassa* that the MT array was able to advance as a unit as the hyphae elongate. The basis for this bulk flow has not been resolved yet.

Actin versus microtubule cytoskeleton

The exact role that MTs play in polarized growth is not yet clear. Surely, depolymerization of MTs leads to a ten-fold reduction of the hyphal growth rate (Horio & Oakley, 2005). In contrast, the actin cytoskeleton is essential and disassembly of F-actin by e.g. latrunculin B causes a fast block in hyphal extension (Sampson & Heath, 2005) (Fig. 7.5). Likewise, deletion of the myosin gene, *myoA*, is lethal (McGoldrick *et al.*, 1995). There are two likely contributions of the actin cytoskeleton to polarized growth. On the one hand the actin–myosin cytoskeleton is used for vesicle transportation and secretion. On the other hand cortical proteins are brought into place by this system in *S. cerevisiae* and guarantee proper attachment of MTs to the cortex (Schuyler & Pellman, 2001b). Because MT attachment sites required for polarized growth seem to be very defined in the apical dome (see above), it is conceivable that the actin cytoskeleton plays a role at this point. However, further experiments are required to unravel the exact mechanisms.

Another element of the actin cytoskeleton is a protein complex localized at the incipient bud of *S. cerevisiae*, which was named the polarisome. This structure is involved in the organization of the actin cytoskeleton and its appearance resembles the Spitzenkörper in filamentous fungi (Sagot *et al.*, 2002). However, there is evidence that this protein complex also exists in filamentous fungi in addition to the Spitzenkörper (Harris & Momany, 2004). The existence of polarisome components in filamentous fungi was shown first in *A. nidulans*. Sharpless and Harris demonstrated that SepA, a homologue of a key component of the yeast polarisome, Bni1, colocalize with the Spitzenkörper (Sharpless & Harris, 2002). Similarly in *A. gossypii*, a filamentous fungus very closely related to *S. cerevisiae* (Wendland & Walther, 2005), a homologue of the *S. cerevisiae* polarisome protein Spa2 was analyzed (Knechtle *et al.*, 2003). This protein was not essential in *A. gossypii* but necessary for fast polarized growth. In the same year, a Spa2 homologue was characterized in *C. albicans* and studied during filamentous growth (Zheng *et al.*, 2003). The protein persistently localized at hyphal tips and deletion caused defects in polarity establishment. Recently, Crampin *et al.* suggested that the polarisome and Spitzenkörper are distinct structures which coexist in hyphae (Crampin *et al.*, 2005; Sagot *et al.*, 2002) (Fig. 7.6). According to this model, filamentous fungal cells employ both the MT and the actin cytoskeleton, and related to these structural elements, the Spitzenkörper as vesicle supply centre and the polarisome for actin organization.

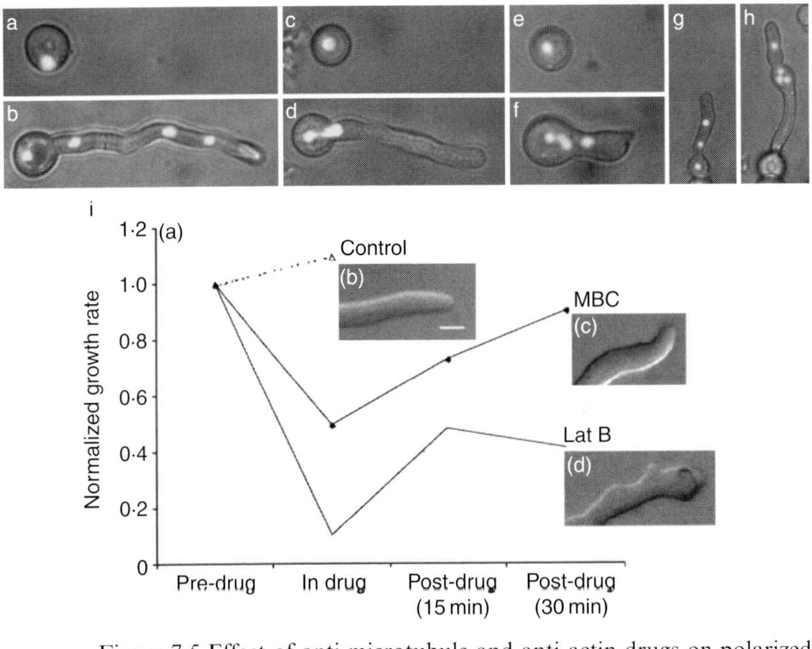

Figure 7.5 Effect of anti-microtubule and anti-actin drugs on polarized growth of *A. nidulans*. (a, c) Conidiospores of a wild-type strain with a GFP-stained nucleus. (b, d) Germlings of the strain in (a, c) grown for 10 h at 30°C in minimal medium (b) or minimal medium supplemented with 1.5 μg/ml benomyl (d). Hyphae elongated, the nucleus divided but nuclei did not migrate into the germtube. (e–h) Germination of a conidiospore in the presence of 50 μg/ml cytochalasin D. Hyphal morphology and polarized growth are affected. (i) Comparison of the growth rate of hyphal tips in the presence of the MT drug MBC or the actin drug latrunculin B (LatB). Taken from Sampson & Heath (2005).

Another aspect of MT function could be the maintenance of straight growth as discussed above. Low concentrations of benomyl caused hyphae to meander (Riquelme *et al.*, 1998). A similar effect was observed in *kipA* mutants in *A. nidulans* and in analogy to the situation in *S. pombe*, it was suggested that components of the MT plus end protein complex in combination with cortical proteins (cell end markers) determine growth directionality (Fig. 7.3). It was demonstrated in *S. pombe* that the kinesin Tea2 transports Tea1 towards MT plus ends, which then hitchhikes with the MT tip to the cortex where it subsequently interacts with Mod5 (Snaith & Sawin, 2003). We have searched the *A. nidulans* genome and identified a protein with similarity to Tea1 and named it TeaA. Although deletion of *teaA* caused a defect in cell polarity, the phenotype was slightly distinct from the *kipA* deletion phenotype (Konzack & Fischer, unpublished results). A protein with similarity to Mod5 could not be identified. The results in *S. pombe* and comparison with *A. nidulans* demonstrate that we are just at the beginning of understanding how growth directionality is achieved in filamentous fungi (Fig. 7.6).

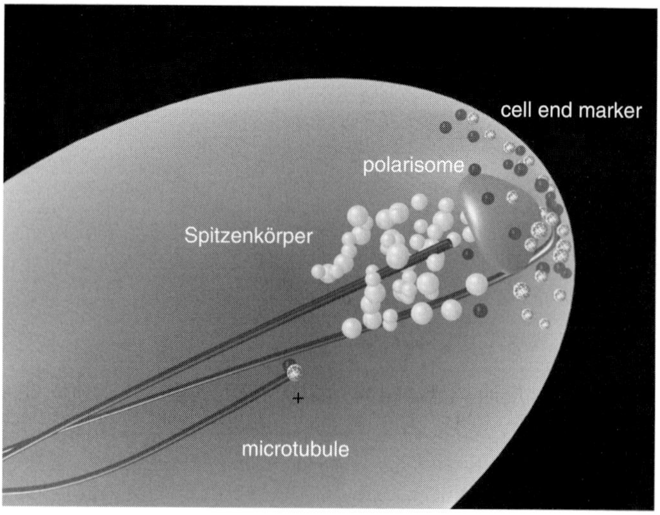

Figure 7.6 Three-dimensional model of a growing hyphal tip with MTs, the Spitzenkörper, the polarisome and cell end markers. Modified after Harris *et al.* (2005). It is not absolutely clear at the moment if the Spitzenkörper and the polarisome are separate structures or whether the components of the polarisome are subunits of the Spitzenkörper.

Conclusion

The last few years have offered many new insights into the arrangement, the formation and function of the MT cytoskeleton and the corresponding motor proteins. It appears that the main function of MTs is to deliver vesicles and cell end markers. The latter function, especially, needs much more attention, because only one putative cell end marker protein has been identified in *A. nidulans* so far. If homologues of *S. pombe* cell end markers exist in filamentous fungi, the question remains what their biochemical functions are, and which downstream events they trigger to allow straight hyphal growth. The publication of several fungal genomes along with continuous improvement of microscopical techniques promise a fruitful future for cytoskeletal research in fungi.

References

Akhmanova, A. & Hoogenraad, C. C. (2005). Microtubule plus-end-tracking proteins: mechanisms and functions. *Current Opinions in Cell Biology*, **17**, 47–54.

Aldaz, H., Rice, L. M., Stearns, T. & Agard, D. A. (2005). Insights into microtubule nucleation from the crystal structure of human gamma-tubulin. *Nature*, **435**, 523–7.

Baas, P. W., Pienkowski, T. P., Cimbalnik, K. A., Toyama, K., Bakalis, S., Ahmad, F. J. & Kosik, K. S. (1994). Tau confers drug stability but not cold stability to microtubules in living cells. *Journal of Cell Science*, **107**, 135–43.

Baas, P. W. & Qiang, L. (2005). Neuronal microtubules: when the MAP is the roadblock. *Trends in Cell Biology*, **15**, 183–7.

Bartnicki-Garcia, S., Bartnicki, D. D., Gierz, G., Lopez-Franco, R. & Bracker, C. E. (1995). Evidence that Spitzenkörper behavior determines the shape of a fungal hypha: a test of the hyphoid model. *Experimental Mycology*, **19**, 153–9.

Browning, H., Hayles, J., Mata, J., Aveline, L., Nurse, P. & McIntosh, J. R. (2000). Tea2p is a kinesin-like protein required to generate polarized growth in fission yeast. *Journal of Cell Biology*, **151**, 15–27.

Browning, H., Hackney, D. D. & Nurse, P. (2003). Targeted movement of cell end factors in fission yeast. *Nature Cell Biology*, **5**, 812–18.

Brunswick, H. (1924). *Untersuchungen über die Geschlechts und Kernverhältnisse bei der Hymenomyzetengattung Corpinus*. Jena: Gustav Fisher.

Busch, K. E., Hayles, J., Nurse, P. & Brunner, D. (2004). Tea2p kinesin is involved in spatial microtubule organization by transporting tip1p on microtubules. *Developmental Cell*, **16**, 831–43.

Carazo-Salas, R., Antony, C. & Nurse, P. (2005). The kinesin Klp2 mediates polarization of interphase microtubules in fission yeast. *Science*, **309**, 297–300.

Carminati, J. L. & Stearns, T. (1997). Microtubules orient the mitotic spindle in yeast through dynein-dependent interactions with the cell cortex. *Journal of Cell Biology*, **138**, 629–41.

Carvalho, P., Gupta, M. L. J., Hoyt, M. A. & Pellman, D. (2004). Cell cycle control of kinesin-mediated transport of Bik1 (CLIP-170) regulates microtubule stability and dynein activation. *Developmental Cell*, **6**, 815–29.

Cassimeris, L. & Spittle, C. (2001). Regulation of microtubule-associated proteins. *International Review of Cytology*, **210**, 163–226.

Crampin, H., Finley, K., Gerami-Nejad, M., Court, H., Gale, C., Berman, J. & Sudbery, P. (2005). *Candida albicans* hyphae have a Spitzenkörper that is distinct from the polarisome found in yeast and pseudohyphae. *Journal of Cell Science*, **118**, 2935–47.

Ding, D. Q., Chikashige, Y., Haraguchi, T. & Hiraoka, Y. (1998). Oscillatory nuclear movement in fission yeast meiotic prophase is driven by astral microtubules, as revealed by continuous observation of chromosomes and microtubules in living cells. *Journal of Cell Science*, **111**, 701–12.

Doxsey, S., McCollum, D. & Theurkauf, W. (2005). Centrosomes in cellular regulation. *Annual Review of Cell and Developmental Biology*, **21**, 411–34.

Efimov, V., Zhang, J. & Xiang, X. (2006). CLIP-170 homologue and NUDE play overlapping roles in NUDF localization in *Aspergillus nidulans*. *Molecular Biology of the Cell*, **17**, 2021–34.

Enos, A. P. & Morris, N. R. (1990). Mutation of a gene that encodes a kinesin-like protein blocks nuclear division in *A. nidulans*. *Cell*, **60**, 1019–27.

Eshel, D., Urrestarazu, L. A., Vissers, S., Jauniaux, J. C., Van Vliet-Reedijk, J. C., Planta, R. J. & Gibbons, I. R. (1993). Cytoplasmic dynein is required for normal nuclear segregation in yeast. *Proceedings of the National Academy of Sciences USA*, **90**, 11172–6.

Fischer, R. & Timberlake, W. E. (1995). *Aspergillus nidulans* apsA (anucleate primary sterigmata) encodes a coiled-coil protein necessary for nuclear positioning and completion of asexual development. *Journal of Cell Biology*, **128**, 485–98.

Freitag, M., Hickey, P. C., Raju, N. B., Selker, E. U. & Read, N. D. (2004). GFP as a tool to analyze the organization, dynamics and function of nuclei and microtubules in *Neurospora crassa*. *Fungal Genetics and Biology*, **41**, 897–910.

Fuchs, F., Prokisch, H., Neupert, W. & Westermann, B. (2002). Interaction of mitochondria with microtubules in the filamentous fungus *Neurospora crassa*. *Journal of Cell Science*, **115**, 1931–7.

Fuchs, F. & Westermann, B. (2005). Role of Unc104/KIF1-related motor proteins in mitochondrial transport in *Neurospora crassa*. *Molecular Biology of the Cell*, **16**, 153–61.

Girbardt, M. (1957). Der Spitzenkörper von Polystictus versicolor. *Planta*, **50**, 47–59.

Hagan, I. M. (1998). The fission yeast microtubule cytoskeleton. *Journal of Cell Science*, **111**, 1603–12.

Han, G., Liu, B., Zhang, J., Zuo, W., Morris, N. R. & Xiang, X. (2001). The *Aspergillus* cytoplasmic dynein heavy chain and NUDF localize to microtubule ends and affect microtubule dynamics. *Current Biology*, **11**, 19–24.

Harris, S. D. & Momany, C. (2004). Polarity in filamentous fungi: moving beyond the yeast paradigm. *Fungal Genetics and Biology*, **41**, 391–400.

Harris, S. D., Read, N. D., Roberson, R. W., Shaw, B., Seiler, S., Plamann, M. & Momany, M. (2005). Polarisome meets Spitzenkörper: Microscopy, genetics, and genomics converge. *Eukaryotic Cell*, **4**, 225–9.

Heath, I. B. (1981). Nucleus-associated organelles in fungi. *International Review of Cytology*, **69**, 191–221.

Hestermann, A., Rehberg, M. & Gräf, R. (2002). Centrosomal microtubule plus end tracking proteins and their role in *Dictyostelium* cell dynamics. *Journal of Muscle Research and Cell Motility*, **23**, 621–30.

Hirokawa, N. (1998). Kinesin and dynein superfamily proteins and the mechanism of organelle transport. *Science*, **279**, 519–26.

Hoepfner, D., Brachat, A. & Philippsen, P. (2000). Time-lapse video microscopy analysis reveals astral microtubule detachment in the yeast spindle pole mutant *cnm67*^{V}. *Molecular Biology of the Cell*, **11**, 1197–211.

Horio, T. & Oakley, B. R. (2005). The role of microtubules in rapid hyphal tip growth of *Aspergillus nidulans*. *Molecular Biology of the Cell*, **16**, 918–26.
Jaspersen, S. L. & Winey, M. (2004). The budding yeast spindle pole body: structure, duplication, and function. *Annual Review of Cell and Developmental Biology*, **20**, 1–28.
Job, D., Valiron, O. & Oakley, B. R. (2003). Microtubule nucleation. *Current Opinion in Cell Biology*, **15**, 111–17.
Jung, M. K., May, G. S. & Oakley, B. R. (1998). Mitosis in wild-type and β-tubulin mutant strains of *Aspergillus nidulans*. *Fungal Genetics and Biology*, **24**, 146–60.
Kapitein, L. C., Peterman, E. J., Kwok, B. H., Kim, J. H., Kapoor, T. M. & Schmidt, C. F. (2005). The bipolar mitotic kinesin Eg5 moves on both microtubules that it crosslinks. *Nature*, **435**, 114–18.
Knechtle, P., Dietrich, F. & Philippsen, P. (2003). Maximal polar growth potential depends on the polarisome component AgSpa2 in the filamentous fungus *Ashbya gossypii*. *Molecular Biology of the Cell*, **14**, 4140–54.
Konzack, S., Rischitor, P., Enke, C. & Fischer, R. (2005). The role of the kinesin motor KipA in microtubule organization and polarized growth of *Aspergillus nidulans*. *Molecular Biology of the Cell*, **16**, 497–506.
Lehmler, C., Steinberg, G., Snetselaar, K. M., Schliwa, M., Kahmann, R. & Bölker, M. (1997). Identification of a motor protein required for filamentous growth in *Ustilago maydis*. *EMBO Journal*, **16**, 3464–73.
Maekawa, H., Usui, T., Knop, M. & Schiebel, E. (2003). Yeast Cdk1 translocates to the plus end of cytoplasmic microtubules to regulate bud cortex interactions. *EMBO Journal*, **22**, 438–49.
Maekawa, H. & Schiebel, E. (2004). Cdk1-Clb4 controls the interaction of astral microtubule plus ends with subdomains of the daughter cell cortex. *Genes and Development*, **18**, 1709–24.
Martin, R., Walther, A. & Wendland, J. (2004). Deletion of the dynein heavy-chain gene DYN1 leads to aberrant nuclear positioning and defective hyphal development in *Candida albicans*. *Eukaryotic Cell*, **3**, 1574–88.
Martin, S. G. & Chang, F. (2003). Cell polarity: a new mod(e) of anchoring. *Current Biology*, **13**, R711–730.
Martin, S. G. & Chang, F. (2005). New end take off: regulating cell polarity during the fission yeast cell cycle. *Cell Cycle*, **4(8)**, 1046–9.
Mata, J. & Nurse, P. (1997). *tea1* and the microtubular cytoskeleton are important for generating global spatial order within the fission yeast cell. *Cell*, **89**, 939–49.
McDaniel, D. P. & Roberson, R. W. (1998). γ-tubulin is a component of the Spitzenkörper and centrosomes in hyphal-tip cells of *Allomyces macrogynus*. *Protoplasma*, **203**, 118–23.
McGoldrick, C. A., Gruver, C. & May, G. S. (1995). *myoA* of *Aspergillus nidulans* encodes an essential myosin I required for secretion and polarized growth. *Journal of Cell Biology*, **128**, 577–87.
Miller, R. K., Heller, K. K., Frisèn, L., Wallack, D. L., Loayza, D., Gammie, A. E. & Rose, M. D. (1998). The kinesin-related proteins, Kip2p and Kip3p, function differently in nuclear migration in yeast. *Molecular Biology of the Cell*, **9**, 2051–68.
Morris, N. R. (1976). Mitotic mutants of *Aspergillus nidulans*. *Genetical Research*, **26**, 237–54.
Mouriño-Pérez, R. R., Roberson, R. W. & Bartnicki-Garcia, S. (2006). Microtubule dynamics and organization during hyphal growth and branching in *Neurospora crassa*. *Fungal Genetics and Biology*, **43(6)**, 389–400.
O'Connell, M. J., Meluh, P. B., Rose, M. D. & Morris, N. R. (1993). Suppression of the *bimC4* mitotic spindle defect by deletion of *klpA*, a gene encoding a *KAR3*-related kinesin-like protein in *Aspergillus nidulans*. *Journal of Cell Biology*, **120**, 153–62.

Oakley, B. R., Oakley, C. E., Yoon, Y. & Jung, K. M. (1990). γ-Tubulin is a component of the spindle pole body in *Aspergillus nidulans*. *Cell*, **61**, 1289–301.

Oakley, B. R. (1995). A nice ring to the centrosome. *Nature*, **378**, 555–6.

Oakley, B. R. (2000). An abundance of tubulins. *Trends in Cell Biology*, **10**, 537–42.

Oakley, B. R. (2004). Tubulins in *Aspergillus nidulans*. *Fungal Genetics and Biology*, **41**, 420–7.

Oakley, C. E. & Oakley, B. R. (1989). Identification of gamma-tubulin, a new member of the tubulin superfamily encoded by *mipA* gene of *Aspergillus nidulans*. *Nature*, **338**, 662–4.

Ovechkina, Y., Maddox, P., Oakley, C. E., Xiang, X., Osmani, S. A., Salomon, E. D. & Oakley, B. R. (2003). Spindle formation in *Aspergillus* is coupled to tubulin movement into the nucleus. *Molecular Biology of the Cell*, **14**, 2192–200.

Pereira, G. & Schiebel, E. (1997). Centrosome-microtubule nucleation. *Journal of Cell Science*, **110**, 295–300.

Philippsen, P., Kaufmann, A. & Schmitz, H.-P. (2005). Homologues of yeast polarity genes control the development of multinucleated hyphae in *Ashbya gossypii*. *Current Opinions in Microbiology*, **8**, 370–7.

Prigozhina, N. L., Walker, R. A., Oakley, C. E. & Oakley, B. R. (2001). Gamma-tubulin and the C-terminal motor domain kinesin-like protein, KLPA, function in the establishment of spindle bipolarity in *Aspergillus nidulans*. *Molecular Biology of the Cell*, **12**, 3161–74.

Pruyne, D., Legesse-Miller, A., Gao, L., Dong, Y. & Bretscher, A. (2004). Mechanisms of polarized growth and organelle segregation in yeast. *Annual Review of Cell and Developmental Biology*, **20**, 559–91.

Requena, N., Alberti-Segui, C., Winzenburg, E., Horn, C., Schliwa, M., Philippsen, P., Liese, R. & Fischer, R. (2001). Genetic evidence for a microtubule-destabilizing effect of conventional kinesin and analysis of its consequences for the control of nuclear distribution in *Aspergillus nidulans*. *Molecular Microbiology*, **42**, 121–32.

Riquelme, M., Reynaga-Peña, C. G., Gierz, G. & Bartnicki-García, S. (1998). What determines growth direction in fungal hyphae? *Fungal Genetics and Biology*, **24**, 101–9.

Riquelme, M., Fischer, R. & Bartnicki-Garcia, S. (2003). Apical growth and mitosis are independent processes in *Aspergillus nidulans*. *Protoplasma*, **222**, 211–15.

Riquelme, M. & Bartnicki-Garcia, S. (2004). Key differenences between lateral and apical branching in hyphae of *Neurospora crassa*. *Fungal Genetics and Biology*, **41**, 842–51.

Rischitor, P., Konzack, S. & Fischer, R. (2004). The Kip3-like kinesin KipB moves along microtubules and determines spindle position during synchronized mitoses in *Aspergillus nidulans* hyphae. *Eukaryotic Cell*, **3**, 632–45.

Sagot, I., Klee, S. K. & Pellman, D. (2002). Yeast formins regulate cell polarity by controlling the assembly of actin cables. *Nature Cell Biology*, **4**, 42–50.

Sampson, K. & Heath, I. B. (2005). The dynamic behaviour of microtubules and their contributions to hyphal tip growth in *Aspergillus nidulans*. *Microbiology*, **151**, 1543–55.

Sawin, K. E., Lourenco, P. C. C. & Snaith, H. A. (2004). Microtubule nucleation at non-spindle pole body microtubule-organizing centers requires fission yeast centrosomin-related protein mod20p. *Current Biology*, **14**, 763–75.

Sawin, K. E. & Snaith, H. A. (2004). Role of microtubules and tea1p in establishment and maintenance of fission yeast cell polarity. *Journal of Cell Science*, **117**, 689–700.

Schliwa, M. & Woehlke, G. (2003). Molecular motors. *Nature*, **422**, 759–65.

Schuyler, S. C. & Pellman, D. (2001a). Search, capture and signal: games microtubules and centrosomes play. *Journal of Cell Science*, **114**, 247–55.

Schuyler, S. C. & Pellman, D. (2001b). Microtubule 'plus-end-tracking proteins': The end is just the beginning. *Cell*, **105**, 421–4.

Seiler, S., Nargang, F. E., Steinberg, G. & Schliwa, M. (1997). Kinesin is essential for cell morphogenesis and polarized secretion in *Neurospora crassa*. *EMBO Journal*, **16**, 3025–34.

Seiler, S., Plamann, M. & Schliwa, M. (1999). Kinesin and dynein mutants provide novel insights into the roles of vesicle traffic during cell morphogenesis in *Neurospora*. *Current Biology*, **9**, 779–85.

Sharpless, K. E. & Harris, S. D. (2002). Functional characterization and localization of the *Aspergillus nidulans* formin SEPA. *Molecular Biology of the Cell*, **13**, 469–79.

Sheeman, B., Carvalho, P., Sagot, I., Geiser, J., Kho, D., Hoyt, M. A. & Pellman, D. (2003). Determinants of *S. cerevisiae* dynein localization and activation: Implications for the mechanism of spindle positioning. *Current Biology*, **13**, 364–72.

Snaith, H. A. & Sawin, K. E. (2003). Fission yeast mod5p regulates polarized growth through anchoring of tea1p at cell tips. *Nature*, **423**, 647–51.

Snell, V. & Nurse, P. (1994). Genetic analysis of cell morphogenesis in fission yeast – a role for casein kinase II in the establishment of polarized growth. *EMBO Journal*, **13**, 2066–74.

Steinberg, G., Wedlich-Söldner, R., Brill, M. & Schulz, I. (2001). Microtubules in the fungal pathogen *Ustilago maydis* are highly dynamic and determine cell polarity. *Journal of Cell Science*, **114**, 609–22.

Straube, A., Enard, W., Berner, A., Wedlich-Söldner, R., Kahmann, R. & Steinberg, G. (2001). A split motor domain in a cytoplasmic dynein. *EMBO Journal*, **20**, 5091–100.

Straube, A., Brill, M., Oakley, B. R., Horio, T. & Steinberg, G. (2003). Microtubule organization requires cell cycle-dependent nucleation at dispersed cytoplasmic sites: polar and perinuclear microtubule organizing centers in the plant pathogen *Ustilago maydis*. *Molecular Biology of the Cell*, **14**, 642–57.

Suelmann, R. & Fischer, R. (2000). Mitochondrial movement and morphology depend on an intact actin cytoskeleton in *Aspergillus nidulans*. *Cell Motility and the Cytoskeleton*, **45**, 42–50.

Veith, D., Scherr, N., Efimov, V. P. & Fischer, R. (2005). Role of the spindle-pole body protein ApsB and the cortex protein ApsA in microtubule organization and nuclear migration in *Aspergillus nidulans*. *Journal of Cell Science*, **118**, 3705–16.

Venkatram, S., Jennings, J. L., Link, A. & Gould, K. L. (2005). Mto2p, a novel fission yeast protein required for cytoplasmic microtubule organization and anchoring of the cytokinetic actin ring. *Molecular Biology of the Cell*, **16**, 3052–63.

Wendland, J. & Walther, A. (2005). *Ashbya gossypii*: a model for fungal developmental biology. *Nature Reviews. Microbiology*, **3**, 421–9.

West, R. R., Malmstrom, T. & McIntosh, J. R. (2002). Kinesins *klp5*$^+$ and *klp6*$^+$ are required for normal chromosome movement in mitosis. *Journal of Cell Science*, **115**, 931–40.

Woehlke, G. & Schliwa, M. (2000). Walking on two heads: The many talents of kinesin. *Nature Reviews. Molecular Cell Biology*, **1**, 50–8.

Wu, Q., Sandrock, T. M., Turgeon, B. G., Yoder, O. C., Wirsel, S. G. & Aist, J. R. (1998). A fungal kinesin required for organelle motility, hyphal growth, and morphogenesis. *Molecular Biology of the Cell*, **9**, 89–101.

Xiang, X., Beckwith, S. M. & Morris, N. R. (1994). Cytoplasmic dynein is involved in nuclear migration in *Aspergillus nidulans*. *Proceedings of the National Academy of Sciences USA*, **91**, 2100–4.

Xiang, X. & Fischer, R. (2004). Nuclear migration and positioning in filamentous fungi. *Fungal Genetics and Biology*, **41**, 411–19.

Yamamoto, A. & Hiraoka, Y. (2003). Cytoplasmic dynein in fungi: insights from nuclear migration. *Journal of Cell Science*, **116**, 4501–12.

Yildiz, A. & Selvin, P. R. (2005). Kinesin: walking, crawling or sliding along? *Trends in Cell Biology*, **15**, 112–20.

Zhang, J., Han, G. & Xiang, X. (2002). Cytoplasmic dynein intermediate chain and heavy chain are dependent upon each other for microtubule end localization in *Aspergillus nidulans*. *Molecular Microbiology*, **44**, 381–92.

Zhang, J., Li, S., Fischer, R. & Xiang, X. (2003). The accumulation of cytoplasmic dynein and dynactin at microtubule plus-ends is kinesin dependent in *Aspergillus nidulans*. *Molecular Biology of the Cell*, **14**, 1479–88.

Zheng, X. D., Wang, Y. M. & Wang, Y. (2003). CaSPA2 is important for polarity establishment and maintenance in *Candida albicans*. *Molecular Microbiology*, **49**, 1391–405.

8

The cellular response to protein unfolding stress

M. SCHRÖDER

Introduction

Yeast and fungi are important for the production of industrial and therapeutic proteins (Tables 8.1 and 8.2). Product titres can reach several g/l, e.g. for human serum albumin expressed in *Pichia pastoris* (~7 g/l) (Kobayashi *et al.*, 2000) or cellobiohydrolase expressed in *Trichoderma reesei* (~20 g/l) (Durand, Clanet & Tiraby, 1988; Nakari-Setälä & Penttilä, 1995). However, in most cases product titres are ~1000-fold lower (Archer, Jeenes & Mackenzie, 1994; Penttilä, 1998; Cereghino & Cregg, 2000). Identification and engineering of the bottleneck in these, usually heterologous, protein production processes will increase their cost efficiency and competitiveness.

The first bottlenecks that were identified were gene copy number (Clare *et al.*, 1991; McGrew *et al.*, 1997; Vassileva *et al.*, 2001) and transcription efficiency (Outchkourov, Stiekema & Jongsma, 2002). Both were overcome by increasing gene dosage (Clare *et al.*, 1991; Parekh, Forrester & Wittrup, 1995; McGrew *et al.*, 1997; Vassileva *et al.*, 2001) and optimization of codon usage (Gouka, Punt & van den Hondel, 1997a; Kraševec, van den Hondel & Komel, 2000; Moralejo *et al.*, 2000; Outchkourov *et al.*, 2002; Cardoza *et al.*, 2003). Reports emerged stating that with increased gene copy number protein production decreased (Parekh *et al.*, 1995). Recombinant protein was shown to remain associated with the cell (Hohenblum, Borth & Mattanovich, 2003) and localized to the endoplasmic reticulum (ER) (Kauffman *et al.*, 2002). Thus, protein folding and posttranslational modification in the ER are a major bottleneck for protein secretion. In this review I will summarize principles of protein folding in the ER and cellular responses to protein unfolding in the ER in yeast and filamentous fungi. I will close by discussing the impact of these

Exploitation of Fungi, eds. G. D. Robson, P. van West and G. M. Gadd. Published by Cambridge University Press.

Table 8.1 *Induction of ER chaperone and foldase gene expression by protein overexpression in yeasts and filamentous fungi. Abbreviations:* bipA – Aspergillus *BiP,* clxA – Aspergillus *calnexin,* pdiA – Aspergillus *protein disulfide isomerase,* prpA – pdiA *related gene, and* tigA – *tunicamycin inducible gene A. Induction is represented by '↑', no change by '–'.*

Protein	Expression Level	Gene	Reference
Aspergillus awamori			
Bovine prochymosin		*prpA* (↑)	(Wang & Ward, 2000)
Endoxylanase II	5–10 times WT	*bipA* (–)	(Hessing *et al.*, 1994)
Single chain antibody (against *Streptococcus sanguis*)		*bipA* (↑)	(Punt *et al.*, 1998)
Single chain antibody (against lysozyme)		*bipA* (↑)	(Punt *et al.*, 1998)
Aspergillus niger			
Glucoamylase	5–10 times WT	*bipA* (↑)	(Verdoes *et al.*, 1993)
Glucoamylase – glucuronidase fusion	5–10 times WT	*bipA* (↑)	(Punt, Veldhuisen & van den Hondel, 1994)
Glucoamylase – human interleukin 6 fusion	2–3 times WT	*bipA* (–)	(Broekhuijsen *et al.*, 1993)
Hen egg white lysozyme	3 mg/l	*bipA* (↑) *pdiA* (↑) *tigA* (↑)	(Ngiam *et al.*, 2000)
	70 mg/l	*bipA* (↑) *pdiA* (↑) *tigA* (↑)	(Ngiam *et al.*, 2000)
Manganese peroxidase		*bipA* (↑) *clxA* (↑)	(Conesa *et al.*, 2002) (Conesa *et al.*, 2002)
Pichia pastoris			
Human trypsinogen		*bip* (↑)	(Hohenblum *et al.*, 2004)
Saccharomyces cerevisiae			
Single chain antibody		*bip* (↑)	(Kauffman *et al.*, 2002)

findings for engineering the next generation of microbial cell factories for the production of recombinant proteins.

Protein folding in the endoplasmic reticulum

To successfully fold a protein a cell has to solve three protein folding problems: (1) aggregation of nascent polypeptide chains, (2) huge discrepancies in the rates of protein folding and synthesis of the polypeptide

Table 8.2 *The effect of engineering of chaperone systems on protein production by yeasts and fungi. Abbreviations:* bipA – Aspergillus *BiP, clxA* – Aspergillus *calnexin,* pdiA – Aspergillus *protein disulfide isomerase,* prpA – pdiA *related gene, and Ubi4p – ubiquitin. Overexpression is represented by '↑', and decreased expression by '↓' under the heading 'manipulation'. No effect on expression of the target protein is represented by '–', increased expression of the target protein by '↑', and decreased expression of the target protein by '↓' in the column 'effect'.*

Protein	Yield	Manipulation	Effect	Reference
Aspergillus awamori				
Acid phosphatase	20 U/mg	*pdiA* ↑	–	(Moralejo *et al.*, 2001)
α-amylase	0.8 U/mg	*pdiA* ↑	–	(Moralejo *et al.*, 2001)
Bovine prochymosin	12 U	*prpA* ↑	–	(Wang & Ward, 2000)
Cutinase (wild-type)	55 mg/l	*bipA* ↑	–	(van Gemeren *et al.*, 1998)
Cutinase (T179Y)	25 mg/l	*bipA* ↑	↓	(van Gemeren *et al.*, 1998)
Glucoamylase – human interleukin 6 fusion		*bipA* ↑	–	(Punt *et al.*, 1998)
Glucoamylase – single chain antibody fusion (against *Streptococcus sanguis*)		*bipA* ↑	–	(Punt *et al.*, 1998)
Thaumatin	5 mg/l	*pdiA* ↑	↑	(Moralejo *et al.*, 2001)
Aspergillus niger				
Glucoamylase	200 mg/l	*pdiA* ↑	–	(Ngiam *et al.*, 2000)
		pdiA ↓	↓	(Ngiam *et al.*, 2000)
Hen egg white lysozyme	60 mg/l	*pdiA* ↑	–	(Ngiam *et al.*, 2000)
Manganese peroxidase		*clxA* ↑	↑	(Conesa *et al.*, 2002)
		bipA ↑	↓	(Conesa *et al.*, 2002)
Kluyveromyces lactis				
Human interleukin 1β		Ubi4p ↑	–	(Bao & Fukuhara, 2001)
		Pdi1p ↑	–	
Human serum albumin		Ubi4p ↑	↑	(Bao & Fukuhara, 2001)
		Pdi1p ↑	↑	
Saccharomyces cerevisiae				
Acid phosphatase (*Schizosaccharomyces pombe*)	2.5 A_{435}/A_{600}	Kar2p ↑	–	(Robinson *et al.*, 1996)
	42 A_{435}/ml/min	Kar2p ↓	↓	(Robinson *et al.*, 1996)
		Pdi1p ↑	↑	(Robinson, Hines & Wittrup, 1994)
Antistasin	0.3 mg/l	Pdi1p ↑	↑	(Schultz *et al.*, 1994)
Bovine pancreatic trypsin inhibitor	5 mg/l	Kar2p ↑	–	(Robinson *et al.*, 1996)
		Kar2p ↓	↓	(Robinson *et al.*, 1996)

Table 8.2 (*cont.*)

Protein	Yield	Manipulation	Effect	Reference
Carboxypeptidase A		Pdi1p ↓	↓	(Tachibana & Stevens, 1992)
Elafin	40 µl/l	Ubi4p ↑	↑	(Chen, Pioli & Piper, 1994)
Erythropoietin		Kar2p ↑	↑	(Gouka, Punt & van den Hondel, 1997b)
Human granulocyte colony-stimulating factor	0.1 mg/l	Kar2p ↑	–	(Robinson *et al.*, 1996)
		Kar2p ↓	↓	(Robinson *et al.*, 1996)
Human lysozyme	0.5 mg/l	Pdi1p ↑	↑	(Hayano, Hirose & Kikuchi, 1995)
Human platelet derived growth factor B homodimer	0.14 mg/l	Pdi1p ↑	↑	(Robinson *et al.*, 1994)
Prochymosin	0.04 g/kg	Kar2p ↑	↑	(Harmsen *et al.*, 1996)
Single-chain antibody fragments		Kar2p ↑	↑	(Shusta *et al.*, 1998)
		Pdi1p ↑	↑	
Thaumatin	0.4 g/kg	Kar2p ↑	–	(Harmsen *et al.*, 1996)
Yeast killer toxin		Pdi1p ↓	↓	(Dunn *et al.*, 1995)

chain by the ribosome, and (3) kinetically unfavoured protein folding reactions. The hydrophobic effect is a major driving force in protein folding because of the hydrophobic nature of many amino acid side chains. The initial event in folding of many proteins is a 'hydrophobic collapse' (Stevens & Argon, 1999) in which hydrophobic amino acid residues form a hydrophobic protein core. These hydrophobic interactions are not restricted to a productive folding pathway and also occur between two proteins resulting in the formation of large unfolded protein aggregates, e.g. inclusion bodies in *Escherichia coli*. Formation of protein secondary structures such as α-helices or β-turns occurs on a µs scale (Dobson, Sali & Karplus, 1998), whereas synthesis of the polypeptide chain by the translating ribosome proceeds at 2–8 amino acids/s (Palmiter, 1975). This requires that regions far apart in the primary structure but interacting in the secondary structure of a protein, e.g. β-sheets or disulphide bonds, are preserved in an unfolded conformation until the interacting sequence has been synthesized. Finally, the rates for two folding reactions, protein disulphide bond formation and *cis-trans* peptidyl prolyl isomerization are too low to allow for productive protein folding without catalysis.

The solution to problems (1) and (2) are molecular chaperones. Enzymatic catalysis of disulphide bond formation by protein disulphide isomerases (PDIs) and *cis-trans* peptidyl prolyl isomerization by *cis-trans* peptidyl prolyl isomerases (PPIs) is the solution to problem (3). PDIs and PPIs are referred to as foldases, an enzyme class that catalyzes protein folding reactions. In contrast, molecular chaperones do not act as enzymes. Molecular chaperones shield unfolded regions of a protein displaying hydrophobic amino acid residues on their surfaces from each other by recognizing these hydrophobic regions (Flynn *et al.*, 1991; Blond-Elguindi *et al.*, 1993). In this way they create a private folding environment for an unfolded protein, also called an Anfinsen cage, to prevent aggregation of unfolded proteins. Further, chaperone interaction with unfolded proteins has two additional consequences: (1) activation of signalling pathways, collectively called the unfolded protein response (UPR), that regulate adaptive responses to protein unfolding in the ER, and (2) retention of slowly folding or folding incompetent proteins in the ER, and targeting of these molecules for destruction by the proteasome.

In yeast, two major chaperone classes operate in the ER. The HSP70 chaperones Kar2p/GRP78/BiP and Lhs1p/Cer1p/Ssi1p, and the lectin chaperone calnexin (Cne1p). HSP70 chaperones shuttle through a cycle of substrate binding and release coupled to ATP hydrolysis (Schröder & Kaufman, 2005a, b). ATP hydrolysis is stimulated by DnaJ-like co-chaperones (Scj1p, Jem1p, and Sec63p) and ADP ATP nucleotide exchange by GrpE-like co-chaperones (Sil1p/Sls1p). Calnexin is a component of the calnexin cycle. In this cycle, *N*-linked oligosaccharides of a folding protein are glucosylated and deglucosylated by action of uridine diphosphate (UDP)-glucose:glycoprotein glucosyl transferase (UGGT) and α-glucosidase II (Ellgaard & Helenius, 2003). CNX recognizes the monoglucosylated form and retains it in the ER, allowing it to continue to fold. UGGT has been proposed to be the gatekeeper of this cycle by sensing the folding status of the client protein. Kre5p may be the *S. cerevisiae* UGGT (Shahinian *et al.*, 1998). BiP can work on completely unfolded substrates, whereas the calnexin cycle works on partially folded substrates. Prolonged interaction of a substrate protein with either chaperone system triggers targeting of the protein for degradation. The lectin Mnl1p/Htm1p/ER degradation-enhancing α-mannosidase-like protein (EDEM) extracts these proteins from the calnexin cycle and targets them for degradation by the proteasome (Molinari *et al.*, 2003; Oda *et al.*, 2003).

Endoplasmic reticulum stress

In homeostasis, the influx of nascent, unfolded polypeptide chains into the ER is in balance with the folding capacity of the ER provided by its chaperone machinery. When the folding demand on the ER exceeds its folding capacity adaptive responses to this stress situation are initiated to return to homeostasis. These adaptive responses include the upregulation of ER chaperone and foldase expression, upregulation of phospholipid biosynthesis to increase ER size and dilute its unfolded protein content, attenuation of influx of nascent, unfolded polypeptide chains into the ER, increased clearance of unfolded proteins by upregulation of ER associated protein degradation (ERAD) by the proteasome, mounting an antioxidant response, and apoptosis. These responses are coordinated by signal transduction pathways collectively termed the unfolded protein response (UPR).

The unfolded protein response in *S. cerevisiae*

The UPR is currently much better understood in *S. cerevisiae* than in filamentous fungi. Figure 8.1 provides an overview of the UPR in yeast.

Mechanism of signal transduction

The ER stress signal is transduced across the ER membrane by the type I transmembrane protein kinase endoribonuclease Ire1p. Ire1p consists of an *N*-terminal ER luminal domain, and a cytosolic, or nucleoplasmic, *C*-terminal protein kinase domain and a domain with strong homology to RNase L (Fig. 8.2A). Despite low sequence conservation of the ER luminal domains between species, metazoan and plant ER luminal domains fused to the cytosolic portion of yeast Ire1p complement an *IRE1* deletion (Liu, Schröder & Kaufman, 2000; Okushima *et al.*, 2002), thus demonstrating functional conservation of the ER luminal domain in all eukaryotes.

The conserved portion of the Ire1p luminal domain has similarity to a β-propeller consisting of four four-stranded β-sheets ('blades') (Ponting, 2000). β-propellers with four blades are the substrate binding domains in extracellular matrix proteases (Fülöp & Jones, 1999; Paoli, 2001), suggesting that this domain may directly be involved in unfolded protein recognition by Ire1p. In unstressed cells the chaperone BiP is associated with the luminal domain of Ire1p. When unfolded proteins accumulate in the ER they sequester BiP from the ER luminal domain of Ire1p (Bertolotti *et al.*, 2000). Ire1p then dimerizes and activates its RNase domain by autophosphorylation in *trans*. Mutation of phosphorylation sites in yeast Ire1p

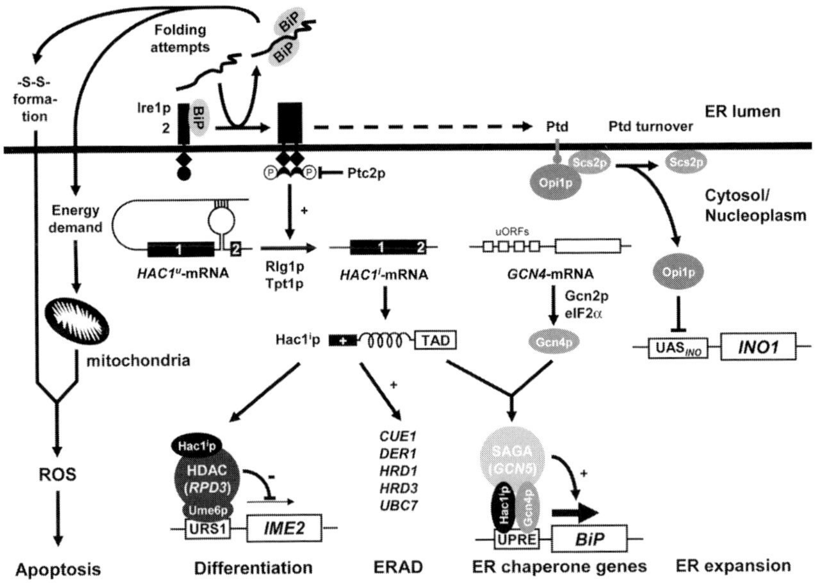

Figure 8.1 The UPR in *S. cerevisiae*. The ER luminal domain of Ire1p is depicted by a rectangle, the kinase domain by a square and the RNase domain by a circle or a half moon representing a conformational change (Papa *et al.*, 2003). The two *HAC1* exons are labelled 1 and 2, and the intron by a hairpin. Base pairing between the 5′ untranslated region (UTR) and the *HAC1* intron is represented by five parallel lines between the 5′ UTR and the intron (Rüegsegger *et al.*, 2001). Abbreviations: BiP – heavy chain binding protein, *CUE1* – coupling of ubiquitin conjugation to ER degradation 1, *DER1* – degradation in the ER 1, eIF2α – eukaryotic initiation factor 2α, ER – endoplasmic reticulum, ERAD – ER associated protein degradation, Gcn2p – general control non-derepressible 2 protein, *GCN4* – general control non-derepressible 4, *GCN5* – general control non-derepressible 5, $HAC1^u$ – homologous to activating transcription factor (ATF)/cAMP response binding protein (CREB), uninduced form, $HAC1^i$ – homologous to activating transcription factor (ATF)/cAMP response binding protein (CREB), induced form, HDAC – histone deacetylase, *HRD1* and *HRD3* – HMG-CoA reductase degradation 1 and 3, Ire1p – inositol requiring 1 protein, P – phosphorylation, Ptc2p – phosphatase two C protein, Ptd – phosphatidic acid, Rlg1 – tRNA ligase, ROS – reactive oxygen species, RPD3 – reduced potassium dependency 3, SAGA – Spt-Ada-Gcn5 acetyltransferase, -S-S- – disulphide bond, TAD – transcriptional activation domain, Tpt1p – tRNA 2′-phosphotransferase, UAS_{INO} – upstream activating site in the *myo*-inositol-1-phosphate synthase gene (*INO1*) promoter, UBC7 – ubiquitin-conjugating enzyme gene 7, uORF – upstream open reading frame, Ume6p – unscheduled meiosis 6 protein, UPRE – unfolded protein response element, and URS1 – upstream repression site 1.

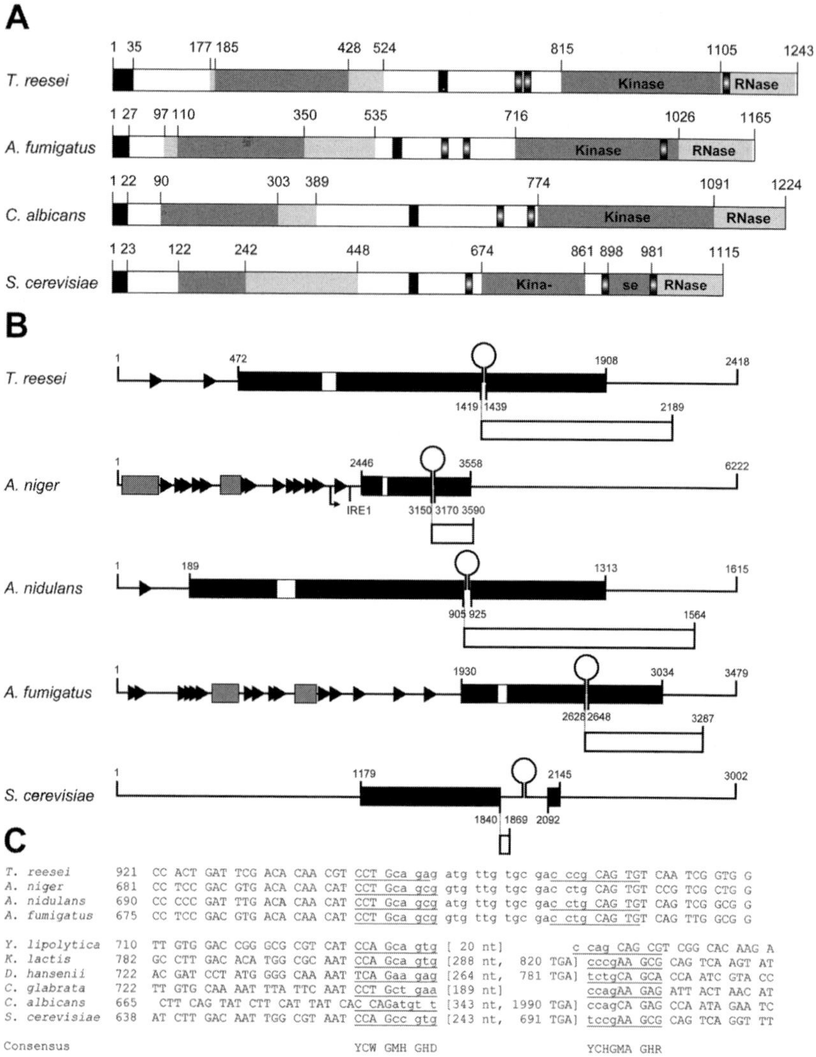

Figure 8.2 Regulatory components of the UPR in yeast and filamentous fungi. (A) Schematic representation of IRE1 protein from *Trichoderma reesei* (accession no. AAP92915), *Aspergillus fumigatus* (accession no. EAL87884), *Candida albicans* (accession no. EAK98566, EAK98642, XP_717532 and XP_7176061), and *Saccharomyces cerevisiae* (accession no. CAA77763, NP_011946 and P32361). Signal sequence and transmembrane domain are depicted by black boxes and nuclear localization sequences by white stars on a black background (Nakai & Horton, 1999). Unlabelled dark grey bars represent homology to the metazoan ER luminal domain (ProDom domain PD152704) and unlabelled light grey bars to yeast ER luminal domains (PD146858 and PD217699). For *S. cerevisiae* Ire1p

(S840 and S841) abolished its RNase activity. Dominant-negative mutations in the RNase domain suggest that an Ire1p dimer is the active form of the enzyme (Tirasophon *et al.*, 2000).

The substrate for the endoribonuclease domain of Ire1p is the mRNA for the basic leucine zipper (bZIP) transcription factor Hac1p. Unspliced yeast *HAC1* mRNA (*HAC1*^{u}, u for uninduced) consists of two exons that are separated by an unusual *C*-terminal intron of 252 nucleotides (nt) (Fig. 8.2B). Activated Ire1p cleaves conserved stem-loop structures at both exon-intron junctions in *HAC1*^{u} mRNA. tRNA ligase (Rlg1p/Trl1p) then joins the two exons. tRNA ligase leaves a 2′ phosphate group at the splice junction which is removed by the NAD^+-dependent phosphatase Tpt1p (Zillmann, Gorovsky & Phizicky, 1991). Temperature sensitive mutations in the spliceosome did not block *HAC1* mRNA splicing (Sidrauski, Cox & Walter, 1996), showing that this unusual splicing reaction bypasses the spliceosome. In yeast, genome-wide screens did not identify any additional substrates for the RNase activity of Ire1p (Niwa *et al.*, 2005). The splicing reaction introduces a new *C*-terminus into Hac1p (Hac1^{i}p, i for induced) with increased transcriptional activation potential (Mori *et al.*, 2000) (Fig. 8.2C). In *S. cerevisiae*, the splicing reaction also removes a translational attenuator residing in the intron of the unspliced mRNA (Chapman & Walter, 1997; Rüegsegger, Leber & Walter, 2001). The subcellular localization of *HAC1* splicing is controversial. tRNA ligase has been localized to the nucleus (Clark & Abelson, 1987), suggesting nuclear splicing of *HAC1* mRNA. In contrast, cytosolic and polyribosomal *HAC1*^{u} mRNA are substrates for Ire1p (Rüegsegger *et al.*, 2001), indicating cytosolic localization of the splicing event.

Fig 8.2 (cont.)
homology to PD217699 is shown. (B) Functional elements of *T. reesei* (accession no. AJ413273), *A. niger* (accession no. AY303684), *A. nidulans* (accession no. 413273), *A. fumigatus* (accession no. NW_876231) and *S. cerevisiae HAC1* (YFL031w) loci. Black bars represent coding sequence for the induced form of HAC1, white bars introns, triangles and grey bars uORF. The intron cleaved by IRE1 is shown as a hairpin. White bars below the locus represent the alternative *C*-terminus for the uninduced form of HAC1. The arrow represents the transcription start site for *A. niger hacA*, and Ire1 an additional potential substrate site in *A. niger hacA*. (C) Alignment of exon intron splice junctions cleaved by IRE1. Sequences are depicted in the translation frame for uninduced HAC1 to illustrate a translational frameshift introduced by removal of the intron. Where the intron encodes a stop codon, the translation frame of induced HAC1 is shown throughout.

Ire1p is dephosphorylated by the protein phosphatase Ptc2p (Welihinda *et al.*, 1998). *HAC1* mRNA splicing and expression of UPRE driven β-galactosidase reporters are elevated in *ptc2Δ* strains and attenuated in cells overexpressing WT, but not catalytically inactive Ptc2p (Welihinda *et al.*, 1998). Thus, Ptc2p is a negative regulator of the UPR.

Regulation of adaptive responses to endoplasmic reticulum stress

Molecular chaperones and foldases

Of the responses regulated by the UPR transcriptional activation of ER chaperone and foldase genes is best understood. In addition, Hac1^{i}p autostimulates its own transcription, which is required for prolonged survival of ER stress (Ogawa & Mori, 2004). The promoters of these genes harbour a partially palindromic sequence, the UPR element (UPRE, consensus sequence CAGCGTG). UPRE is the DNA binding site for Hac1^{i}p. Based on genetic data it was proposed that Hac1^{i}p binds UPRE as a homodimer. More recent evidence suggests that UPRE is bound by a heterodimer of Hac1^{i}p with the bZIP transcription factor Gcn4p (Patil, Li & Walter, 2004). Transcriptional activation by Hac1^{i}p is abolished in cells deleted for components of the Spt-Ada-Gcn5 acetyltransferase (SAGA) and the acetyltransferase SALSA (SAGA altered, Spt8 absent), e.g. Ada5p (Welihinda *et al.*, 1997). Further, Hac1^{i}p interacted with Ada3p and Ada5p in vitro (Welihinda, Tirasophon & Kaufman, 2000). These data suggest that Hac1^{i}p recruits SAGA to UPRE to acetylate the *N*-terminal tails of the nucleosomal core histone to promote transcription.

ER associated protein degradation

Transcriptional induction of genes involved in ER associated protein degradation (ERAD), Ubc7p, Cue1p and Hrd1p, requires both *IRE1* and *HAC1* (Friedlander *et al.*, 2000; Travers *et al.*, 2000). A UPRE or common promoter element has not yet been identified in the promoters of these genes. Additional promoter elements on which Hac1^{i}p activates transcription are UPRE-2 (TACGTG), and UPRE-3 (AGGACAAC) in ER chaperone and foldases gene promoters (Patil *et al.*, 2004), and a subtelomeric activating transcription factor (ATF)/cAMP response element binding protein (CREB) GTA variant element (SACE, ATGGTA TCAT) (Spode *et al.*, 2002). These elements may be involved in regulation of ERAD genes. In addition, modest transcriptional activation of core promoters in response to ER stress has been shown to be independent of *IRE1* and *HAC1* (Schröder, Clark & Kaufman, 2003).

Phospholipid biosynthesis

INO1, encoding inositol-1-phosphate synthase, is induced during ER stress in an *IRE1* and *HAC1* dependent manner (Cox, Chapman & Walter, 1997; Travers *et al.*, 2000). *ire1*Δ and *hac1*Δ strains are *myo*-inositol auxotrophs (Nikawa & Yamashita, 1992; Nikawa *et al.*, 1996). These observations support the idea that the UPR increases ER size through upregulation of phospholipid synthesis to dilute the unfolded protein load of the ER (Cox *et al.*, 1997). Initially, it was thought that Hac1p is directly involved in the transcriptional regulation of *INO1* (Cox *et al.*, 1997). More recent reports have shown that alternative carbon sources, e.g. oleic acid (Stroobants *et al.*, 1999), and mutations causing an overproduction of inositol (Opi^-) phenotype (Chang, Jones & Henry, 2002) bypass the requirement for *IRE1* and *HAC1* in transcription of *INO1*. The key transcriptional regulator of *INO1* is the bZIP protein Opi1p (Carman & Henry, 1989). Opi1p is tethered to the ER membrane by binding to phosphatidic acid (Loewen *et al.*, 2004) and held there in an inactive complex with Scs2p (Brickner & Walter, 2004; Loewen *et al.*, 2004). Addition of *myo*-inositol to cells converts phosphatidic acid to phosphatidylinositol, releases Opi1p from the ER membrane, and represses *INO1*. During activation the *INO1* locus relocates from the nucleoplasm to the nuclear membrane (Brickner & Walter, 2004). *ire1*Δ and *hac1*Δ cells display decreased *INO1* mRNA levels and decreased phosphatidic acid and phosphatidylinositol levels during prolonged inositol starvation (Chang *et al.*, 2002). Opi^- cells, in which the requirement for *IRE1* and *HAC1* to maintain high levels of *INO1* mRNA during *myo*-inositol starvation is bypassed, displayed elevated phosphatidic acid levels (Chang *et al.*, 2002). This observation supports the idea that *IRE1* and *HAC1* promote accumulation of phosphatidic acid in the ER membrane to relieve repression of *INO1*.

Attenuation of the influx of nascent, unfolded polypeptide chains into the endoplasmic reticulum

In contrast to all other eukaryotes, mechanisms for decreasing the influx of nascent, unfolded polypeptide chains into the ER during periods of ER stress have not yet been observed. Transient stimulation of *GCN4* translation required the eukaryotic initiation factor 2α (eIF2α) kinase Gcn2p during ER stress (Patil *et al.*, 2004). Phosphorylation of eIF2α at

S51 by Gcn2p attenuates general translation, but also promotes skipping of short upstream open reading frames (uORF) and translation of downstream open reading frames, as in the case for *GCN4*. Thus, transient attenuation of general translation to decrease the unfolded protein burden of the ER may exist after all in yeast.

Apoptosis

Accumulation of unfolded proteins in the ER triggers production of reactive oxygen species (ROS), externalization of phosphatidylserine to the outer leaflet of the plasma membrane, and DNA cleavage in ERAD deficient cells (Haynes, Titus & Cooper, 2004). These are hallmarks of apoptosis in eukaryotes, including yeast (Madeo, Fröhlich & Fröhlich, 1997; Madeo *et al.*, 1999). Reactive oxygen species are sufficient and required to trigger apoptosis in yeast (Madeo *et al.*, 1999) and originate from oxidative protein folding in the ER, especially disulphide bond formation, and ATP generation in mitochondria (Haynes *et al.*, 2004). *ire1*Δ and *hac1*Δ cells are protected from ROS accumulation and DNA cleavage (Haynes *et al.*, 2004). Repeated attempts to fold unfolded proteins are associated with disulphide bond isomerization and ATP consumption by molecular chaperones. Reactive oxygen species accumulation in ERAD deficient cells may arise from uncoupling two protective responses regulated by the UPR, chaperone and foldases expression and degradation of unfolded proteins, in these cells. Alternatively, a direct regulatory link between the UPR and apoptosis awaits discovery in yeast.

The unfolded protein response as a nutrient sensor

The informative content of UPR signalling extends beyond reporting on the balance between the folding burden and folding capacity of the ER. The UPR also informs on the overall nutritional state of the cell, as has been shown by investigation of diploid yeast cells defective in the UPR. Diploid yeasts have two differentiation responses to severe nitrogen starvation, pseudohyphal growth and meiosis (Gimeno *et al.*, 1992; Mitchell, 1994). Pseudohyphal growth is an alternative growth form characterized by an elongated cell shape, cells sticking together after cell division has been completed, synchronization of cell division in mother and daughter cells, and a change from a bipolar to unipolar budding pattern (Kron *et al.*, 1994). Meiosis produces an ascus harbouring four haploid spores. Pseudohyphal growth was derepressed in *ire1*Δ/*ire1*Δ and *hac1*Δ/*hac1*Δ cells (Schröder, Chang & Kaufman, 2000). Inhibition of protein folding in the ER or a ten-fold increase in the ammonium sulphate concentration in the medium inhibited pseudohyphal growth in an *IRE1* and *HAC1* dependent manner (Schröder *et al.*, 2000). Overexpression of Hac1^{i}p under nitrogen limitation repressed pseudohyphal growth in WT cells (Schröder *et al.*, 2000). A basal level of *HAC1* mRNA splicing was observed in exponentially growing yeast cells. Depletion of nitrogen rapidly (<15 min) inhibited the splicing reaction, and addition of ammonium salts to nitrogen starved cells was sufficient to rapidly (<15 min) reactivate *HAC1* splicing to levels comparable prior to induction of nitrogen starvation (Schröder *et al.*, 2000). Hac1^{i}p has a very short half-life (1–2 min) (Kawahara *et al.*, 1997). Thus, Hac1^{i}p rapidly depletes in nitrogen starved cells. In summary, the UPR represses pseudohyphal growth in a nitrogen-rich environment (Schröder *et al.*, 2000).

The UPR also regulates entry into meiosis. Meiosis is governed by a transcriptional cascade consisting of *IME1*, early, middle, and late meiotic genes (Kupiec *et al.*, 1997). *IME1* is a transcriptional activator of the early meiotic genes (EMGs), e.g. *HOP1*, *IME2* and *SPO13* (Kupiec *et al.*, 1997). Hac1^{i}p interfered with activation of EMGs during starvation, and EMG mRNA levels were derepressed in *hac1Δ* cells (Schröder *et al.*, 2000). Overexpression or deletion of *HAC1* had no effects on *IME1* mRNA levels. In contrast, *ire1Δ* cells were defective in activation of *IME1* and EMGs. Expression of Hac1^{i}p in *ire1Δ* cells did not restore *IME1* transcription, suggesting that Ire1p regulates *IME1* independent of Hac1p (Schröder *et al.*, 2000).

The observation that Hac1^{i}p interferes with activation of EMGs makes Hac1p the first bZIP transcription factor in yeast with both positive and negative roles in transcription. All EMGs contain the regulatory promoter element URS1 (TCGGCGGCT) in their promoters (Mitchell, 1994). Mutation of URS1 in endogenous promoters of the EMGs *DMC1* and *REC104* abolished negative regulation by Hac1^{i}p (Schröder *et al.*, 2004). URS1 is the DNA binding site for the transcriptional regulator Ume6p (Strich *et al.*, 1994). *ume6Δ* cells are defective in negative regulation of EMGs by Hac1^{i}p (Schröder *et al.*, 2004). In nitrogen-rich conditions Ume6p recruits two transcriptional repression complexes to URS1, the ISW2 chromatin remodelling complex (Goldmark *et al.*, 2000) and the Rpd3p-Sin3p histone deacetylase (HDAC) (Kadosh & Struhl, 1997; Washburn & Esposito, 2001). In nitrogen starvation, the transcriptional activator Ime1p is expressed, binds to Ume6p, and the Ume6p–Ime1p complex activates transcription of EMGs (Bowdish, Yuan & Mitchell, 1995; Rubin-Bejerano *et al.*, 1996). Epistasis analysis established that the ISW2 chromatin remodelling complex is dispensable and that the

Rpd3p-Sin3p HDAC is required for repression by Hac1^{i}p (Schröder *et al.*, 2004). Catalytically inactivating point mutations in Rpd3p that do not interfere with assembly of the multiprotein HDAC (Kadosh & Struhl, 1998) also abolished repression of EMGs by Hac1^{i}p (Schröder *et al.*, 2004). Hac1^{i}p co-immunoprecipitated with Rpd3p and Sin3p (Schröder *et al.*, 2004). Transcriptional repression by the HDAC was partially intact in *hac1*Δ cells, arguing that Hac1^{i}p is a peripheral or regulatory component of the HDAC (Schröder *et al.*, 2004). These data show that Hac1^{i}p associates with the HDAC in a nitrogen-rich environment to increase the transcriptional repression potential of the HDAC. This interaction is predicted to (a) induce structural changes in the HDAC, and (b) alter the catalytic activity or the substrate spectrum of the HDAC. Antibodies directed against Rpd3p co-immunoprecipitate Sin3p with Rpd3p more efficiently in wild-type (WT) than *hac1*Δ cells, showing that such a structural change exists (Schröder *et al.*, unpublished). In summary, Hac1^{i}p transduces a nitrogen signal from the ER to the HDAC to repress entry into meiosis in nitrogen-rich conditions, defining a role for the UPR in nutrient sensing and control of starvation programmes.

The unfolded protein response in filamentous fungi

Protein folding in the ER and the cellular response to protein unfolding in this organelle are very similar in filamentous fungi. Homologues for major chaperones and foldases have been cloned from several filamentous fungi (Tables 8.1 and 8.2). HSP70 co-chaperones, UGGT and α-glucosidase II homologues have not yet been cloned from these organisms. A few notable distinctions in the UPR of filamentous fungi have been identified. These are discussed in the following paragraphs.

Mechanism of signal transduction

IRE1 has been cloned from *Trichoderma reesei* (*Hypocrea jecorina*) (Valkonen, Penttilä & Saloheimo, 2004) (accession no. AY328083), and is present in the *Aspergillus fumigatus* genome (Fig. 8.2A). Sequence conservation of the bZIP transcription factor Hac1p is limited to its DNA binding region, and the stem loop structures at both exon-intron junctions that are recognized by IRE1. Functional homologues have been cloned from *Aspergillus nidulans* (HACA) (Saloheimo, Valkonen & Penttilä, 2003), *Aspergillus niger* (HACA) (Mulder *et al.*, 2004), and *T. reesei* (HAC1) (Saloheimo *et al.*, 2003) (Fig. 8.2B). Homologues of tRNA ligase and the NAD^+ dependent 2′ phosphatase have not been cloned from filamentous fungi. The intron in *A. nidulans hacA* (Saloheimo *et al.*, 2003),

A. niger hacA (Mulder *et al.*, 2004), and *T. reesei hac1* (Saloheimo *et al.*, 2003) mRNA is only 20 nt long. Splicing of these mRNAs also introduces an alternative *C*-terminus (Saloheimo *et al.*, 2003) (Fig. 8.2C). It has not yet been shown that these alternative *C*-termini encode more potent transcriptional activation domains. Several short uORFs reside in the 5′ mRNA ends of these mRNAs (Saloheimo *et al.*, 2003) (Fig. 8.2B), suggesting that these mRNAs are under translational control. All three mRNAs undergo a terminal shortening of their 5′ ends of 150, 230 or 250 nt, respectively (Saloheimo *et al.*, 2003; Mulder *et al.*, 2004). Removal of the 5′ mRNA end from *T. reesei HAC1* mRNA improved translational efficiency of this mRNA (Saloheimo *et al.*, 2003). Thus, truncation of these mRNAs removes sequences that inhibit translation of *hacA* and *hac1* mRNAs. It is not known if strong secondary structure elements reside in these 5′ ends that interfere with scanning of the untruncated mRNA by the 43 S preinitiation complex of the ribosome in analogy to yeast *HAC1* mRNA, or if translational repression is due to the presence of small uORFs present in the 5′ untranslated regions of these mRNAs. Truncated clones isolated from cDNA libraries do not contain the 20 nt intron, suggesting rapid processing of the exon-intron junctions by IRE1 or posttranscriptional truncation by an endonuclease system. A putative IRE1 cleavage site (CUUG↓CAGUG) overlapping with the start of the truncated *hacA*[i] transcript from *A. niger* has been identified (Mulder *et al.*, 2004). This site corresponds well to the consensus sequence for IRE1 cleavage (BCNGMNGHR; B = A, C, G; H = A, C, U; M = A, C; N = A, C, G, U; R = A, G) (Gonzalez *et al.*, 1999; Mulder *et al.*, 2004). How this, presumably uncapped, mRNA is translated and protected from degradation remains to be determined.

Regulation of adaptive responses

As in yeast, the UPR induces transcription of ER chaperone genes, e.g. *pdi1* in *T. reesei* (Saloheimo, Lund & Penttilä *et al.*, 1999) and *bipA* in *A. nidulans* (Valkonen *et al.*, 2003b). HACA also autostimulates its own transcription in *A. niger* (Mulder *et al.*, 2004). Much less is known about the regulation of ERAD, phospholipid synthesis or apoptosis by the UPR in these organisms. In contrast to yeast a mechanism, termed repression under secretion stress (RESS), downregulates the influx of nascent, unfolded polypeptide chains into the ER when protein folding in the ER is inhibited. Levels of several transcripts encoding secretory proteins, e.g. cellobiohydrolases I and II (*cbh1* and *cbh2*), endoglucanases I and II (*egl1* and *egl2*), hydrophonin (*hfb2*), and xylanase (*xyn1*) were downregulated after treatment of cells with DTT or the Ca^{2+} ionophore A23187 (Pakula *et al.*, 2003). Both treatments inhibit protein folding in the ER by reducing disulphide bonds or depleting ER luminal Ca^{2+} and inhibiting chaperone function. The promoter of the *cbh1* gene conferred this downregulation to a reporter gene (Pakula *et al.*, 2003). Similar observations were made with the *glaA* gene encoding glucoamylase in *A. niger* (Al-Sheikh *et al.*, 2004). Antisense inhibition of *pdiA* interfered with accumulation of *glaA* transcripts, but did not activate *hacA* mRNA splicing to significant levels (Al-Sheikh *et al.*, 2004). These observations suggest that RESS may be independent of spliced HACA. Promoter regions in the *cbh1* promoter that mediate RESS have been identified (Pakula *et al.*, 2004). Genome-wide studies will be necessary to show that RESS extends to, and is limited to, the majority of secretory proteins. Likewise, the molecular mechanism for RESS remains to be identified.

Engineering of the unfolded protein response to increase recombinant protein productivity

Expression of recombinant proteins induces the UPR as indicated by increased chaperone mRNA levels in most systems (Table 8.1). The conclusion that engineering of cells with increased chaperone expression will increase recombinant protein secretion turned out to be true for some, but not all, proteins of interest (Table 8.2). Further, in some instances increased chaperone levels resulted in decreased secretion of the recombinant protein (Table 8.2). Chaperone–client protein interactions will be different for each chaperone–client pair studied, even if chaperones like BiP or Pdi1p are quite general players. An important factor may be the time a chaperone stays associated with its client proteins. Prolonged interactions will result in targeting of the protein for degradation. Careful optimization of chaperone levels and activities may be necessary to circumvent this problem. Alternatively, several chaperones may work together on one client protein (Steel *et al.*, 2004), or co-chaperone or ER luminal ATP levels may become limiting for chaperone function. Thus, manipulation of individual chaperone levels may have little, if any, effect on productive folding of a client protein.

Engineering of the UPR, e.g. by expression of spliced Hac1p, is an excellent candidate to circumvent the latter problem. Indeed, overexpression of yeast $Hac1^{i}p$ or *T. reesei* spliced HAC1 increased secretion of *Bacillus amyloliquefaciens* α-amylase by *S. cerevisiae* (Valkonen, Penttilä & Saloheimo, 2003a). In *A. niger*, overexpression of spliced HACA increased secretion of *Trametes versicolor* laccase and of bovine preprochymosin

(Valkonen *et al.*, 2003b). However, overexpression of yeast Hac1^{i}p had no effect on secretion of *T. reesei* endoglucanase EGI by yeast (Valkonen *et al.*, 2003a). In conclusion, engineering of the UPR holds promise to increase productivity of eukaryotic host organisms for secretion of recombinant proteins. It should be kept in mind that nature invented the UPR to ensure survival of a cell, or a population. Additional responses are under control of the UPR, which may not be beneficial to recombinant protein production. A more detailed molecular understanding of how the UPR coordinates these responses will be crucial for the development of the next generation of production organisms.

References

Al-Sheikh, H., Watson, A. J., Lacey, G. A., Punt, P. J., MacKenzie, D. A., Jeenes, D. J., Pakula, T., Penttilä, M., Alcocer, M. J. & Archer, D. B. (2004). Endoplasmic reticulum stress leads to the selective transcriptional downregulation of the glucoamylase gene in *Aspergillus niger*. *Molecular Microbiology*, **53**, 1731–42.

Archer, D. B., Jeenes, D. J. & Mackenzie, D. A. (1994). Strategies for improving heterologous protein production from filamentous fungi. *Antonie Van Leeuwenhoek*, **65**, 245–50.

Bao, W. G. & Fukuhara, H. (2001). Secretion of human proteins from yeast: stimulation by duplication of polyubiquitin and protein disulfide isomerase genes in *Kluyveromyces lactis*. *Gene*, **272**, 103–10.

Bertolotti, A., Zhang, Y., Hendershot, L. M., Harding, H. P. & Ron, D. (2000). Dynamic interaction of BiP and ER stress transducers in the unfolded-protein response. *Nature Cell Biology*, **2**, 326–32.

Blond-Elguindi, S., Cwirla, S. E., Dower, W. J., Lipshutz, R. J., Sprang, S. R., Sambrook, J. F. & Gething, M.-J. (1993). Affinity panning of a library of peptides displayed on bacteriophages reveals the binding specificity of BiP. *Cell*, **75**, 717–28.

Bowdish, K. S., Yuan, H. E. & Mitchell, A. P. (1995). Positive control of yeast meiotic genes by the negative regulator UME6. *Molecular and Cellular Biology*, **15**, 2955–61.

Brickner, J. H. & Walter, P. (2004). Gene recruitment of the activated *INO1* locus to the nuclear membrane. *Public Library of Science Biology*, **2**, 1–11.

Broekhuijsen, M. P., Mattern, I. E., Contreras, R., Kinghorn, J. R. & van den Hondel, C. A. M. J. J. (1993). Secretion of heterologous proteins by *Aspergillus niger*: production of active human interleukin-6 in a protease-deficient mutant by KEX2-like processing of a glucoamylase-hIL6 fusion protein. *Journal of Biotechnology*, **31**, 135–45.

Cardoza, R. E., Gutierrez, S., Ortega, N., Colina, A., Casqueiro, J. & Martin, J. F. (2003). Expression of a synthetic copy of the bovine chymosin gene in *Aspergillus awamori* from constitutive and pH-regulated promoters and secretion using two different pre-pro sequences. *Biotechnology and Bioengineering*, **83**, 249–59.

Carman, G. M. & Henry, S. A. (1989). Phospholipid biosynthesis in yeast. *Annual Review of Biochemistry*, **58**, 635–69.

Cereghino, J. L. & Cregg, J. M. (2000). Heterologous protein expression in the methylotrophic yeast *Pichia pastoris*. *FEMS Microbiology Reviews*, **24**, 45–66.

Chang, H. J., Jones, E. W. & Henry, S. A. (2002). Role of the unfolded protein response pathway in regulation of *INO1* and in the *sec14* bypass mechanism in *Saccharomyces cerevisiae*. *Genetics*, **162**, 29–43.

Chapman, R. E. & Walter, P. (1997). Translational attenuation mediated by an mRNA intron. *Current Biology*, **7**, 850–9.

Chen, Y., Pioli, D. & Piper, P. W. (1994). Overexpression of the gene for polyubiquitin in yeast confers increased secretion of a human leucocyte protease inhibitor. *Biotechnology (NY)*, **12**, 819–23.

Clare, J. J., Romanos, M. A., Rayment, F. B., Rowedder, J. E., Smith, M. A., Payne, M. M., Sreekrishna, K. & Henwood, C. A. (1991). Production of mouse epidermal growth factor in yeast: high-level secretion using *Pichia pastoris* strains containing multiple gene copies. *Gene*, **105**, 205–12.

Clark, M. W. & Abelson, J. (1987). The subnuclear localization of tRNA ligase in yeast. *Journal of Cell Biology*, **105**, 1515–26.

Conesa, A., Jeenes, D., Archer, D. B., van den Hondel, C. A. M. J. J. & Punt, P. J. (2002). Calnexin overexpression increases manganese peroxidase production in *Aspergillus niger*. *Applied and Environmental Microbiology*, **68**, 846–51.

Cox, J. S., Chapman, R. E. & Walter, P. (1997). The unfolded protein response coordinates the production of endoplasmic reticulum protein and endoplasmic reticulum membrane. *Molecular Biology of the Cell*, **8**, 1805–14.

Dobson, C. M., Sali, A. & Karplus, M. (1998). Protein folding: a perspective from theory and experiment. *Angewandte Chemie International Edition in English*, **37**, 868–93.

Dunn, A., Luz, J. M., Natalia, D., Gamble, J. A., Freedman, R. B. & Tuite, M. F. (1995). Protein disulphide isomerase (PDI) is required for the secretion of a native disulphide-bonded protein from *Saccharomyces cerevisiae*. *Biochemical Society Transactions*, **23**, 78S.

Durand, H., Clanet, M. & Tiraby, G. (1988). Genetic improvement of *Trichoderma reesei* for large-scale cellulase production. *Enzyme and Microbial Technology*, **10**, 341–6.

Ellgaard, L. & Helenius, A. (2003). Quality control in the endoplasmic reticulum. *Nature Reviews Molecular and Cellular Biology*, **4**, 181–91.

Flynn, G. C., Pohl, J., Flocco, M. T. & Rothman, J. E. (1991). Peptide-binding specificity of the molecular chaperone BiP. *Nature*, **353**, 726–30.

Friedlander, R., Jarosch, E., Urban, J., Volkwein, C. & Sommer, T. (2000). A regulatory link between ER-associated protein degradation and the unfolded-protein response. *Nature Cell Biology*, **2**, 379–84.

Fülöp, V. & Jones, D. T. (1999). b propellers: structural rigidity and functional diversity. *Current Opinion in Structural Biology*, **9**, 715–21.

Gimeno, C. J., Ljungdahl, P. O., Styles, C. A. & Fink, G. R. (1992). Unipolar cell divisions in the yeast *S. cerevisiae* lead to filamentous growth: regulation by starvation and RAS. *Cell*, **68**, 1077–90.

Goldmark, J. P., Fazzio, T. G., Estep, P. W., Church, G. M. & Tsukiyama, T. (2000). The Isw2 chromatin remodeling complex represses early meiotic genes upon recruitment by Ume6p. *Cell*, **103**, 423–33.

Gonzalez, T. N., Sidrauski, C., Dörfler, S. & Walter, P. (1999). Mechanism of non-spliceosomal mRNA splicing in the unfolded protein response pathway. *EMBO Journal*, **18**, 3119–32.

Gouka, R. J., Punt, P. J. & van den Hondel, C. A. M. J. J. (1997a). Glucoamylase gene fusions alleviate limitations for protein production in *Aspergillus awamori* at the transcriptional and (post) translational levels. *Applied and Environmental Microbiology*, **63**, 488–97.

Gouka, R. J., Punt, P. J. & van den Hondel, C. A. M. J. J. (1997b). Efficient production of secreted proteins by *Aspergillus*: progress, limitations and prospects. *Applied Microbiology and Biotechnology*, **47**, 1–11.

Harmsen, M. M., Bruyne, M. I., Raue, H. A. & Maat, J. (1996). Overexpression of binding protein and disruption of the PMR1 gene synergistically stimulate secretion of bovine prochymosin but not plant thaumatin in yeast. *Applied Microbiology and Biotechnology*, **46**, 365–70.

Hayano, T., Hirose, M. & Kikuchi, M. (1995). Protein disulfide isomerase mutant lacking its isomerase activity accelerates protein folding in the cell. *FEBS Letters*, **377**, 505–11.

Haynes, C. M., Titus, E. A. & Cooper, A. A. (2004). Degradation of misfolded proteins prevents ER-derived oxidative stress and cell death. *Molecular Cell*, **15**, 767–76.

Hessing, J. G. M., van Rotterdam, C., Verbakel, J. M., Roza, M., Maat, J., van Gorcom, R. F. M. & van den Hondel, C. A. M. J. J. (1994). Isolation and characterization of a 1,4-β-endoxylanase gene of *A. awamori*. *Current Genetics*, **26**, 228–32.

Hohenblum, H., Borth, N. & Mattanovich, D. (2003). Assessing viability and cell-associated product of recombinant protein producing *Pichia pastoris* with flow cytometry. *Journal of Biotechnology*, **102**, 281–90.

Hohenblum, H., Gasser, B., Maurer, M., Borth, N. & Mattanovich, D. (2004). Effects of gene dosage, promoters, and substrates on unfolded protein stress of recombinant *Pichia pastoris*. *Biotechnology and Bioengineering*, **85**, 367–75.

Kadosh, D. & Struhl, K. (1997). Repression by Ume6 involves recruitment of a complex containing Sin3 corepressor and Rpd3 histone deacetylase to target promoters. *Cell*, **89**, 365–71.

Kadosh, D. & Struhl, K. (1998). Histone deacetylase activity of Rpd3 is important for transcriptional repression in vivo. *Genes and Development*, **12**, 797–805.

Kauffman, K. J., Pridgen, E. M., Doyle, F. J., 3rd, Dhurjati, P. S. & Robinson, A. S. (2002). Decreased protein expression and intermittent recoveries in BiP levels result from cellular stress during heterologous protein expression in *Saccharomyces cerevisiae*. *Biotechnology Progress*, **18**, 942–50.

Kawahara, T., Yanagi, H., Yura, T. & Mori, K. (1997). Endoplasmic reticulum stress-induced mRNA splicing permits synthesis of transcription factor Hac1p/Ern4p that activates the unfolded protein response. *Molecular Biology of the Cell*, **8**, 1845–62.

Kobayashi, K., Kuwae, S., Ohya, T., Ohda, T., Ohyama, M. & Tomomitsu, K. (2000). High level secretion of recombinant human serum albumin by fed-batch fermentation of the methylotrophic yeast, *Pichia pastoris*, based on optimal methanol feeding strategy. *Journal of Bioscience and Bioengineering*, **90**, 280–8.

Krasevec, N., van den Hondel, C. A. M. J. J. & Komel, R. (2000). Expression of human lymphotoxin alpha in *Aspergillus niger*. *Pflugers Arch*, **440**, R83–5.

Kron, S. J., Styles, C. A. & Fink, G. R. (1994). Symmetric cell division in pseudohyphae of the yeast *Saccharomyces cerevisiae*. *Molecular Biology of the Cell*, **5**, 1003–22.

Kupiec, M., Byers, B., Esposito, R. E. & Mitchell, A. P. (1997). Meiosis and sporulation in *Saccharomyces cerevisiae*. In *The Molecular and Cellular Biology of the Yeast Saccharomyces*, Vol. 3, ed. J. R. Pringle, J. R. Broach & E. W. Jones. Plainview, NY: Cold Spring Harbor Laboratory Press, pp. 889–1036.

Liu, C. Y., Schröder, M. & Kaufman, R. J. (2000). Ligand-independent dimerization activates the stress response kinases IRE1 and PERK in the lumen of the endoplasmic reticulum. *Journal of Biological Chemistry*, **275**, 24881–5.

Loewen, C. J., Gaspar, M. L., Jesch, S. A., Delon, C., Ktistakis, N. T., Henry, S. A. & Levine, T. P. (2004). Phospholipid metabolism regulated by a transcription factor sensing phosphatidic acid. *Science*, **304**, 1644–7.

Madeo, F., Fröhlich, E. & Fröhlich, K.-U. (1997). A yeast mutant showing diagnostic markers of early and late apoptosis. *Journal of Cell Biology*, **139**, 729–34.

Madeo, F., Fröhlich, E., Ligr, M., Grey, M., Sigrist, S. J., Wolf, D. H. & Fröhlich, K.-U. (1999). Oxygen stress: a regulator of apoptosis in yeast. *Journal of Cell Biology*, **145**, 757–67.

McGrew, J. T., Leiske, D., Dell, B., Klinke, R., Krasts, D., Wee, S. F., Abbott, N., Armitage, R. & Harrington, K. (1997). Expression of trimeric CD40 ligand in *Pichia pastoris*: use of a rapid method to detect high-level expressing transformants. *Gene*, **187**, 193–200.

Mitchell, A. P. (1994). Control of meiotic gene expression in *Saccharomyces cerevisiae*. *Microbiological Reviews*, **58**, 56–70.

Molinari, M., Calanca, V., Galli, C., Lucca, P. & Paganetti, P. (2003). Role of EDEM in the release of misfolded glycoproteins from the calnexin cycle. *Science*, **299**, 1397–400.

Moralejo, F. J., Cardoza, R. E., Gutierrez, S., Sisniega, H., Faus, I. & Martin, J. F. (2000). Overexpression and lack of degradation of thaumatin in an aspergillopepsin A-defective mutant of *Aspergillus awamori* containing an insertion in the *pepA* gene. *Applied Microbiology and Biotechnology*, **54**, 772–7.

Moralejo, F. J., Watson, A. J., Jeenes, D. J., Archer, D. B. & Martín, J. F. (2001). A defined level of protein disulfide isomerase expression is required for optimal secretion of thaumatin by *Aspegillus awamori*. *Molecular Genetics and Genomics*, **266**, 246–53.

Mori, K., Ogawa, N., Kawahara, T., Yanagi, H. & Yura, T. (2000). mRNA splicing-mediated C-terminal replacement of transcription factor Hac1p is required for efficient activation of the unfolded protein response. *Proceedings of the National Academy of Sciences of the United States of America*, **97**, 4660–5.

Mulder, H. J., Saloheimo, M., Penttilä, M. & Madrid, S. M. (2004). The transcription factor HACA mediates the unfolded protein response in *Aspergillus niger*, and up-regulates its own transcription. *Molecular Genetics and Genomics*, **271**, 130–40.

Nakai, K. & Horton, P. (1999). PSORT: a program for detecting sorting signals in proteins and predicting their subcellular localization. *Trends in Biochemical Sciences*, **24**, 34–6.

Nakari-Setälä, T. & Penttilä, M. (1995). Production of *Trichoderma reesei* cellulases on glucose-containing media. *Applied and Environmental Microbiology*, **61**, 3650–5.

Ngiam, C., Jeenes, D. J., Punt, P. J., van den Hondel, C. A. M. J. J. & Archer, D. B. (2000). Characterization of a foldase, protein disulfide isomerase A, in the protein secretory pathway of *Aspergillus niger*. *Applied and Environmental Microbiology*, **66**, 775–82.

Nikawa, J., Akiyoshi, M., Hirata, S. & Fukuda, T. (1996). *Saccharomyces cerevisiae IRE2/HAC1* is involved in *IRE1*-mediated *KAR2* expression. *Nucleic Acids Research*, **24**, 4222–6.

Nikawa, J. & Yamashita, S. (1992). IRE1 encodes a putative protein kinase containing a membrane-spanning domain and is required for inositol phototrophy in *Saccharomyces cerevisiae*. *Molecular Microbiology*, **6**, 1441–6.

Niwa, M., Patil, C. K., DeRisi, J. & Walter, P. (2005). Genome-scale approaches for discovering novel nonconventional splicing substrates of the Ire1 nuclease. *Genome Biology*, **6**, R3.

Oda, Y., Hosokawa, N., Wada, I. & Nagata, K. (2003). EDEM as an acceptor of terminally misfolded glycoproteins released from calnexin. *Science*, **299**, 1394–7.

Ogawa, N. & Mori, K. (2004). Autoregulation of the *HAC1* gene is required for sustained activation of the yeast unfolded protein response. *Genes to Cells*, **9**, 95–104.

Okushima, Y., Koizumi, N., Yamaguchi, Y., Kimata, Y., Kohno, K. & Sano, H. (2002). Isolation and characterization of a putative transducer of endoplasmic reticulum stress in *Oryza sativa*. *Plant Cellular Physiology*, **43**, 532–9.

Outchkourov, N. S., Stiekema, W. J. & Jongsma, M. A. (2002). Optimization of the expression of equistatin in *Pichia pastoris*. *Protein Expression and Purification*, **24**, 18–24.

Pakula, T., Saloheimo, M., Uusitalo, J., Huuskonen, A., Watson, A., Jeenes, D., Archer, D. & Pentilla, M. (2004). Method for production of secreted proteins in fungi. Sughrue Mion, PLLC: Canada. CA 2,438,356 (A1), 1–58.

Pakula, T. M., Laxell, M., Huuskonen, A., Uusitalo, J., Saloheimo, M. & Penttilä, M. (2003). The effects of drugs inhibiting protein secretion in the filamentous fungus

Trichoderma reesei. Evidence for down-regulation of genes that encode secreted proteins in the stressed cells. *Journal of Biological Chemistry*, **278**, 45011–20.

Palmiter, R. D. (1975). Quantitation of parameters that determine the rate of ovalbumin synthesis. *Cell*, **4**, 189–97.

Paoli, M. (2001). Protein folds propelled by diversity. *Progress in Biophysics and Molecular Biology*, **76**, 103–30.

Papa, F. R., Zhang, C., Shokat, K. & Walter, P. (2003). Bypassing a kinase activity with an ATP-competitive drug. *Science*, **302**, 1533–7.

Parekh, R., Forrester, K. & Wittrup, D. (1995). Multicopy overexpression of bovine pancreatic trypsin inhibitor saturates the protein folding and secretory capacity of *Saccharomyces cerevisiae*. *Protein Expression and Purification*, **6**, 537–45.

Patil, C. K., Li, H. & Walter, P. (2004). Gcn4p and novel upstream activating sequences regulate targets of the unfolded protein response. *Public Library of Science Biology*, **2**, 1208–23.

Penttilä, M. (1998). Heterologous protein production in *Trichoderma*. In *Trichoderma And Gliocladium*, Vol. 2, ed. H. G. E. & C. P. Kubicek. London: Taylor & Francis, pp. 365–82.

Ponting, C. P. (2000). Proteins of the endoplasmic-reticulum-associated degradation pathway: domain detection and function prediction. *Biochemical Journal*, **351**, 527–35.

Punt, P. J., van Gemeren, I. A., Drint-Kuijvenhoven, J., Hessing, J. G. M., van Muijlwijk-Harteveld, G. M., Beijersbergen, A., Verrips, C. T. & van den Hondel, C. A. M. J. J. (1998). Analysis of the role of the gene *bipA*, encoding the major endoplasmic reticulum chaperone protein in the secretion of homologous and heterologous proteins in black *Aspergilli*. *Applied Microbiology and Biotechnology*, **50**, 447–54.

Punt, P. J., Veldhuisen, G. & van den Hondel, C. A. M. J. J. (1994). Protein targeting and secretion in filamentous fungi. A progress report. *Antonie Van Leeuwenhoek*, **65**, 211–16.

Robinson, A. S., Bockhaus, J. A., Voegler, A. C. & Wittrup, K. D. (1996). Reduction of BiP levels decreases heterologous protein secretion in *Saccharomyces cerevisiae*. *Journal of Biological Chemistry*, **271**, 10017–22.

Robinson, A. S., Hines, V. & Wittrup, K. D. (1994). Protein disulfide isomerase overexpression increases secretion of foreign proteins in *Saccharomyces cerevisiae*. *Biotechnology (NY)*, **12**, 381–4.

Rubin-Bejerano, I., Mandel, S., Robzyk, K. & Kassir, Y. (1996). Induction of meiosis in *Saccharomyces cerevisiae* depends on conversion of the transcriptional represssor Ume6 to a positive regulator by its regulated association with the transcriptional activator Ime1. *Molecular and Cellular Biology*, **16**, 2518–26.

Rüegsegger, U., Leber, J. H. & Walter, P. (2001). Block of HAC1 mRNA translation by long-range base pairing is released by cytoplasmic splicing upon induction of the unfolded protein response. *Cell*, **107**, 103–14.

Saloheimo, M., Lund, M. & Penttilä, M. E. (1999). The protein disulphide isomerase gene of the fungus *Trichoderma reesei* is induced by endoplasmic reticulum stress and regulated by the carbon source. *Molecular and General Genetics*, **262**, 35–45.

Saloheimo, M., Valkonen, M. & Penttilä, M. (2003). Activation mechanisms of the HAC1-mediated unfolded protein response in filamentous fungi. *Molecular Microbiology*, **47**, 1149–61.

Schröder, M., Chang, J. S. & Kaufman, R. J. (2000). The unfolded protein response represses nitrogen-starvation induced developmental differentiation in yeast. *Genes and Development*, **14**, 2962–75.

Schröder, M., Clark, R. & Kaufman, R. J. (2003). *IRE1*- and *HAC1*-independent transcriptional regulation in the unfolded protein response of yeast. *Molecular Microbiology*, **49**, 591–606.

Schröder, M., Clark, R., Liu, C. Y. & Kaufman, R. J. (2004). The unfolded protein response represses differentiation through the *RPD3-SIN3* histone deacetylase. *EMBO Journal*, **23**, 2281–92.

Schröder, M. & Kaufman, R. J. (2005a). ER stress and the unfolded protein response. *Mutation Research*, **569**, 29–63.

Schröder, M. & Kaufman, R. J. (2005b). The mammalian unfolded protein response. *Annual Review of Biochemistry*, **74**, 739–89.

Schultz, L. D., Markus, H. Z., Hofmann, K. J., Montgomery, D. L., Dunwiddie, C. T., Kniskern, P. J., Freedman, R. B., Ellis, R. W. & Tuite, M. F. (1994). Using molecular genetics to improve the production of recombinant proteins by the yeast *Saccharomyces cerevisiae*. *Annals of the New York Academy of Sciences*, **721**, 148–57.

Shahinian, S., Dijkgraaf, G. J., Sdicu, A. M., Thomas, D. Y., Jakob, C. A., Aebi, M. & Bussey, H. (1998). Involvement of protein N-glycosyl chain glucosylation and processing in the biosynthesis of cell wall β-1,6-glucan of *Saccharomyces cerevisiae*. *Genetics*, **149**, 843–56.

Shusta, E. V., Raines, R. T., Pluckthun, A. & Wittrup, K. D. (1998). Increasing the secretory capacity of *Saccharomyces cerevisiae* for production of single-chain antibody fragments. *Nature Biotechnology*, **16**, 773–7.

Sidrauski, C., Cox, J. S. & Walter, P. (1996). tRNA ligase is required for regulated mRNA splicing in the unfolded protein response. *Cell*, **87**, 405–13.

Spode, I., Maiwald, D., Hollenberg, C. P. & Suckow, M. (2002). ATF/CREB sites present in sub-telomeric regions of *Saccharomyces cerevisiae* chromosomes are part of promoters and act as UAS/URS of highly conserved *COS* genes. *Journal of Molecular Biology*, **319**, 407–20.

Steel, G. J., Fullerton, D. M., Tyson, J. R. & Stirling, C. J. (2004). Coordinated activation of Hsp70 chaperones. *Science*, **303**, 98–101.

Stevens, F. J. & Argon, Y. (1999). Protein folding in the ER. *Seminars in Cell and Developmental Biology*, **10**, 443–54.

Strich, R., Surosky, R. T., Steber, C., Dubois, E., Messenguy, F. & Esposito, R. E. (1994). UME6 is a key regulator of nitrogen repression and meiotic development. *Genes and Development*, **8**, 796–810.

Stroobants, A. K., Hettema, E. H., van den Berg, M. & Tabak, H. F. (1999). Enlargement of the endoplasmic reticulum membrane in *Saccharomyces cerevisiae* is not necessarily linked to the unfolded protein response via Ire1p. *FEBS Letters*, **453**, 210–14.

Tachibana, C. & Stevens, T. H. (1992). The yeast *EUG1* gene encodes an endoplasmic reticulum protein that is functionally related to protein disulfide isomerase. *Molecular and Cellular Biology*, **12**, 4601–11.

Tirasophon, W., Lee, K., Callaghan, B., Welihinda, A. & Kaufman, R. J. (2000). The endoribonuclease activity of mammalian IRE1 autoregulates its mRNA and is required for the unfolded protein response. *Genes and Development*, **14**, 2725–36.

Travers, K. J., Patil, C. K., Wodicka, L., Lockhart, D. J., Weissman, J. S. & Walter, P. (2000). Functional and genomic analyses reveal an essential coordination between the unfolded protein response and ER-associated degradation. *Cell*, **101**, 249–58.

Valkonen, M., Penttilä, M. & Saloheimo, M. (2003a). Effects of inactivation and constitutive expression of the unfolded-protein response pathway on protein production in the yeast *Saccharomyces cerevisiae*. *Applied and Environmental Microbiology*, **69**, 2065–72.

Valkonen, M., Penttilä, M. & Saloheimo, M. (2004). The ire1 and ptc2 genes involved in the unfolded protein response pathway in the filamentous fungus *Trichoderma reesei*. *Molecular Genetics and Genomics*, **272**, 443–51.

Valkonen, M., Ward, M., Wang, H., Penttilä, M. & Saloheimo, M. (2003b). Improvement of foreign-protein production in *Aspergillus niger* var. *awamori* by constitutive induction of the unfolded-protein response. *Applied and Environmental Microbiology*, **69**, 6979–86.

van Gemeren, I. A., Beijersbergen, A., van den Hondel, C. A. M. J. J. & Verrips, C. T. (1998). Expression and secretion of defined cutinase variants by *Aspergillus awamori*. *Applied and Environmental Microbiology*, **64**, 2794–9.

Vassileva, A., Chugh, D. A., Swaminathan, S. & Khanna, N. (2001). Effect of copy number on the expression levels of hepatitis B surface antigen in the methylotrophic yeast *Pichia pastoris*. *Protein Expression and Purification*, **21**, 71–80.

Verdoes, J. C., Punt, P. J., Schrickx, J. M., van Verseveld, H. W., Stouthamer, A. H. & van den Hondel, C. A. M. J. J. (1993). Glucoamylase overexpression in *Aspergillus niger*: molecular genetic analysis of strains containing multiple copies of the *glaA* gene. *Transgenic Research*, **2**, 84–92.

Wang, H. & Ward, M. (2000). Molecular characterization of a PDI-related gene *prpA* in *Aspergillus niger* var. *awamori*. *Current Genetics*, **37**, 57–64.

Washburn, B. K. & Esposito, R. E. (2001). Identification of the Sin3-binding site in Ume6 defines a two-step process for conversion of Ume6 from a transcriptional repressor to an activator in yeast. *Molecular and Cellular Biology*, **21**, 2057–69.

Welihinda, A. A., Tirasophon, W., Green, S. R. & Kaufman, R. J. (1997). Gene induction in response to unfolded protein in the endoplasmic reticulum is mediated through Ire1p kinase interaction with a transcriptional coactivator complex containing Ada5p. *Proceedings of the National Academy of Sciences of the United States of America*, **94**, 4289–94.

Welihinda, A. A., Tirasophon, W., Green, S. R. & Kaufman, R. J. (1998). Protein serine/threonine phosphatase Ptc2p negatively regulates the unfolded-protein response by dephosphorylating Ire1p kinase. *Molecular and Cellular Biology*, **18**, 1967–77.

Welihinda, A. A., Tirasophon, W. & Kaufman, R. J. (2000). The transcriptional co-activator *ADA5* is required for *HAC1* mRNA processing *in vivo*. *Journal of Biological Chemistry*, **275**, 3377–81.

Zillmann, M., Gorovsky, M. A. & Phizicky, E. M. (1991). Conserved mechanism of tRNA splicing in eukaryotes. *Molecular and Cellular Biology*, **11**, 5410–16.

9

Protein secretion and associated stresses in *Aspergillus*: a genomic perspective

D. B. ARCHER, S. E. BARNES AND T. GUILLEMETTE

Introduction

Aspergillus spp. are filamentous fungi of which several have been studied in relation to the secretion of proteins. This chapter will only discuss those species for which there is currently genome sequence information even though the secretion of proteins is probably an important aspect of the lifestyles of the more than 180 known species of *Aspergillus*. Therefore, the emphasis is with a common saprophyte and human pathogen (*Aspergillus fumigatus*), a sexual species and one that has been developed as a convenient laboratory species for basic studies (*Aspergillus nidulans*), and two species that are widely exploited commercially for their secreted enzymes (*Aspergillus niger* and *Aspergillus oryzae*). The genome sequence data have been recently acquired and made available to the scientific community so it is particularly timely that the annotated sequences serve as a framework for this chapter on protein secretion. That will also avoid a substantial amount of repetition because there are several excellent reviews on protein secretion by *Aspergillus* that are primarily pre-genomic in their outlook (Conesa *et al.*, 2001; Punt *et al.*, 2002; MacKenzie *et al.*, 2004) even though extensive use of gene sequences and expression data has been made in the reviewed literature. More recent reviews have been able to take a more genomic view from the outset (Archer & Dyer, 2004; Archer & Turner, 2005) but, even then, the full annotated genome sequences were not available. Thus, this area is moving rapidly and current knowledge on protein secretion is founded on a wealth of pre-genomic and some emerging post-genomic data.

This chapter will aim to review the new genomic-based information on the capacity of *Aspergillus* spp. to secrete proteins and also to emphasize the developments in understanding the unfolded protein response and

Exploitation of Fungi, eds. G. D. Robson, P. van West and G. M. Gadd. Published by Cambridge University Press.

stresses associated with the secretion of proteins (most noticeable with heterologous proteins).

Aspergillus genomics

At the time of writing, manuscripts that describe the genome sequences and annotation for *A. nidulans* (Galagan *et al.*, 2005), *A. fumigatus* (Nierman *et al.*, 2005) and *A. oryzae* (Machida *et al.*, 2005) have been submitted for publication. The genome of *A. niger* has been sequenced (see Archer & Dyer, 2004) and the DSM sequence (www.dsm.com/dfs/innovation/genomics/) is being annotated by DSM and an invited group of experts. The intention is that a description of the annotated sequence will be published in 2006. In addition, the genome of another strain of *A. niger* is being sequenced by the US Department of Energy (www.microbialgenome.org/announcement/seq2005.shtml) and the genome of *A. flavus* is also being sequenced and the annotated version will soon be accessible (www.aspergillusflavus.org/genomics.html). A useful website with links to all the *Aspergillus* genome data is www.aspergillus.man.ac.uk. Other useful websites are referred to in Archer and Dyer (2004).

A brief summary of the genome sizes and predicted gene numbers for the four sequenced *Aspergillus* spp. is provided in Table 9.1. The numbers in tables of this sort change as the quality of the information improves over time, but the data are approximately correct. All species have similar (c. 50 mole %) G + C content. The genome sizes of *A. fumigatus* and *A. nidulans* are similar to each other and are smaller than the genomes of *A. niger* and *A. oryzae*. *A. oryzae* is closely related to, and often considered to be a 'domesticated' version of, *A. flavus* (Montiel *et al.*, 2003). Preliminary data indicate that the *A. flavus* genome, like that of *A. oryzae*, is larger than those of *A. nidulans* and *A. fumigatus* (Pritchard *et al.*, 2005). In the absence of available annotated data for *A. niger*, it is interesting to note that the *A. oryzae* genome appears to have expanded to accommodate additional genes for secretory hydrolytic enzymes, metabolism (including secondary metabolites) and uptake transporters for sugars and amino acids (Machida *et al.*, 2005). That increase in size (which may also be true of its close relative *A. flavus*) was suggested to have occurred by acquisition of genes rather than gene or chromosomal duplication. That is not to suggest that *A. oryzae* has the genome of, say, *A. fumigatus* and more because *A. fumigatus* itself has c. 500 genes that appear not to have orthologues in *A. nidulans* or *A. oryzae* (Nierman *et al.*, 2005). The annotation of these genomes in relation to the secretion of proteins and the impact that has on their respective lifestyles is discussed below.

Table 9.1 *Summary of information derived from the genome sequences of* Aspergillus *spp. The data were modified from Archer and Dyer (2004) using information from Galagan* et al. *(2005), Nierman* et al. *(2005) and Machida* et al. *(2005). Data for A. niger were updated by DSM (H. Stam and N. van Peij).*

	A. fumigatus	*A. nidulans*	*A. niger*	*A. oryzae*
Genome size (Mb)	29.2	30.1	35.9	37.6
GC content (mole %)	49	50	50	48
Predicted genes	10 034	9541	14 165	12 108

Genome sequences provide a source for mining for genes that are predicted to encode secreted proteins. That is achieved by sequence comparisons with genes known to encode proteins with enzyme activities that are mainly associated with the extra-cellular environment (e.g. cellulases, hemicellulases, amylases) and by the prediction of a signal sequence at the N-terminus of the encoded protein. The signal sequence, if cleaved, would direct the protein to the exterior of the cell, or at least as far as the outside of the plasma-membrane. A predicted signal sequence, which is not predicted to cleave (but has the other structural/charge attributes of the signal sequence), is termed a signal anchor because it directs the protein to the secretory system of the cell but the protein remains anchored to a membrane. These predictions can accept different confidence limits and, at a confidence level of 95%, it was predicted that c. 8% of the genes in *A. fumigatus* encode secreted proteins (Robson *et al.*, 2005). It is not appropriate to reiterate here detailed information on predicted secreted enzymes that is already catalogued elsewhere, and especially for *A. fumigatus* (Nierman *et al.*, 2005; Robson *et al.*, 2005) and *A. oryzae* (Machida, 2002; Kitamoto, 2002: Machida *et al.*, 2005). Even so, it can be summarized that *A. oryzae*, *A. nidulans* and *A. fumigatus* all encode a multiplicity of glycosyl hydrolases, pectin-degrading enzymes and proteases (Robson *et al.*, 2005; Machida *et al.*, 2005; Table 9.2). It should be mentioned too that not all oligosaccharidase enzymes appear to be extracellular, indicating a capacity for at least some fungi to degrade oligosaccharides intracellularly (de Vries *et al.*, 2005a). The catalogue of predicted genes and encoded products (Table 9.2) does not provide any information about differential expression of the genes under prevailing environmental conditions or exposure of the fungi to particular substrates. Those studies rely upon detailed molecular and biochemical analyses, which can be aided by

Table 9.2 *Numbers of hydrolases and proteases predicted from the genome sequences of* A. fumigatus, A. nidulans *and* A. oryzae. *The data were taken from Machida* et al. *(2005) and provided by Michio Takaeuchi, University of Tokyo.*

Note that there are other hydrolases without obvious predicted binding modules (Robson *et al.*, 2005).

Enzyme	*A. fumigatus*	*A. nidulans*	*A. oryzae*
Pectin methylesterase	4	3	5
Pectin/pectate lyase	9	13	9
Polygalacturonase	11	9	20
Glycosyl hydrolase + CBM[a]	14	7	5
Glycosyl hydrolase + SBM[b]	3	4	1
Glycosyl hydrolase + ChBM[c]	5	14	6
Secreted exopeptidase	22	17	26
Secreted endopeptidase	19	14	26

Notes:
[a] CBM: cellulose binding module
[b] SBM: starch binding module
[c] ChBM: chitin binding module

the availability of gene arrays. In addition, the proteome can also be analyzed to provide information on the secreted proteins detected under different conditions and the partial amino acid sequences can be compared to the genome sequence to provide full sequence. This type of analysis has been achieved, for example, in *Phanerochaete chrysosporium* (van den Wymelenberg *et al.*, 2005). Although conditions that have a direct effect upon the induction or repression of transcription of genes encoding secreted proteins will clearly affect the proteins that are secreted, conditions that affect the growth rate of the fungus also affect the rates of both protein synthesis and secretion (Pakula *et al.*, 2005).

In addition to the prediction of genes encoding secreted enzymes, the genome sequences can also be used to predict the presence of genes responsible both for components of the secretory system and also the cellular responses to secretion stress (discussed further below). That has been achieved with *A. nidulans* (Galagan *et al.*, 2005; Sims *et al.*, 2005), *A.oryzae* (Machida *et al.*, 2005) and *A. fumigatus* (Nierman *et al.*, 2005; Robson *et al.*, 2005).

To move beyond the analysis of genome sequences relies upon methods for assay of gene expression and for manipulation of the genomes in order to analyze the functionality of genes. This is the move from genomics to post-genomics that is so crucial in understanding the lifestyles of organisms and their utility in biotechnological applications. Whilst the availability of complete gene arrays for *Aspergillus* spp. would facilitate the genome-wide analysis of gene expression under a variety of conditions, methods are not yet in place, in contrast to the situation in the yeast *Saccharomyces cerevisiae*, for easy genome-wide analysis of gene function. Gene arrays of varying design and completeness have been produced from *Aspergillus* spp. The most complete arrays have been fabricated by Affymetrix using oligonucleotides designed from the *A. niger* genome sequence, but these are currently only available through DSM (www.dsm.com/dfs/innovation/genomics/). Other arrays based on ESTs (expressed sequence tags) have proven to be useful for transcriptomic studies in *A. nidulans* (Sims *et al.*, 2004, 2005) and *A. fumigatus* arrays comprised of c. $10\,k \times 70$ mer oligonucleotides are now available (http://pfgrc.tigr.org/desc.shtml). The tools for transcriptomic analyses of *Aspergillus* are therefore emerging even if they remain far from complete, especially for high-throughput genome-wide studies. The deficiencies and promise in these areas have recently been discussed elsewhere (Archer & Dyer, 2004; Archer & Turner, 2005).

Proteins secreted by *Aspergillus*

Many species of *Aspergillus* are noted for their widespread distribution. They are highly successful saprobes and their success is explained by their capacity to secrete a range of enzymes that degrade a variety of organic substrates. The range of enzymes secreted is responsive to the environment, indicating a capacity to both sense the environment outside of the cell and then to respond to it by producing the most useful enzymes at that time. In contrast to the other aspergilli discussed here, *A. fumigatus* is thermotolerant, being able to grow at temperatures beyond 50° C and to survive even higher temperatures for extended periods. Its saprophytic capacity and thermotolerance combine to make *A. fumigatus* the major fungal colonizer of compost heaps (Beffa *et al.*, 1998). Its conidia are common in the atmosphere and they are therefore inhaled. Although that does not usually present a health risk, *A. fumigatus* can cause allergic responses, disease in poultry and other animals and is a causal agent of invasive aspergillosis in immuno-compromised humans (Latge, 1999). Other *Aspergillus* spp. have also been associated with invasive aspergillosis but *A. fumigatus* is the most common species. That leads to the issue of what features of *A. fumigatus* are key to its pathogenicity and, while it appears to be complex and most probably multifactorial, the secretion of enzymes is an undoubted contributory factor (Robson *et al.*, 2005).

The data in Table 9.2 have already highlighted the multiplicity of enzymic activities that are predicted to be secreted by *Aspergillus* spp. The ongoing annotation of the *A. niger* genome will undoubtedly also confirm that it is another *Aspergillus* spp. with the potential for degradation of a wide range of organic materials. It is clear that cellulose, pectins, hemicellulases, xylans, starch, lipids and proteins are all potential targets for degradation by *Aspergillus* spp. It was suggested by Robson *et al.* (2005) that the relatively large number of predicted glycosyl hydrolases that include a cellulose binding module might be important in its cellulose-rich natural environment. As a human pathogen, it is also possible that other secreted enzymes such as (phospho)lipases and proteases contribute to invasion of tissues by *A. fumigatus*.

The genome of *A. oryzae* is larger than those of *A. fumigatus* and *A. nidulans* (Table 9.1). It has been argued by Machida *et al.* (2005) that this difference in size is most probably due to acquisition of sequence (and not chromosomal duplication) by *A. oryzae* rather than loss by *A. fumigatus* and *A. nidulans*. The additional sequence is enriched for regions encoding enzymes involved in both primary and secondary metabolism, but also to provide enrichment in some secreted enzymes (e.g. Table 9.2). When the annotation of the *A. flavus* and *A. niger* genomes are published, it will be interesting to note whether they have similar features to those associated with *A. oryzae*.

The capacity of the *Aspergillus* spp. to secrete a wide diversity of enzymes has led to the non-pathogenic and non-toxic species to be exploited commercially as enzyme factories. *A. niger* and *A. oryzae* are prominent commercial species for production of enzymes used either extracellularly (e.g. glucoamylase from *A. niger*) or as cell-associated activities as part of a process (e.g. *A. oryzae* in fermented soy products) (Archer, 2000). Several secreted enzymes are important in the initial degradation of soy and other materials that eventually lead to a finished product such as soy sauce or sake. A full audit of all the enzymes involved is not known even if many have been detected and their function assessed (Sakaguchi *et al.*, 1992; Kitamoto, 2002) but, even so, the genome has a far larger capacity for secretion of potentially useful enzymes. Some initial expression analysis of *A. oryzae* grown under different conditions, including solid state, has been reported (Biesebeke *et al.*, 2005). The use of complete *Aspergillus* genome arrays to study expression patterns has not yet been reported, although the availability of *A. niger* Affymetrix GeneChips (see above) would facilitate such analysis. Some arrays have been constructed and used for expression profiling (Archer & Dyer, 2004), for example with *A. nidulans* (Sims *et al.*, 2004) and a study focused on expression profiling of *A. niger* pectinolytic genes (de Vries *et al.*, 2002). The disparity between any list of commercially available enzymes from *Aspergillus* spp. and the range of enzymes predicted to be encoded by *Aspergillus* spp. is immediately obvious, demonstrating that there is a large but hitherto untapped resource of enzyme activities available. Many commercial enzyme preparations are actually mixtures of activities that have use in a particular application. Others are enriched for particular enzymes or may be further purified to be essentially one component.

The arrival of the annotated genome sequences permits a new approach for identification of new fungal-derived enzymes, which is to express new activities either in the same host or in alternative expression hosts. For example, many of the genes that are predicted to encode particular enzymic activities may be either expressed at a very low level or not at all under convenient growth conditions. In those cases, it is possible to clone the gene and express it under the control of a strong promoter with which there is commercial experience. This can be done systematically, especially so because of the increasing utility of cloning vectors adapted for use with fungi (Archer & Turner, 2005), and in a variety of host backgrounds. It is therefore to be expected that new fungus-derived enzymes with new applications will be offered in the foreseeable future.

Filamentous fungi, including *Aspergillus*, possess what might be classified as a typically eukaryotic secretory system but one that is marked by the peculiarity of the hypha, highly polarized tip growth that requires delivery of enzymes to the tip, and a mix of some components that might be viewed as variations on a eukaryotic theme. Until very recently, the characteristics of the secretion pathway that might define the filamentous fungi have been based on a limited number of studies that have been highly focused (recently reviewed by Archer & Turner, 2005). Global analysis of the secretion pathway from fungal genome sequences (Sims *et al.*, 2005; Robson *et al.*, 2005; Machida *et al.*, 2005; Diener *et al.*, 2004) has recently begun.

Aspergillus spp. continue to be exploited as hosts for the production of heterologous enzymes, i.e. enzymes that are encoded by non-native genes. *A. niger* and *A. oryzae* are the most commonly used species for that purpose although others may also be used, e.g. *Aspergillus vadensis* (de Vries *et al.*, 2004, 2005b) or *A. sojae* (Heerikhuisen *et al.*, 2004). There is an increasing list of heterologous proteins expressed by *Aspergillus* spp. (MacKenzie *et al.*, 2004) and some of these proteins are produced commercially by expression in *Aspergillus*. It is commonly experienced that the yields of target heterologous proteins are less than desired, and there are a number of recent reviews that discuss the underlying reasons for this (Punt *et al.*, 2002; Conesa *et al.*, 2001; MacKenzie *et al.*, 2004). The secretion of the heterologous proteins often presents a difficulty and the key issue then is that a heterologous protein is perceived by the cell as being foreign. That is probably because the secretory system within the *Aspergillus* has not co-evolved with proteins having the properties of the target heterologous proteins. Stress responses are then initiated within the cell and it is understanding those responses and learning from the new knowledge that holds promise in improving the secreted yields of heterologous proteins from *Aspergillus*.

Protein secretion stress in *Aspergillus*

The process of protein secretion is largely conserved amongst all eukaryotic cells. A key element of the secretion process is achieving the correct fold of proteins in the lumen of the endoplasmic reticulum (ER) so that the cell's natural quality control systems are satisfied, and then the folded proteins can continue the secretion process. The natural quality control systems within the secretory pathway of eukaryotic cells ensure that only properly folded protein is secreted (Ellgaard & Helenius, 2003). How cells discriminate between properly folded versus misfolded protein, and how they selectively retain and target the latter for degradation is a very challenging question that remains to be answered, although more information is now becoming available. Improperly folded proteins (i.e. those that have not attained their native fold) induce the unfolded protein response (UPR) and other ER stress responses in yeasts and filamentous fungi (Patil & Walter, 2001; Travers *et al.*, 2000; Ngiam *et al.*, 2000; Al-Sheikh *et al.*, 2004). The UPR mediates the transcriptional upregulation of genes encoding ER-resident chaperones and foldases, and upregulates the ER-associated degradation (ERAD) following retro-translocation of proteins from the ER to the cytoplasm, where those proteins are tagged by ubiquitin and targeted for degradation by the proteasome. The UPR is mediated by the bZIP transcription factor HacA in *Aspergillus* (Hac1p in *S. cerevisiae*) although some ER stress responses appear to be independent of Hac1p and the ER membrane protein Ire1p that senses unfolded proteins and initiates the signal from the ER (Schröder *et al.*, 2003). Hac1p is the translation product from a spliced HAC1 mRNA in yeast (reviewed in Patil & Walter, 2001) and analogous sensing by Ire and mRNA processing occurs in *Aspergillus*, albeit with differences in the detail (Saloheimo *et al.*, 2003). Sensing of

ER stress in eukaryotes has been summarized recently (Zhang & Kaufman, 2004; Rutkowski & Kaufman, 2004; Schröder & Kaufman, 2005) and it is clear that the UPR has a significant role in several aspects of normal cellular function. There is further control over the UPR, for example by nitrogen availability, that influences the UPR and there is an interface between the two bZip transcription factors Hac1p and Gcn4p in yeast (Patil *et al.*, 2004; Kaufman *et al.*, 2002). It is becoming apparent that the UPR, and probably ER stresses more generally, are important aspects of the biology of the cell and are not just factors in maintaining cellular homeostasis under abnormal stress conditions. There are alternative possible fates for degradation of misfolded proteins depending upon which checkpoints within the secretion pathway have been passed. On that basis, properly folded proteins are those that have passed scrutiny by a series of checkpoints (Vashist & Ng, 2004).

Secretion of at least some heterologous proteins induce the UPR in *Aspergillus* (e.g. Wiebe *et al.*, 2001; Mulder *et al.*, 2004) and, as with studies in yeast, chemicals such as dithiothreitol (DTT) and tunicamycin also induce the UPR in *Aspergillus*, even though their responses need not be the same. DTT in particular induces a redox stress, may arrest growth, and therefore has wide-ranging effects. The genome sequences, together with gene chips/arrays, provides the opportunity to assess the global transcriptional responses to ER stress in *Aspergillus*. This has been achieved already with *S. cerevisiae* where there is a platform for comparative studies with *Aspergillus*. In *S. cerevisiae* nearly 400 genes were transcriptionally upregulated due directly or indirectly to UPR (Travers *et al.*, 2000) indicating that the effect is broad and has an impact on many aspects of the cellular metabolism. Although UPR leads to a selective upregulation of transcription of genes encoding proteins that combat stress due to malfolded proteins, it may also be beneficial to homeostasis if genes that encode secretory proteins are selectively transcriptionally downregulated, thus reducing the protein traffic in the secretory pathway. It has recently been shown that ER stress may lead to the transcriptional downregulation of genes encoding at least some secreted proteins, but not non-secreted proteins, in filamentous fungi (Pakula *et al.*, 2003; Al-Sheikh *et al.*, 2004) and probably in plants (Martinez & Chrispeels, 2003). This response is called RESS (repression under secretion stress) and it appears to be initiated independently of the UPR (Al-Sheikh *et al.*, 2004) although any possible interface has not been explored in detail. The response may compensate for the apparent lack of translational attenuation initiated by ER stress in fungi; in mammalian cells, translational attenuation occurs and is mediated through the ER trans-membrane protein PERK (reviewed recently by Kaufman, 2004).

The Hac1p transcription factor of yeast (HacA in *A. niger*) is essential for the UPR; Hac1p binds to the UPR element in sensitive promoters and leads to the UPR-mediated transcriptional upregulation of genes encoding chaperones and foldases. Rüegsegger *et al.* (2001) reported translational regulation of *HAC1* mRNA upon induction of the UPR in *S. cerevisiae*. They showed that the unspliced *HAC1* mRNA is associated with polyribosomes even though it does not lead to the synthesis of full-length protein. The expression of the unspliced *HAC1* mRNA was already known to be post-transcriptionally regulated by an intron (and this block was removed upon excision of the intron by a non-conventional splicing event) but the mechanism was not fully appreciated. Rüegsegger *et al.* (2001) showed that the intron base-paired with a region of the 5′ UTR to cause stalling. So, despite being associated with polyribosomes no Hac1p was produced. Saloheimo *et al.* (2003) have confirmed some RNA secondary structures in the *hac* mRNAs from *A. nidulans* and *T. reesei* that may regulate translation in these fungi. It has also been confirmed that HacA mediates the UPR in *A. niger* and that it upregulates its own transcription (Mulder *et al.*, 2004). It is possible to delete the *hacA* gene from *A. niger* (H. Mulder, personal communication) to facilitate studies on HacA dependency (as was used, for example, in *S. cerevisiae* by Travers *et al.*, 2000) but deletion of *ireA* (encoding the trans-membrane kinase that mediates the UPR) has not been achieved. The constitutive upregulation of *hacA* enhances the secreted yield of at least two heterologous proteins by *A. niger* (Valkonen *et al.*, 2003). Other studies indicate that there may be an optimal level of individual foldases and chaperones for maximized secretion of any given heterologous protein (Moralejo *et al.*, 2001; Lombraña *et al.*, 2004) by *Aspergillus* spp.

In a recent study, Al-Sheikh *et al.* (2004) showed that transcriptional downregulation of the gene (*glaA*) encoding glucoamylase (a major secreted protein) in *A. niger* occurs in response to DTT-induced stress. Nuclear run-on studies were a part of that study, and showed that the effect was transcriptional rather than due to stability of mRNA. Although the UPR was also induced under the DTT stress conditions, it was also shown that a strain of *A. niger* in which the *pdiA* gene was downregulated by antisense exhibited the same downregulation of *glaA* without induction of the UPR. RESS operating on *glaA* is mediated by a region of the *glaA* promoter between 1 and 2 kb upstream of the translational start, but it is not known how it is mediated. There is no evidence for RESS occurring in yeast judged by the database available in Travers *et al.* (2000).

A gene array study under stress conditions has recently been published with *A. nidulans* where c. 30% of the predicted genes were represented on the array (Sims *et al.*, 2005). There were no reports in this analysis of genes affected by RESS but the transcriptional responses of several genes, encoding proteins that are implicated in protein secretion, to both DTT stress and the stress of secreting a heterologous protein were documented. A summary of some of these genes, and their transcriptional responses to DTT, together with the likely locations in the secretory system of some of the encoded proteins (Fig. 9.1) is provided in Table 9.3. Included in Table 9.3 is some analogous data with *A. niger* (Guillemette, van Peij, Goosen, Lanthaler, Robson, van den Hondel, Stam and Archer, unpublished) using the DSM Affymetrix GeneChips. Note that both the *A. nidulans* and *A. niger* studies noted that DTT does not induce the same response as secretion of a heterologous protein, but it is also likely that different heterologous proteins will induce different responses.

The genes listed in Table 9.3 are a secretion-relevant or stress-relevant subset of those mentioned by Sims *et al.* (2005). DTT was used to stress both *A. nidulans* (Sims *et al.*, 2005) and *A. niger* (Guillemette *et al.*, unpublished) using similar, if not identical, conditions. For the most part, the two datasets are similar but there are differences that cannot be immediately

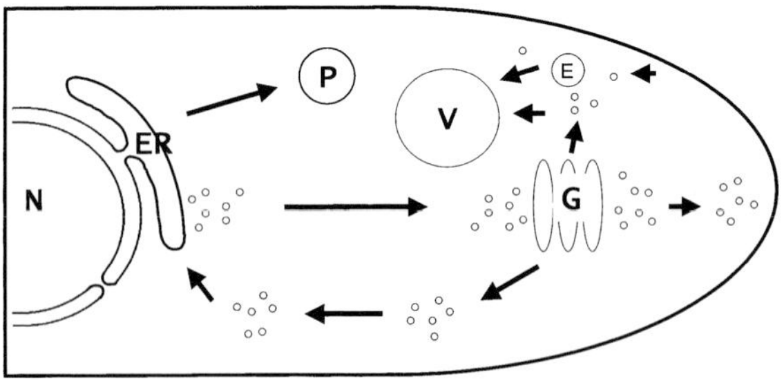

Figure 9.1 Some key compartments in the secretion of proteins by the fungal hypha are shown. N = nucleus, ER = endoplasmic reticulum, P = proteasome, V = vacuole, G = Golgi equivalent, E = endosome. All components shown, including small vesicles, are membrane bound apart from the proteasome. The data shown in Table 9.3 map on to the compartments in this figure. Translocation is the movement of proteins from the cytosol to the lumen of the ER. ERAD is ER-associated degradation and involves the dislocation of proteins from the ER to the cytosol and their degradation by the proteasome, the chaperones and foldases listed in Table 9.3 are located in the lumen of the ER whereas the heat shock proteins are cytosolic.

Table 9.3 *The data compare the transcriptional regulation of selected genes in* A. nidulans *(Sims* et al., *2005) and* A. niger *(Guillemette* et al., *unpublished) by DTT. The genes chosen for this analysis are those already mentioned and tabulated by Sims* et al. *(2005). The treatments with 20 mM DTT were for 1 h with* A. nidulans *and 2 h with* A. niger. *The numbers refer to fold changes compared to water-treated controls. NC = no change (within +/− 1.5 fold). ND = no data (i.e. probably not represented on the* A. nidulans *gene array). The location of gene products are: translocation (from cytosol to ER lumen); ERAD (ER to proteasome); chaperones, foldases (ER-resident); heat shock proteins and peptidyl prolyl isomerases (cytosol, mitochondrial (ECM10) or nuclear (PINA)); cellular export and secretion (various locations but mainly involved in vesicular biogenesis, transport or fusion); transcription (ER membrane (IreA) and cytosol to nucleus (HacA)). The* A. nidulans *ORF reference numbers can be found at www.broad.mit.edu and the* A. niger *numbers are from DSM: www.dsm.com/dfs/innovation/genomics/.*

Note that known *A. nidulans* genes are described in lower case whereas the yeast genes are described in upper case. For example, with the GTPase-encoding genes that are implicated in vesicle transport most genes are described in both *Aspergillus* and yeast but some mammalian-type GTPases found in *Aspergillus* (Punt et al., 2001) are not known in yeast.

A. nidulans ORF	*A. nidulans* + DTT	*A. niger* ORF	*A. niger* + DTT
Translocation			
SEC61/AN7721.2	ND	An00g08106	+1.5
SEC63/AN0834.2	+5.5	An00g11030	+1.7
srpA/SRP54/AN8246.2	ND	An00g03561	NC
ERAD			
HDR3/AN0810.2	+1.5	An00g06489	NC
UBA1/AN2174.2	+1.9	An00g12015	NC
Chaperones and foldases			
bipA/KAR2/AN2062.2	+3.9	An00g00175	+4.7
pdiA/PDI1/AN7436.2	NC	An00g03066	+2.2
tigA/AN0075.2	NC	An00g03070	+1.5
prpA/AN0248.2	NC	An00g03054	+4.0
clxA/CNE1/AN3592.2	NC	An00g07353	+3.2
Heat shock proteins			
hscA/SSB2/X98931	+2.0	An00g08012	−4.0
SSA1/AN5129.2	+1.5	An00g08563	NC
SSB1/AN4616.2	ND	An00g08562	−2.5
ECM10/AN6010.2	NC	An00g08064	NC
cypA/CPR3/AN3814.2	NC	An00g00169	NC
FPR1/AN3598.2	NC	An00g10232	NC
PIN1/AN6145.2	+2.0	An00g10225	+1.9

Table 9.3 (cont.)

A. nidulans ORF	*A. nidulans* + DTT	*A. niger* ORF	*A. niger* + DTT
Membrane and cell wall biogenesis			
INO1/AN7625.2	NC	An00g09942	+2.0
ECM3/AN0930.2	NC	An00g00116	NC
SPF1/AN3146.2	NC	An00g11666	NC
Vesicle trafficking, vacuolar protein sorting			
srgA/*SEC4*/AN6974.2	NC	An00g03554	−1.7
srgB/*YPT1*/AN4281.2	NC	An00g03555	NC
srgC/*YPT6*/AN7602.2	ND	An00g08527	NC
srgD/AN5106.2	NC	An00g03556	NC
srgE/*YPT31*/AN0347.2	+2.3	An00g08525	NC
srgF/*YPT7*/AN0089.2	ND	An00g10819	NC
srgG/*YPT51*/AN4915.2	NC	An00g11056	+1.9
srgH/*YPT52*/AN3842.2	NC	An00g08531	+2.2
srgL/AN0069.2	ND	An00g10476	NC
sarA/*SAR1*/AN0411.2	NC	An00g03557	−1.9
SNX41/AN6351.2	NC	An00g09223	NC
VPS41/AN4876.2	NC	An00g12071	NC
sagA/*END3*/AN1023.2	NC	An00g10935	NC
nsfA/*SEC18*/AN3098.2	NC	An00g03558	+2.5
SHR3/AN1845.2	NC	An00g04133	+1.6
SEC27/AN5972.2	NC	An00g07562	NC
ssoA/*SSO1*/AN3416.2	NC	An00g11215	NC
Transcription			
hacA/AN9397.2	+1.7	An00g05764	NC
ireA/AN0235.2	NC	An00g10665	NC
Glycosylation			
WBP1/AN4683.2	ND	An00g08015	NC
OST2/AN4031.2	NC	An00g11344	NC
ALG11/AN5725.2	NC	An00g10484	NC
ALG2/AN6874.2	−5.6	An00g09668	+1.7
MNN4/AN4030.2	NC	An00g04020	+2.7
PSA1/AN1911.2	ND	An00g09662	+2.1

explained even if the experimental set-ups might go some way towards providing a reason for the differences. The transcriptional upregulation of foldase-encoding genes (*pdiA*, *tigA* and *prpA*) and the calnexin-encoding gene (*clxA*) is consistent with published data in *A. niger* but none of these genes appear to be upregulated in *A. nidulans* (Sims *et al.*, 2005). One other clear anomaly is the transcriptional downregulation of the *A. nidulans* homologue of the yeast *ALG2* gene whereas it appears to be upregulated in *A. niger*. In *S. cerevisiae*, *ALG2* is an essential gene presumed to encode a mannosyltransferase involved in the N-linked glycosylation pathway.

Use of the *A. niger* Affymetrix GeneChips has also been extended to examination of UPR by DTT and tunicamycin, as well as study of a strain that secretes the heterologous protein tissue plasminogen activator (t-PA) (Guillemette *et al.*, unpublished). Although the detailed results have yet to be published, it is clear that a large number of genes appear to be transcriptionally affected (either up or down) by each of the conditions, and although there is overlap between conditions each condition affects a unique gene set (Fig. 9.2). It is the challenge now to confirm the transcriptional effects, explore selected genes that might prove to be informative in terms of unravelling mechanisms of regulation and then to capitalize on this information by improving the secreted yields of proteins secreted by commercial species of *Aspergillus*. The information may also be applicable in providing antifungal targets in the treatment of invasive aspergillosis.

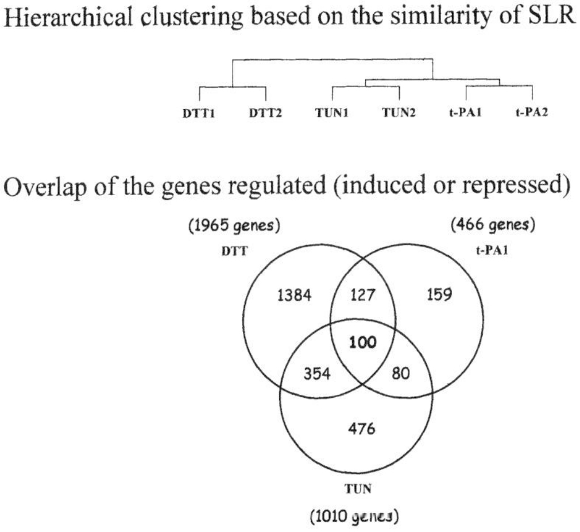

Figure 9.2 *A. niger* was exposed to dithiothreitol (DTT) or tunicamycin (TUN) and the effect of these treatments on transcript levels was assessed globally using Affymetrix GeneChips (Guillemette *et al.*, unpublished). In addition, a strain secreting the heterologous protein tissue plasminogen activator (t-PA) (Wiebe *et al.*, 2001), with the gene expressed using the constitutive *gpdA* promoter, was also assessed using the Chips. The studies were performed in duplicate (e.g. DTT1, DTT2, etc.) and scored for change in transcript level (either + or −). The results show that the three conditions elicit changes which include both commonality but also changes that are specific to the particular condition. SLR = signal log ratio.

Summary

Aspergillus spp. are widely distributed in nature. While some species are causal agents of disease, others are safe and exploitable fungi used commercially as factories for secreted production of both native and heterologous enzymes, as well as some metabolites. The recent availability of genome sequences for important *Aspergillus* spp. has provided a wealth of information that should underpin studies aimed at gaining a better understanding of the biological processes that are relevant to *Aspergillus* lifestyles. The future studies should also be able to identify factors that limit the further exploitation of *Aspergillus* spp. commercially. It must be recognized, however, that it is not yet routine in *Aspergillus* to undertake gene manipulation on anything approaching a global scale. When that is achieved, studies with *Aspergillus* will have properly entered the post-genomic era.

Acknowledgements

The authors acknowledge funding for their research from DSM and the Biotechnology and Biological Sciences Research Council.

References

Al-Sheikh, H. M., Watson, A. J., Lacey, G. A., Punt, P., MacKenzie, D. A., Jeenes, D. J., Pakula, T., Penttilä, M., Alcocer, M. J. C. & Archer, D. B. (2004). Endoplasmic reticulum stress leads to the selective transcriptional downregulation of the glucoamylase gene in *Aspergillus niger*. *Molecular Microbiology*, **53**, 1731–42.

Archer, D. B. (2000). Filamentous fungi as microbial cell factories for food use. *Current Opinion in Biotechnology*, **11**, 478–83.

Archer, D. B. & Dyer, P. S. (2004). From genomics to post-genomics in *Aspergillus niger*. *Current Opinion in Microbiology*, **7**, 499–504.

Archer, D. B. & Turner, G. (2005). Genomics of protein secretion and hyphal growth in *Aspergillus*. In *The Mycota XIII, Fungal Genomics*, ed. A. J. Brown. Berlin: Springer-Verlag.

Beffa, T., Staib, F., Fischer, J. L., Lyon, P. F., Gumowski, P., Marfenina, O. E., Dunoyer-Geindre, S., Georgen, F., Roch-Susuki, R., Gallaz, L. & Latge, J.-P. (1998). Mycological control and surveillance of biological waste and compost. *Medical Mycology*, **36** (suppl. I), 137–45.

Biesebeke, R. T., Levin, A., Sagt, C., Bartels, J., Goosen, T., Ram, A., van den Hondel, C. A. M. J. J. & Punt, P. (2005). Identification of growth phenotype-related genes in *Aspergillus oryzae* by heterologous macroarray and suppression subtractive hybridization. *Molecular Genetics and Genomics*, **272**, 33–42.

Conesa, A., Punt, P. J., van Luijk, N. & van den Hondel, C. A. M. J. J. (2001). The secretion pathway in filamentous fungi: a biotechnological view. *Fungal Genetics and Biology*, **33**, 155–71.

de Vries, R. P., Jansen, J., Aguilar, G., Parenicova, L., Joosten, V., Wulfert, F., Benen, J. A. E. & Visser, J. (2002). Expression profiling of pectinolytic genes from *Aspergillus niger*. *FEBS Letters*, **530**, 41–7.

de Vries, R. P., Burgers, K., van de Vondervoort, P. J. I., Frisvad, J. C., Samson, R. A. & Visser, J. (2004). A new black *Aspergillus* species, *A. vadensis*, is a promising host for homologous and heterologous protein production. *Applied and Environmental Microbiology*, **70**, 3954–9.

de Vries, R. P., van Grieken, C., van Kuyk, P. A. & Wosten, H. A. B. (2005a). The value of genome sequences in the rapid identification of novel genes encoding specific plant cell wall degrading enzymes. *Current Genomics*, **6**, 157–87.

de Vries, R. P., Frisvad, J. C., van de Vondervoort, P. J. I., Burgers, K., Kuijpers, A. F., Samson, R. A. & Visser, J. (2005b). *Aspergillus vadensis*, a new species of the group of black Aspergilli. *Antonie van Leeuwenhoek International Journal of General and Molecular Microbiology*, **87**, 195–203.

Diener, S. E., Dunn-Coleman, N., Foreman, P., Houfek, T. D., Teunissen, P. J., van Solingen, P., Dankmeyer, L., Mitchell, T. K., Ward, M. & Dean, R. A. (2004). Characterization of the protein processing and secretion pathways in a comprehensive set of expressed sequence tags from *Trichoderma reseei*. *FEMS Microbiology Letters*, **230**, 275–82.

Ellgaard, L. & Helenius, A. (2003). Quality control in the endoplasmic reticulum. *Nature Reviews in Molecular Cell Biology*, **4**, 181–91.

Galagan, J. E., Calvo, S. E., Cuomo, C. *et al.* (2005). Sequencing of *Aspergillus nidulans* and comparative analysis with *A. fumigatus* and *A. oryzae*. *Nature*, **438**, 1105–15.

Heerikhuisen, M., van den Hondel, C. & Punt, P. (2004). *Aspergillus sojae*. In *Production of Recombinant Proteins: Novel Microbial and Eukaryotic Expression Systems*, ed. G. Gellissen. The Netherlands: Wiley, pp. 191–214.

Kaufman, R. J. (2004). Regulation of mRNA translation by protein folding in the endoplasmic reticulum. *Trends in Biochemical Sciences*, **29**, 152–8.

Kaufman, R. J., Scheuner, D., Schröder, M., Shen, X., Lee, K., Liu, C. Y. & Arnold, S. M. (2002). The unfolded protein response in nutrient sensing and differentiation. *Nature Reviews in Molecular Cell Biology*, **3**, 411–21.

Kitamoto, K. (2002). Molecular biology of the Koji moulds. *Advances in Applied Microbiology*, **51**, 129–53.

Latge, J.-P., (1999). *Aspergillus fumigatus* and aspergillosis. *Clinical Microbiology Reviews*, **12**, 310–50.

Lombraña, M., Moralejo, F. J., Pinto, R. & Martin, J. F. (2004). Modulation of thaumatin secretion by *Aspergillus awamori* by modification of *bipA* gene expression. *Applied and Environmental Microbiology*, **70**, 5145–52.

Machida, M. (2002). Progress of *Aspergillus oryzae* genomics. *Advances in Applied Microbiology*, **51**, 81–106.

Machida, M., Asai, K., Sano, M., *et al.* (2005). Genome sequencing and analysis of *Aspergillus oryzae*. *Nature*, **438**, 1157–61.

MacKenzie, D. A., Jeenes, D. J. & Archer, D. B. (2004). Filamentous fungi as expression systems for heterologous proteins. In *The Mycota II*, Ch.15, ed. U. Kück. Springer, pp. 289–315.

Martinez, I. M. & Chrispeels, M. J. (2003). Genomic analysis of the unfolded protein response in *Arabidopsis* shows its connection to important cellular processes. *Plant Cell*, **15**, 561–76.

Montiel, D., Dickinson, M. J., Lee, H. A., Dyer, P. S., Jeenes, D. J., Roberts, I. N., James, S., Fuller, S. J., Matsushima, K. & Archer, D. B. (2003). Genetic differentiation of the *Aspergillus* Section *Flavi* complex using AFLP fingerprints. *Mycological Research* **107**, 1427–34.

Moralejo, F. J., Watson, A. J., Jeenes, D. J., Archer, D. B. & Martín, J. F. (2001). A defined level of protein disulfide isomerase expression is required for optimal secretion of thaumatin by *Aspergillus awamori*. *Molecular Genetics and Genomics*, **266**, 246–53.

Mulder, H. J., Saloheimo, M., Penttilä, M. & Madrid, S. (2004). The transcription factor HACA mediates the unfolded protein response in *Aspergillus niger*, and up-regulates its own transcription. *Molecular Genetics and Genomics*, **271**, 130–40.

Ngiam, C., Jeenes, D. J., Punt, P. J., van den Hondel, C. A. & Archer, D. B. (2000). Characterization of a foldase, protein disulfide isomerase A, in the protein secretory pathway of *Aspergillus niger*. *Applied and Environmental Microbiology*, **66**, 775–82.

Nierman, W., Pain, A., Anderson, M. J., *et al.* (2005). Genomic sequence of the pathogenic and allergenic filamentous fungus *Aspergillus fumigatus*. *Nature*, **438**, 1151–6.

Pakula, T. M., Laxell, M., Huuskonen, A., Uusitalo, J., Saloheimo, M. & Penttilä, M. (2003). The effects of drugs inhibiting protein secretion in the filamentous fungus *Trichoderma reesei*: evidence for down-regulation of genes that encode secreted proteins in the stressed cells. *Journal of Biological Chemistry*, **278**, 45011–20.

Pakula, T. M., Salonen, K., Uusitalo, J. & Penttilä, M. (2005). The effect of specific growth rate on protein synthesis and secretion in the filamentous fungus *Trichoderma reesei*. *Microbiology*, **151**, 135–43.

Patil, C. & Walter, P. (2001). Intracellular signaling from the endoplasmic reticulum to the nucleus: the unfolded protein response in yeast and mammals. *Current Opinion in Cell Biology*, **13**, 349–56.

Patil, C., Li, H. & Walter, P. (2004). Gcn4p and novel upstream activating sequences regulate targets of the unfolded protein response. *Plos Biology*, **2**, 1208–23.

Pritchard, B. L., Yu, J., Nierman, W., Dean, R. A., Bhatnagar, D., Cleveland, T. E. & Payne, G. A. (2005). *Aspergillus flavus* genome sequence: status and release of sequence data. *Proceedings of Plant and Animal Genomes XIII Conference*, San Diego, USA, P037.

Punt, P. J., Seiboth, B., Weenink, X. O., van Zeijl, C., Lenders, M., Konetschny, C., Ram, A. F. J., Montijn, R., Kubicek, C. P. & van den Hondel, C. A. M. J. J. (2001). Identification and characterization of a family of secretion-related small GTPase-encoding genes from the filamentous fungus *Aspergillus niger*: a putative SEC4 homologue is not essential for growth. *Molecular Microbiology*, **41**, 513–25.

Punt, P. J., van Biezen, N., Conesa, A., Albers, A., Mangnus, J. & van den Hondel, C. A. M. J. J. (2002). Filamentous fungi as cell factories for heterologous protein production. *Trends in Biotechnology*, **20**, 200–6.

Robson, G. E., Huang., J., Wortman, J. & Archer, D. B. (2005). A preliminary analysis of the process of protein secretion, and the diversity of putative secreted hydrolases encoded in *Aspergillus fumigatus*: insights from the genome. *Medical Mycology* Supplement 1, **43**, S41–S47.

Rüegsegger, U., Leber, J. H. & Walter, P. (2001). Block of *HAC1* mRNA translation by long-range base pairing is released by cytoplasmic splicing upon induction of the unfolded protein response. *Cell*, **107**, 103–14.

Rutkowski, D. T. & Kaufman, R. J. (2004). A trip to the ER: coping with stress. *Trends in Cell Biology*, **14**, 20–8.

Sakaguchi, K., Takagi, M., Horiuchi, H. & Gomi, K. (1992). Fungal enzymes used in oriental food and beverage industries. In *Applied Molecular Genetics of Filamentous Fungi*, eds. J. R. Kinghorn & G. Turner. London: Blackie, pp. 54–99.

Saloheimo, M., Valkonen, M. & Penttilä, M. (2003). Activation mechanisms of the HACI-mediated unfolded protein response in filamentous fungi. *Molecular Microbiology*, **47**, 1149–61.

Schröder, M., Clark, R. & Kaufman, R. J. (2003). *IRE1*- and *HAC1*-independent transcriptional regulation in the unfolded protein response of yeast. *Molecular Microbiology*, **49**, 591–606.

Schröder, M. & Kaufman, R. J. (2005). ER stress and the unfolded protein response. *Mutation Research*, **569**, 29–63.

Sims, A. H., Gent, M. E., Robson, G. D., Dunn-Coleman, N. S. & Oliver, S. G. (2004). Combining transcriptome data with genomic and cDNA sequence alignments to make confident functional assignments for *Aspergillus nidulans* genes. *Mycological Research*, **108**, 853–7.

Sims, A. H., Gent, M. E., Lanthaler, K., Dunn-Coleman, N. S., Oliver, S. G. & Robson, G. D. (2005). Transcriptome analysis of recombinant protein secretion by *Aspergillus nidulans* and the unfolded protein response *in vivo*. *Applied and Environmental Microbiology*, **71**, 2737–47.

Travers, K. J., Patil, C. K., Wodicka, L., Lockhart, D. J., Weissman, J. S. & Walter, P. (2000). Functional and genomic analyses reveal an essential coordination between the unfolded protein response and ER-associated degradation. *Cell*, **101**, 249–58.

Valkonen, M., Ward, M., Wang, H., Penttilä, M. & Saloheimo, M. (2003). Improvement of foreign protein production in *Aspergillus niger* var. *awamori* by constitutive induction of the unfolded protein response. *Applied and Environmental Microbiology*, **69**, 6979–86.

van den Wymelenberg, A., Sabat, G., Martinez, D., Rajangam, A. S., Teeri, T. T., Gaskell, J., Kersten, P. J. & Cullen. D. (2005). The *Phanerochaete chrysosporium* secretome: database predictions and initial mass spectrometry peptide identifications in cellulose-grown medium. *Journal of Biotechnology*, **118**, 17–34.

Vashist, S. & Ng, D. T. W. (2004). Misfolded proteins are sorted by a sequential checkpoint mechanism of ER quality control. *Journal of Cell Biology*, **165**, 41–52.

Wiebe, M. G., Karandikar, A., Robson, G. D., Trinci, A. P. J., Candia, J. L. F., Trappe, S., Wallis, G., Rinas, U., Derkx, P. M. F., Madrid, S., Sisniega, H., Faus, I., Montijn, R., van den Hondel, C. A. M. J. J. & Punt, P. J. (2001). Production of tissue plasminogen activator (t-PA) in *Aspergillus niger*. *Biotechnology and Bioengineering*, **76**, 164–74.

Zhang, K. & Kaufman, R. J. (2004). Signaling the unfolded protein response from the endoplasmic reticulum. *Journal of Biological Chemistry*, **279**, 25935–8.

IV
Fungal bioremediation

10

Metal stress and the single yeast cell: Berkeley Award Lecture

S. V. AVERY

Introduction

Overview

This paper explores the mechanisms of metal toxicity towards cells, specifically some advances being made in this area through work with the unicellular fungus, *Saccharomyces cerevisiae*. The two principal questions addressed here are:

1. Is oxidative damage the cause of cellular metal toxicity?
2. Why do individual cells exhibit widely differing metal resistances?

Several powerful experimental tools are unique to *S. cerevisiae* among eukaryotes, and are being exploited to help elucidate the mechanism(s) of metal toxicity. Furthermore, in conjunction with its unicellular morphology, *S. cerevisiae* provides an ideal system with which to explore the topical problem of cell individuality, applied here to metal toxicity. This chapter provides an overview of these fields, illustrated with key findings from the author's laboratory.

Metals in the environment and relevance to fungi

A wide range of industrial activities give rise to metal pollutants, which continue to be released into the environment at potentially harmful levels. Localized concentration of certain metals may also arise naturally. For example, toxic levels of the biologically essential metal copper are often associated with certain mineral ores as well as industrial or agricultural discharges. Cadmium is used widely in electroplating and galvanizing industries, as a colour pigment in paints and in batteries, and as a by-product of zinc and lead mining and smelting. Zinc, lead and other metals also may be released from similar types of sources. Concern over

Exploitation of Fungi, eds. G. D. Robson, P. van West and G. M. Gadd. Published by Cambridge University Press.

these activities is exacerbated by reports that pollution with metals (e.g. cadmium) and associated toxicity in natural populations can be considerably higher than previously thought (Larison *et al.*, 2000).

The interactions of yeasts and other microorganisms with metals have been of scientific interest for some time. Free metals and compounds that contain complexed metals, such as copper, have been used widely in fungicides (Gadd, 1993). Furthermore, metal complexation is being used to enhance the effectiveness of certain antifungal drugs (Nicoletti *et al.*, 1999; Mastrolorenzo *et al.*, 2000). Organisms subject to routine metal-exposure in their natural environments generally have had to develop resistance mechanisms. In an applied context, the ability of certain yeasts and filamentous fungi to resist metal toxicity, combined with their generally high metal-absorptive capacities, has been exploited in the development of certain biological metal-removal processes (White *et al.*, 1998; Gadd, 2000; Malik, 2004). Furthermore, to address the need to monitor environmental levels of toxic metals, yeasts such as *S. cerevisiae* are being used in efforts to develop novel metal-ion biosensors (Lehmann *et al.*, 2000; Verma & Singh, 2005).

S. cerevisiae and other fungi may also serve as useful models for characterizing certain aspects of metal toxicity in higher organisms. In humans, metals are implicated in a wide range of degenerative conditions, including neurotoxicity, nephrotoxicity, genotoxicity and carcinogenesis (Halliwell & Gutteridge, 1990; Stohs & Bagchi, 1995; Llanos & Mercer, 2002; Leonard *et al.*, 2004). The requirement to maintain an appropriate homeostatic balance of essential metals such as copper is reflected by the prevalence of Menkes syndrome, associated with copper deficiency, and Wilson's disease, associated with excess copper accumulation (Harris & Gitlin, 1996). Obviously, metal toxicology in higher eukaryotes is an issue of significant concern and there has been considerable effort to characterize the effects of metals that are manifest at the tissue, organ or whole-organism level. Unfortunately, these foci have tended to divert attention from mechanisms of metal toxicity acting at the cellular or molecular level, which ultimately lead to tissue and organ damage. One reason for the current lack of understanding of these aspects is the difficulty of studying them in higher organisms, which has led to a partial reliance on information from potentially non-extrapolatable in vitro work. *S. cerevisiae* and other yeasts provide alternative systems for modelling the cellular effects of metals in vivo.

The yeast model applied to metal toxicology

The ease with which *S. cerevisiae* can be manipulated facilitates detailed elucidation of many aspects of eukaryotic cell physiology and/or toxicology that would not be convenient with filamentous fungi and higher eukaryotic systems. The genetics of *S. cerevisiae* are very well understood and the completion of its genome-sequence in 1996 has facilitated strides in functional genomics technologies applied to *S. cerevisiae* over the last few years (Winzeler *et al.*, 1999; Hughes *et al.*, 2000; Uetz *et al.*, 2000; Allen *et al.*, 2003; Lum *et al.*, 2004). Furthermore, there has been striking conservation of cellular function through evolution between *S. cerevisiae* and mammalian cells, and the similarities are exploited heavily to provide insight into cellular processes that occur in humans and other organisms. With regard to metal toxicity, several additional specific advantages of the yeast model have been highlighted (Avery, 2001). As well as the pertinence of yeast/fungal–metal interactions in an applied context, *S. cerevisiae* provides a well-described eukaryotic framework on which to build a detailed understanding of cellular metal toxicology.

Is oxidative damage the cause of cellular metal toxicity?

Metal-induced generation of reactive oxygen species and oxidative damage

It has been known for some time that many metals can promote the generation of highly damaging free radicals or reactive oxygen species (ROS) in biological systems. However, there is general uncertainty as to whether or not the ROS generated during metal exposure are the cause of metal toxicity. Major cellular macromolecules such as lipids, proteins and DNA are highly susceptible to oxidative attack and degradation (ROS attack on carbohydrates is comparatively slow) (Fig. 10.1).

Metal-induced ROS formation could occur via several mechanisms. The Fenton or Haber-Weiss reactions are catalyzed by redox-active metals (e.g. Cu, Fe, Cr, V) and generate the highly reactive hydroxyl ($OH^{\bullet}$) radical from H_2O_2 and superoxide ($O_2^{\bullet -}$) substrates (the latter being formed during respiration) (Halliwell & Gutteridge, 1999). Cu and Fe may also initiate oxidative damage directly, e.g. by promoting the propagation reactions of lipid peroxidation. Several indirect mechanisms are considered to account for the action of redox-inactive metals such as cadmium, mercury, nickel and lead. For example, these metals might displace redox-active metals from cellular binding sites (Casalino *et al.*, 1997), similar to the proposed action of $O_2^{\bullet -}$ in releasing Fenton catalysts from certain iron-sulphur [4Fe-4S] enzymes (Srinivasan *et al.*, 2000). Further potential indirect mechanisms by which metals may promote ROS damage include inhibition of antioxidant defence enzymes (Sarkar *et al.*, 1997; Murakami & Yoshino, 1999) and depletion of protein-bound sulphhydryl groups and glutathione (Stohs & Bagchi, 1995). The result of each of these processes would be a tendency towards elevated ROS formation and/or increased lipid, protein and DNA oxidation. Indeed, in support of a role for free radical reactions in metal toxicity, there are close similarities between the gross effects of oxidative-damage and metal-damage on cellular macromolecules.

The net result of lipid peroxidation is conversion of unsaturated lipids to polar lipid hydroperoxides that can cause increased membrane fluidity due to disturbance of hydrophobic phospholipids (Dix & Aikens, 1993; Van Ginkel & Sevanian, 1994). Metal stress is also associated with increased membrane fluidity (Assmann *et al.*, 1996; Howlett & Avery, 1997a). Such effects can profoundly influence membrane functions. For example, lipid peroxidation, like metal stress, may be accompanied by membrane permeabilization with efflux of cytosolic solutes (Van Ginkel & Sevanian, 1994; Howlett & Avery, 1997a,b). Extensive lipid peroxidation can lead to the ultimate disintegration of membrane integrity and possible cell death.

Proteins display differing susceptibilities to oxidative attack, which may be linked to variable compositions of sulphydryl groups, Fe-S clusters, reduced haem moieties and Cu prosthetic groups (Liochev & Fridovich, 1999; Shringarpure *et al.*, 2001). Certain amino acids (e.g., Cys, Met, His)

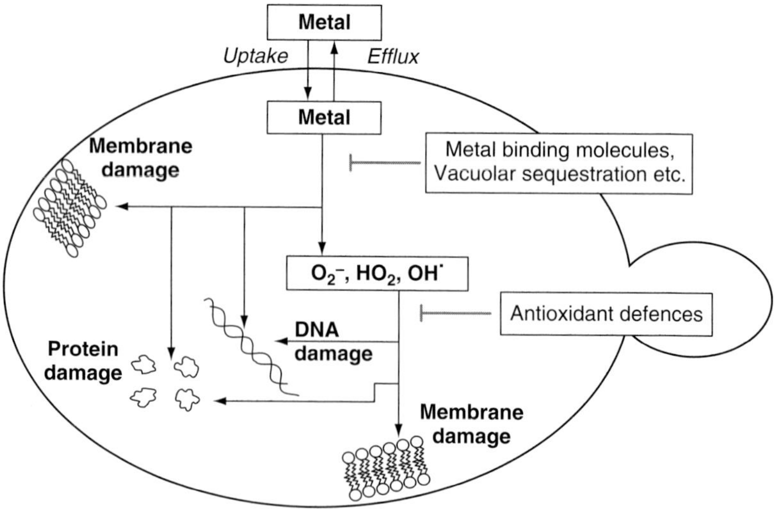

Figure 10.1 Generalized scheme showing principal metal targets and protective mechanisms in yeast cells. Adapted from Avery (2001) with permission.

are also particularly prone to oxidation. Accumulation of oxidized amino acid residues leads to protein conformational changes and inactivation (Stadtman & Berlett, 1997), similar to the reported effects of metals (Hughes & Poole, 1991; Jungmann *et al.*, 1993). A probable role of oxidative damage in metal effects on proteins is reflected in the sensitivity of certain proteins to both oxidative- and metal-stress (Watabe *et al.*, 1995; Shanmuganathan *et al.*, 2004; Sumner *et al.*, 2005). Such effects may result from site-specific ROS damage at sites of metal (e.g. Fe, Cu) binding (Requena *et al.*, 2001). Non-specific metal-induced ROS generation may also affect proteins, for example by causing side-chain modifications (Shringarpure *et al.*, 2001). In addition, membrane proteins may be oxidized by products of (metal-induced) lipid peroxidation.

Oxidative damage to DNA includes base modifications, DNA-protein cross-links, depurination and strand scission (Beckman & Ames, 1997; Cooke *et al.*, 2003). These effects are very similar to those observed during metal-exposure (Brennan & Schiestl, 1996; Lloyd *et al.*, 1997; Kasprzak, 2002). Mutagenicity arises from erratic replication and/or repair of the damaged DNA. When cellular DNA-repair mechanisms are overwhelmed, extensive DNA damage can lead to cell death.

General evidence for involvement of ROS in cellular metal toxicity

At the macroorganism level, there has been growing evidence for a key involvement of oxidative stress in metal-related damage and disease (for reviews, see Halliwell & Gutteridge, 1990; Rotilio *et al.*, 1995; Hippeli & Elstner, 1999; Avery, 2001). Thus, elevated oxidation of cellular components (e.g. lipid peroxidation) is commonly reported in metal-exposed cells and animals (e.g. De Vos *et al.*, 1993; Sarkar *et al.*, 1997; Baryla *et al.*, 2000; Chen *et al.*, 2002; Hepburn *et al.*, 2003) and protection may be conferred by antioxidants (e.g. Sarkar *et al.*, 1997; Pourahmad & O'Brien, 2000). The cumulative evidence shows that many toxic metals promote ROS formation in cells. However, in most cases it has not been clearly resolved whether ROS-mediated effects are responsible for metal-induced cell injury, or if they are merely incidental.

This issue is complicated further by a number of factors. Firstly, certain potentially toxic metals also have antioxidant properties. For example, zinc deficiency is considered to enhance the susceptibility of plants to attack by ROS (Cakmak, 2000) and zinc suppresses cadmium-induced apoptosis in higher cells, which is believed to be in part due to the antioxidant action of zinc (Szuster-Ciesielska *et al.*, 2000). Secondly, redox-active metals such as Fe(II) and Cu(I) may be toxic under both aerobic and anaerobic conditions (Strain & Culotta, 1996; Dunning *et al.*, 1998). Thirdly, ROS-independent toxicity mechanisms are known even for metals that are prime candidates for ROS-dependent toxicity. For example, in addition to effects dependent on the redox-activity of copper, Cu toxicity could be mediated through inappropriate binding of the metal to N, O and S ligands in biomolecules, thereby inactivating enzymes and disrupting organism function (Karlstrom & Levine, 1991). It has been demonstrated that several metals can replace Zn in the zinc finger DNA binding domain of the human estrogen receptor, resulting in a protein that no longer can bind to its target sequence (Predki & Sarkar, 1992). Furthermore, Cs toxicity is known to arise through displacement of essential K^+ ions (Avery, 1995).

Metal toxicity in *S. cerevisiae* and the role of oxidative stress

There is an extensive literature on general aspects of metal uptake, homeostasis and resistance in yeast and fungal cells, coverage of which is beyond the scope of this paper. For detailed reviews on those aspects the reader is referred to Avery (2001) and Rutherford and Bird (2004). Of course a detailed knowledge of homeostasis and resistance mechanisms can, in turn, contribute to our understanding of toxicity. For example, owing to the efficiency of cellular Cu chaperones, it has been estimated that the total cytoplasmic free copper content of *S. cerevisiae* is many orders of magnitude less than one atom per cell (Rae *et al.*, 1999). This underscores the difficulty of interpreting total cellular metal contents in toxicological terms, as also applies to extracellular metals where bioavailability may be diminished by ligand binding and precipitation (Hughes & Poole, 1991; Avery & Tobin, 1993; Gadd, 1993). This finding also underscores the potency of metals such as copper.

Regarding the question of whether oxidative damage is the cause of metal toxicity in *S. cerevisiae*, there have been fewer studies. Moreover, much of this evidence is limited by the non-specificity of the tools used in discriminating oxidative from other mechanisms. The following summarizes some key examples of such evidence (see Avery (2001) for a more complete account).

In principle, any of the biological effects of metals proposed to occur at the molecular level (Gadd, 1993; Avery, 2001) could either promote or be attributable to ROS formation. Evidence for a role of ROS in metal-damage to yeasts includes increased metal-tolerance during anaerobicity (Greco *et al.*, 1990; Galiazzo *et al.*, 1991), protection exerted by certain free radical scavengers (Brennan & Schiestl, 1996) and the many overlaps in the molecular mechanisms used by yeasts to cope with oxidative and metal stress. However, much of the evidence is ambiguous. For example, superoxide scavenging by the Cu, Zn-superoxide dismutase (Sod1p) of *S. cerevisiae* was originally considered to account for Sod1p-dependent Cu resistance. However, Culotta *et al.* (1995) demonstrated that Sod1p promotes Cu resistance through its capacity to bind and buffer cellular Cu. Similarly the antioxidant molecule glutathione (GSH) additionally has metal-binding activity (Lin *et al.*, 1993; Li *et al.*, 1997) and, in the case of arsenic, GSH is required to regenerate active Acr2p reductase that is required for the As detoxification pathway (Mukhopadhyay *et al.*, 2000). It is known that the Cd-binding activity of GSH is responsible for GSH-dependent Cd-resistance, via Ycf1p-mediated co-transport of Cd with glutathione derivatives into the vacuole (Wemmie *et al.*, 1994; Li *et al.*, 1997). The copper-binding metallothionein also has dual metal binding and antioxidant functions (Tamai *et al.*, 1993; Viarengo *et al.*, 2000). As mentioned earlier, the antioxidant properties of certain metals (e.g. Mn, Cu, Zn) adds further to the complexity of linking metal toxicity with free radical damage in *S. cerevisiae* and oxidants can also alter metal metabolism and toxicity (Chiang *et al.*, 2000).

There are reports that the presence of oxygen enhances Cu toxicity towards *S. cerevisiae* (Greco *et al.*, 1990; Galiazzo *et al.*, 1991) suggesting a role for ROS in toxicity, but oxygen availability has non-specific effects on yeast metabolism and on intracellular Cu accumulation (Strain & Culotta, 1996). Also, oxygen-independent mechanisms of free radical propagation might still potentiate metal damage. There is evidence that significant Cu-induced lipid peroxidation in *S. cerevisiae* may occur independently of mitochondrial ($O_2^{\bullet -}$-generating) function and in the presence of various $OH^{\bullet}$ scavengers (Avery *et al.*, unpublished results). Such observations may be related to the propagation reactions of lipid peroxidation, which Cu can promote directly and independently of Fenton reaction catalysis, unlike lipid peroxidation initiation (Halliwell & Gutteridge, 1999).

Metal-induced membrane damage

One of the most widely observed effects of metal toxicity towards yeasts and other organisms is a sudden increase in membrane permeability (Gadd & Mowll, 1983; Ohsumi *et al.*, 1988; Avery *et al.*, 1996). Membrane perturbation is considered a key potential mechanism of metal toxicity (Gadd, 1993; Avery, 2001). Although effects such as trans-membrane cation loss have been commonly used as indices of metal toxicity, the precise mechanisms underlying loss of membrane integrity and subtler effects on membrane physical properties are still poorly understood.

The first clue to help address this issue comes from the fact that metabolic activity is usually a prerequisite for metal-induced membrane leakage (Ohsumi *et al.*, 1988; Avery *et al.*, 1996), which indicates that a physical metal-binding process cannot be the direct cause. It is also known that, as in higher organisms, many metals promote lipid peroxidation in yeasts (Lee *et al.*, 1996; Howlett & Avery, 1997a; Basu *et al.*, 2004). However, to demonstrate that lipid peroxidation is the *cause* of metal-induced membrane damage is more of a challenge. The timescales of metal-dependent lipid peroxidation and membrane damage may be correlated (Howlett & Avery, 1997a), but such evidence is indirect, particularly since indices used for measuring membrane physical properties, permeabilization and lipid peroxidation are unlikely to be linearly related. Even different methods for measuring lipid peroxidation can give quite different results (Howlett & Avery, 1997a).

Membrane-lipid peroxidation is a cause of cadmium toxicity

PHYSIOLOGICAL STUDIES The principal approach adopted in the author's laboratory to relate gross membrane damage to lipid peroxidation has been specifically to manipulate susceptibility to lipid peroxidation and then to test for corresponding changes in susceptibility to membrane damage and toxicity. This has been accomplished using both physiological and genetic approaches. The amenability of *S. cerevisiae* to manipulation of membrane fatty acid composition was exploited first. Thus, cells either enriched with or lacking polyunsaturated fatty acids (PUFAs) were compared for metal resistance. The oxidizability of PUFAs means that PUFA-rich membranes are more susceptible to lipid peroxidation than are membranes that comprise predominantly saturated or monounsaturated fatty acids (Dix & Aikens, 1993). *S. cerevisiae* readily incorporates PUFAs supplied in the medium to greater than 60% of total membrane fatty acid composition (Fig. 10.2A), with no obvious deleterious effects on growth under non-stress conditions. Moreover, PUFA-rich *S. cerevisiae* were found to exhibit a marked sensitivity to Cu (Avery *et al.*, 1996) and to Cd (Howlett & Avery, 1997a). Thus, metal-induced K^+ efflux from cells increased with the degree of membrane fatty acid unsaturation (Avery *et al.*, 1996; Howlett & Avery, 1997a), as did metal-dependent loss of cell viability (Fig. 10.2B). Furthermore, these effects were correlated closely with elevated levels of metal-induced lipid peroxidation in PUFA-rich cells. For example, Cd-dependent conjugated-diene formation as an index of lipid peroxidation was highest in cells enriched with the tri-unsaturated fatty acid linolenate (18:3) and lowest in cells with a native fatty acid composition (Fig. 10.2C). Conjugated-diene formation also increased with Cd concentration. As well as K^+ loss and lipid peroxidation, the findings were also correlated with relative effects on membrane fluidity, a marker of membrane physical status, of differently-enriched cells (Howlett & Avery, 1997a). These findings strongly indicated that

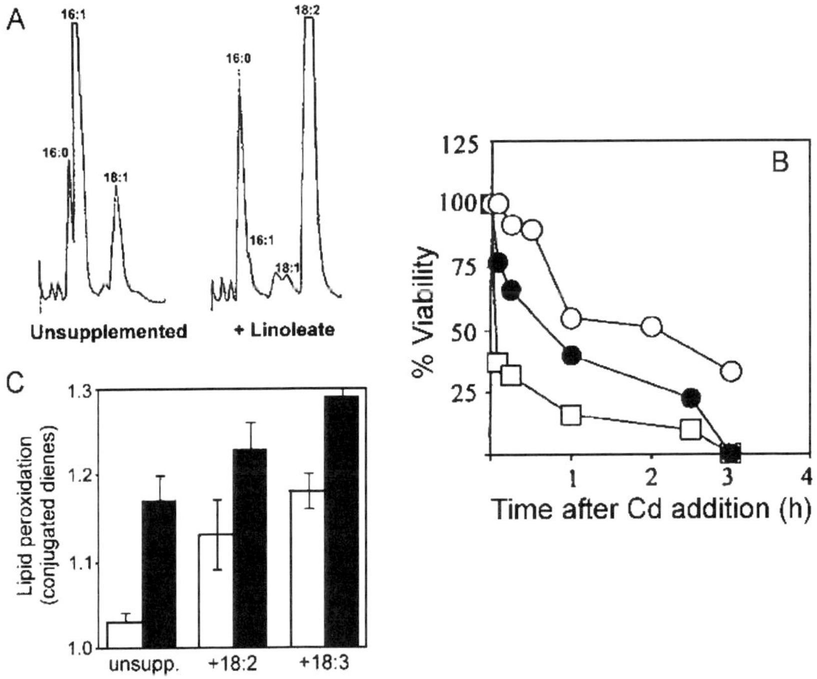

Figure 10.2 Influence of cellular fatty acid composition on cadmium-induced lipid peroxidation and toxicity. (A) Gas-liquid chromatograph traces of the lipid fatty acid composition of *S. cerevisiae* cultured in unsupplemented and linoleate (18:2)-supplemented YEPD medium. (B) Cells were grown in unsupplemented (○), linoleate-supplemented (●) or linolenate (18:3)-supplemented (□) medium and then exposed to 50 μM $Cd(NO_3)_2$. Viability (colony forming ability) was assessed at intervals. (C) Cadmium-induced conjugated-diene formation in unsupplemented or PUFA-supplemented medium. Cells were treated as above then analyzed for conjugated dienes after 1 h exposure to 50 μM (□) or 200 μM (■) $Cd(NO_3)_2$. Adapted from Howlett and Avery (1997a) with permission.

lipid peroxidation is the cause of elevated Cd sensitivity in PUFA-rich *S. cerevisiae* and that lipid peroxidation is a major mechanism of Cd toxicity.

GENETIC STUDIES To substantiate the above findings, a genetic approach was used to modify the susceptibility of *S. cerevisiae* to lipid peroxidation. The glutathione peroxidases (GPxs) are a major but diverse group of antioxidant enzymes. Among these, phospholipid hydroperoxide GPxs (PHGPxs) uniquely can reduce lipid hydroperoxides esterified to biomembranes. Thus, PHGPx is often considered the principal cellular enzyme capable of repairing membrane lipid peroxidation. *S. cerevisiae* expresses three *GPX* genes (Inoue *et al.*, 1999) all of which have been found to encode PHGPx-like proteins, Gpx1, Gpx2 and Gpx3 (Avery & Avery, 2001). Single *gpx3Δ* and triple *gpx1Δ/gpx2Δ/gpx3Δ* *S. cerevisiae* mutants exhibit sensitivity to agents which have a lipid peroxidation-dependent mode of action (Avery & Avery, 2001; Avery *et al.*, 2004; Sumner *et al.*, 2005). However, the phenotypes evident in these mutants could be related to the (PH)GPx enzymes' generic peroxidase activities (Inoue *et al.*, 1999) rather than phospholipid hydroperoxidase activity. In addition, Delaunay *et al.* (2002) have identified a novel role for the major PHGPx-like enzyme of *S. cerevisiae*, Gpx3, as a sensor and transducer of the stress response to hydrogen peroxide; this occurs via Gpx3p-mediated oxidation of the Yap1p transcription factor. These pleiotropic functions of yeast Gpx3p complicate the interpretation of sensitivity phenotypes in the relevant mutant strains, for example a marked sensitization to Cd that was observed in *gpx3Δ* and *gpx1Δ/gpx2Δ/gpx3Δ* mutants versus wild-type cells (Fig. 10.3).

To test the hypothesis that Gpx3-dependent Cd resistance of *S. cerevisiae* was due to the enzyme's phospholipid hydroperoxidase activity specifically, a novel classical glutathione peroxidase (cGPx)-like yeast enzyme – 'cGpx3' – was constructed using the yeast Gpx3 as a template. cGPxs have generic peroxidase activity but lack phospholipid-specific peroxidase activity. Yeast cGpx3 was constructed by introducing consensus cGPx subunit-interaction sequences to the appropriate sites of the yeast enzyme (Avery *et al.*, 2004). Using SDS-PAGE, non-denaturing gel filtration chromatography and activity assays from yeast-cell lysates, these manipulations were confirmed to have been successful in generating a cGPx-like enzyme in yeast, with strong generic-peroxidase activity but lacking phospholipid hydroperoxidase activity (Avery *et al.*, 2004). Unlike the native Gpx3p, expression of cGpx3p in a *gpx1Δ/gpx2Δ/gpx3Δ* triple mutant failed to complement the Cd sensitivity phenotype of this mutant (Fig. 10.3), indicating that the non-phospholipid peroxidase activity of Gpx3p is not important for Cd resistance. To discriminate further between the phospholipid hydroperoxidase and signal transduction activities of Gpx3p in mediating Cd resistance, a Gpx3^{C82S} mutant protein that is defective only for peroxidase activity (Delaunay *et al.*, 2002) was tested.

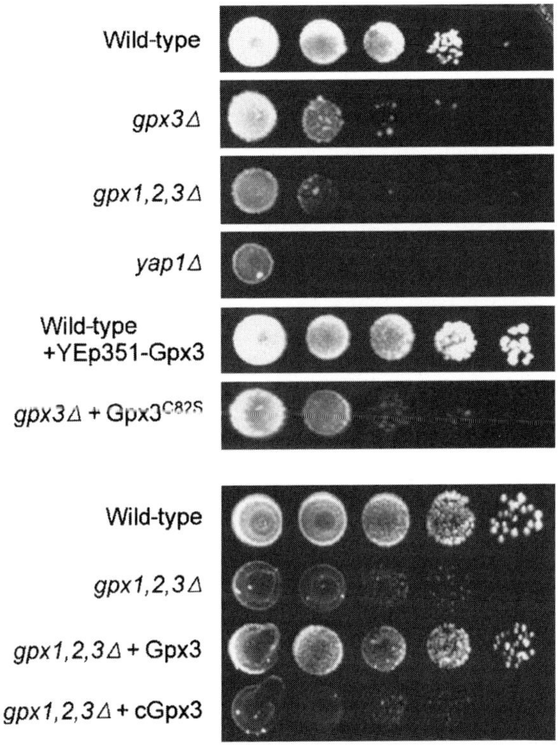

Figure 10.3 Phospholipid-hydroperoxidase activity confers cadmium resistance in *S. cerevisiae*. *S. cerevisiae* BY4741 and the isogenic mutant strains shown were harvested during the exponential phase of growth and dilutions of decreasing concentration were inoculated in 5 μl spots on YEPD agar from left to right on each plate for each strain. Growth in the presence of 50 μM (top panel) or 40 μM (lower panel) $Cd(NO_3)_2$ was examined after incubation for 4 d at 30°C. Control growth in the absence of $Cd(NO_3)_2$ (not shown) was similar for all strains to that shown for the wild-type in the lower panel. '+ YEp351-Gpx3', '+ Gpx3^{C82S}', '+ Gpx3' and '+ cGpx3' indicates strains transformed with the plasmids YEp351-Gpx3, pRS316-HA-Gpx3^{C82S}, YEp351-Gpx3 and YEp351-cGpx3, respectively. Each panel shows typical results from one of at least three independent experiments. Adapted from Avery *et al.* (2004) with permission.

The Cd sensitivity of a *gpx3Δ* mutant was not suppressed by ectopic expression of Gpx3^{C82S} (Fig. 10.3), indicating that the Yap1p-mediated signalling activity of Gpx3p also is not important for Cd resistance. This left phospholipid hydroperoxidase as the critical function.

Combined, the physiological and genetic strategies described above establish that lipid peroxidation-dependent processes are a principal cause of cellular cadmium toxicity. In contrast to Cd, yeast *gpx* mutants are not sensitive to Cu. Consistent with this evidence, while Cu was more toxic at the whole-cell level than Cd, Cd generated the greater extent of plasma membrane permeabilization (data not shown), suggesting that membrane permeabilization makes a more important contribution to Cd toxicity than to Cu toxicity in *S. cerevisiae*. The relative importance of lipid peroxidation as a mechanism of metal toxicity is likely to be greater in organisms with a membrane lipid composition enriched in unsaturated fatty acids.

Metal-induced protein damage

Cadmium activates ubiquitin-dependent proteolysis in *S. cerevisiae* and proteolysis-deficient mutants are hypersensitive to Cd (Jungmann *et al.*, 1993). Metal-induced formation of abnormal proteins could be a major reason for toxicity, particularly if such proteins are not efficiently degraded. Alternatively, essential proteins may be targeted by metals, causing loss of viability. Recently, DNA damage (and concurrent toxicity) resulting from cellular cadmium exposure was shown to be indirect, arising as a consequence of a direct action of Cd on protein-mediated DNA-mismatch repair activity (Jin *et al.*, 2003). There have been several reports of specific protein activities of yeast that are inhibited by metal stress (Fortuniak *et al.*, 1996; Murakami & Yoshino, 1999). However, causal links between these effects, in particular oxidation-dependence, and the toxicology of metals have not been established.

Protein (methionine) oxidation is a cause of chromium toxicity

The approach used in the author's laboratory to establish a dependency of metal toxicity on protein oxidation was to use an oxidation-repair activity directed against proteins. Thus, *S. cerevisiae* cells defective for the peptide methionine sulphoxide reductases MsrA and MsrB were tested. These MSR enzymes have complementary activities, reducing different stereoisomers of oxidized methionine (Met) residues in proteins (L-methionine-S-sulphoxide and L-methionine-R-sulphoxide, respectively) (Kryukov *et al.*, 2002). MSRs provide the only characterized protein oxidation repair activity of cells. Single *msraΔ* and *msrbΔ* mutant strains exhibited marked sensitivity to Cr compared to wild-type cells, and this phenotype was accentuated in a double *msraΔ/msrbΔ* mutant (Fig. 10.4A). These mutants did not exhibit sensitivity to other metals tested, Cd and Cu. Also, Cr did not give a phenotype in *gpx* mutants, unlike Cd (see above), or in *ogg1Δ/rad30Δ* mutants of *S. cerevisiae*, indicating that neither lipid peroxidation nor DNA oxidation are critical for Cr toxicity (Sumner *et al.*, 2005). To substantiate the role for protein (Met) oxidation in Cr toxicity, the *MSRA* and *MSRB* genes were over-expressed: whereas gene deletion can lower the threshold of resistance to

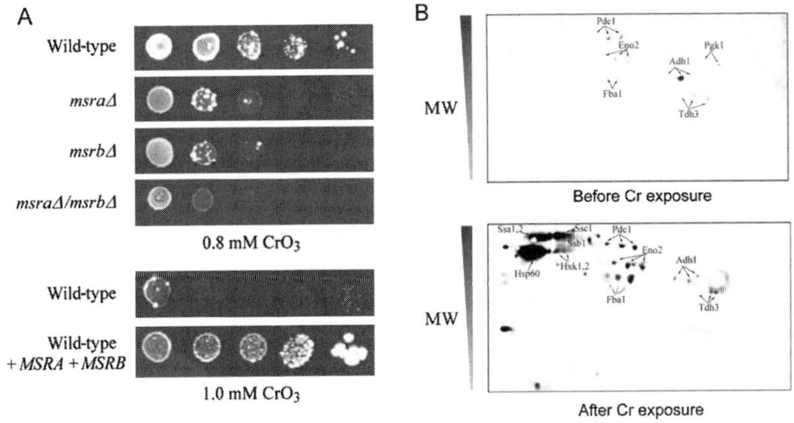

Figure 10.4 Protein oxidation causes Cr toxicity. (A) Dilutions of decreasing cell concentration of exponential phase *S. cerevisiae* were inoculated in 4 μl spots from left to right onto YEPD agar, either supplemented or not with 0.8 or 1.0 mM CrO_3 as indicated. Control growth in the absence of CrO_3 (not shown) was similar for all strains to that shown for the wild-type in the top panel. Plates were inoculated with *S. cerevisiae* BY4741 (wild-type) and either derivative *msraΔ*, *msrbΔ* or *msraΔ/msrbΔ* mutants. Additional plates were inoculated with wild-type cells overexpressing the *MSRA* and *MSRB* genes on multicopy plasmids ('+*MSRA*+*MSRB*'). Growth was assessed after 4 d incubation. Typical results from one of two independent experiments are shown. (B) Protein extracts were prepared from *S. cerevisiae* BY4741 at intervals during treatment with 0.5 mM CrO_3. Two-dimensional western blots were probed with anti-dinitrophenyl antibody and visualized with a chemiluminescent substrate to identify carbonylated (oxidized) proteins. Representative two-dimensional arrays of cellular proteins stained with SYPRO-ruby and corresponding western blots are shown for protein extracts sampled at 0 min (top panel) and 60 min (lower panel) during CrO_3 treatment. The positions of oxidation-susceptible proteins are referenced. Adapted from Sumner *et al.* (2005) with permission.

an agent by sensitizing a new principal cellular target to that agent, only genes that help to protect the normal target(s) of toxicity (or that directly encode that target) should raise the lower resistance-threshold (Avery *et al.*, 2004). Overexpression of *MSRA* and *MSRB* on multicopy plasmids gave markedly enhanced growth of *S. cerevisiae* at a normally inhibitory Cr concentration (Fig. 10.4A). Overexpression of the *MSR* genes also complemented the Cr sensitivity of a *sod1*Δ mutant, defective for the Cu, Zn-superoxide dismutase (data not shown). Note that the Cr sensitivity of the *sod1*Δ mutant has been confirmed to be due to loss of the antioxidant function of Sod1p by testing a variant construct, $Sod1^{H46C}$, which lacks superoxide dismutase catalytic activity but is minimally affected for metal binding (Lu *et al.*, 1996; Wei *et al.*, 2001), as well as incubations under anaerobic conditions (Sumner *et al.*, 2005).

Identification of protein targets of chromium toxicity

To test whether particular protein targets may become highly oxidized during Cr treatment, proteins from cells that had been exposed to Cr for different times were extracted and separated according to their pI and mass using 2D-gels. These were then probed with antibodies specific for derivatized carbonyl residues on proteins, a hallmark of oxidation (Shanmuganathan *et al.*, 2004). Proteins of interest were identified by MALDI–mass spectrometry. Cr-dependent protein oxidation was evident from the increased intensity and numbers of carbonylated proteins in immunoblots derived from extracts of chromium-treated cells (Fig. 10.4B, lower panel). Whereas the total carbonylation induced by Cr was relatively high, Cr targeted a similar range of yeast proteins as found previously for Cu (Shanmuganathan *et al.*, 2004) and other pro-oxidants (Cabiscol *et al.*, 2000; Costa *et al.*, 2002). Enzymes involved in glycolysis or in the fermentation of pyruvate, as well as heat shock proteins, appeared to be particularly susceptible to oxidative modification. These results are consistent with the hypothesis that the glycolytic pathway may become inactivated during (Cr-induced) oxidative stress, so promoting the production of glucose equivalents and of reducing power in the form of $NADPH_2$ within the pentose phosphate pathway (Cabiscol *et al.*, 2000; Costa *et al.*, 2002; Shenton & Grant, 2003).

Since the above results establish protein (Met) oxidation as a primary cause of Cr toxicity, the 2D-analyses presented could suggest some preliminary candidates for specific protein target(s) of cellular Cr toxicity. However, if targeting of a particular protein is the cause of an agent's toxicity, then that protein is likely to be an essential protein (alternatively, it may be a protein that is converted to a toxic form). For example, Naganuma *et al.* (2000) identified the essential enzyme L-glutamine: D-fructose-6-phosphate amidotransferase (GFAT) as the specific critical target of methylmercury (MeHg) toxicity in *S. cerevisiae*. Unfortunately, the majority of the abundant proteins oxidized during Cr exposure that were identified above were non-essential.

The principle of haploinsufficiency provides an alternative approach to identify essential protein targets of metal action. Among the various functional genomics tools that have become available in recent years is a complete collection of heterozygous deletion mutants for *S. cerevisiae*. The heterozygosity means that strains defective for essential genes are represented in the collection. At the same time, since a protein encoded by a gene represented on only one out of two homologous chromosomes is expressed at a lower level than in a corresponding homozygous strain, that level may become (haplo-)insufficient for growth if the protein is simultaneously targeted by a metal or other agent. This principle has been exploited in yeast to characterize essential protein targets of various inhibitory compounds. With microarray technology, the competitive growth of the >6000 heterozygous *S. cerevisiae* strains can be monitored simultaneously under different conditions, allowing identification of haploinsufficiency phenotypes (Lum *et al.*, 2004; Deutschbauer *et al.*, 2005).

Why do individual cells exhibit widely differing metal resistances?

One recently introduced strategy for furthering our understanding of metal toxicity has been to explore phenotypic heterogeneity in yeast metal sensitivity, i.e. the variable metal-susceptibilities of individual cells within yeast cultures. Phenotypic heterogeneity is manifest in all organisms, and in *S cerevisiae* and other microorganisms is readily apparent among individual cells even within isogenic populations. The benefits of such non-genotype-derived heterogeneity for an organism are likely to lie in enhanced adaptability and hardiness of the population as a whole; promoting the chances that better-adapted phenotypic variants may be able to persist following perturbation and seed new (heterogeneous) populations (Kaern *et al.*, 2005; Avery, 2005a,b). Consistent with this, simulations predict that heterogeneous populations can indeed outgrow homogeneous populations during growth in fluctuating conditions (Thattai & van Oudenaarden, 2004), and there is a growing body of evidence that such variability is an evolvable trait (Fraser *et al.*, 2004; Raser & O'Shea, 2004). Consequently, there is currently great interest in unravelling the molecular bases of heterogeneity development. At the level of gene transcription and translation, stochastic variation between cells has been increasingly documented in recent years (Elowitz *et al.*, 2002; Ozbudak *et al.*, 2002; Blake *et al.*, 2003; Raser & O'Shea, 2004). Alternatively, more deterministic parameters that vary between cells, such as the cell cycle, rhythmicity and aging, could potentially drive heterogeneity in gene expression and cellular phenotypes (Davey & Kell, 1996; Sumner & Avery, 2002). Differential metal sensitivity within isogenic yeast cultures is observed frequently, although until recently there had been no serious attempt to provide an explanation for it.

The basis for heterogeneous Cu resistance in *S. cerevisiae*

Studies in the author's laboratory have focused mainly on Cu resistance in individual yeast cells as a model of phenotypic heterogeneity. Much of this work has relied on the ability to separate Cu-resistant and Cu-sensitive cell subpopulations by flow activated cell sorting (FACS) and to analyze them independently for other discriminating properties that might potentially account for their differing Cu sensitivities. Thus, Howlett & Avery (1999) initially showed that variable Cu sensitivity in exponential phase *S. cerevisiae* is cell cycle-dependent and, therefore, is not a stochastic phenomenon. To ascertain this, Cu-sensitive and -resistant cell subpopulations were sorted, permeabilized and stained with a fluorescent probe that reports on cellular DNA content. Reanalysis of the sorted and stained cells by flow cytometry revealed that the cells predisposed to the toxic effects of copper were primarily at later (post-DNA synthesis) stages of the cell cycle, whereas the Cu-resistant cell subpopulation comprised predominantly cells at early cell cycle stages (Sumner *et al.*, 2003). This dependence on cell cycle stage was linked to the initial oxidant status of the cells (Howlett & Avery, 1999) and it hinged on the activity of a single gene product, the Cu, Zn-superoxide dismutase (Sod1). Cell cycle-dependent Cu resistance was abolished in a deletion mutant lacking Sod1p activity (Sumner *et al.*, 2003). The activity of the Sod1 protein was shown to oscillate around five-fold during the yeast cell cycle, in approximate phase with peaks of Cu resistance. The above data were all validated using independent strategies (Sumner *et al.*, 2003). Using flow cytometry, a role has also been established for the replicative age of individual yeast cells in influencing their Cu resistances, which diminish as cells progress through successive rounds of cell division after their 'birth' (Fig. 10.5A). A similar dependency on replicative age has been observed in yeast cells treated with other inhibitory agents (Kale & Jazwinski, 1996; Drakulic *et al.*, 2005). As with cell cycle stage, cell age-dependent Cu resistance was found to require the activity of specific gene products, in particular Sod1p (Fig. 10.5B). Deletion of the *CUP1* gene which encodes the Cu-metallothionein was also found partly to suppress age-dependent Cu resistance (data not shown). It is noteworthy that *CUP1* deletion was not required to eliminate fully age-dependent Cu resistance since this was effectively accomplished by *SOD1* deletion alone, suggesting that the contribution of Cup1p may be Sod1p-dependent. This hypothesis was supported by a ~30% diminishment of Sod1p activity evident in *cup1*Δ cells (data not shown).

The above findings predict elevations in cellular Cu resistance at discrete phases of cell cycle progression and ageing, creating phenotypically heterogeneous populations. It is of particular interest that *SOD1* was the principal gene required for driving heterogeneity in Cu resistance in this study, since Sod1p confers protection against a variety of stresses (Gralla, 1997; Park *et al.*, 1998; Pereira *et al.*, 2003). Such heterogeneity may provide an important insurance mechanism for organisms; creating subpopulations that are pre-equipped for varied activities as needs may arise

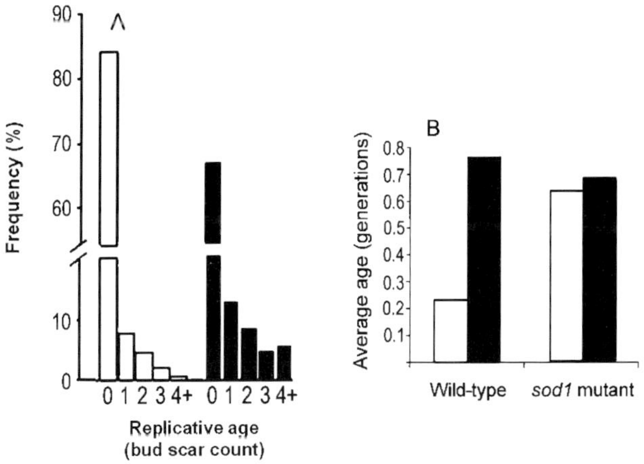

Figure 10.5 The influence of replicative age on single-cell Cu resistance, and dependency on Sod1p. (A) Replicative age profiles of copper-resistant and copper-sensitive subpopulations. Exponential phase *S. cerevisiae* BY1471 was treated with $Cu(NO_3)_2$ for 10 min and sorted with FACS into Cu-resistant (□) and Cu-sensitive (■) subpopulations. Cells were stained with calcofluor white and scored for bud scar (generation) number. (B) Mean ages derived from (A) for wild-type cells, and from a parallel experiment with a *sod1*Δ mutant. At least 400 cells were scored for each subpopulation. Typical results from one of three independent experiments are shown. Adapted from Sumner *et al.* (2003) with permission.

(e.g. when faced with stress), but without the permanent metabolic costs of constitutive expression.

Certain gene functions may act to suppress heterogeneity in metal resistance

In the above example, heterogeneous expression of *SOD1* was a major driver of heterogeneity. It can be easily envisaged how this type of heterogeneity can provide an advantage to a cell population in fluctuating environmental conditions. However, heterogeneity driven by the cell cycle and/or cell age could be considered energetically wasteful in a stable habitat. In this case, perpetual generation of heterogeneous subpopulations via cyclical synthesis and degradation of proteins such as Sod1p and Cup1p might serve no advantage. A better system might be one where there was a mechanism of releasing heterogeneity only when needed, e.g. as soon as an environmental change is sensed. Bergman and Siegal (2003) have put forward one such mechanism. They proposed that the stability of gene expression in *S. cerevisiae* is maintained under non-stress conditions by maintenance of an intact genome/proteome. Thus, individual gene products suppress, or 'buffer', the variable expression of other genes. In this model, environmental stress may cause loss of function of certain gene products, so diminishing the strength of buffering and releasing previously-hidden phenotypic variation when it is needed (i.e. during stress). Theoretical results support this model (Bergman & Siegal, 2003).

Work in the author's lab has revealed a number of gene functions that fit the model of Bergman & Siegal (2003), applied to the problem of heterogeneous metal resistance. Thus, in contrast to the effect of *SOD1* deletion in diminishing heterogeneity, proteins required for vacuolar function have been found to suppress heterogeneity in metal resistance; the relevant deletion mutants exhibit markedly increased cell-to-cell heterogeneity compared with the wild-type strain (Avery *et al.*, unpublished data). Recent studies have indicated similar phenomena in certain yeast mutants treated with pro-oxidants (Drakulic *et al.*, 2005). It remains to be confirmed experimentally that such release of 'hidden' phenotypic variation can indeed provide a benefit to yeast populations during harsh environmental stress.

Under what conditions are particular types of heterogeneity in metal resistance advantageous?

The above account describes some of the sources of heterogeneity that may underlie cell-to-cell variability in resistance to toxic metals. It is likely that these different sources may provide different advantages under different conditions (Tolker-Nielsen *et al.*, 1998). Thus, conventional genotypic variants (mutants) would be likely to help ensure survival when environmental changes are permanent, and the favoured phenotype can become fixed in the population. Non-genotypic variants that arise routinely under standard non-stress conditions (e.g. from cell cycle-/age-dependent Sod1p expression) are more likely to help ensure survival when environmental changes are sudden and non-permanent – continued generation of heterogeneity in the survivors should protect against future environmental changes including reversion to the original conditions. Finally, induction of non-genotypic heterogeneity (and of metal response genes) should help to ensure survival when changes in metal concentration are slow – so allowing time for heterogeneity to be induced – and non-permanent.

Conclusions

Metal toxicity continues to be a major problem and there is a pressing need for the deleterious effects of metals at the cellular and molecular level to be understood. The interactions of yeasts with toxic metals are of considerable interest not only because of the commercial and medical importance of these organisms, but also because they serve as excellent eukaryotic models for delineating the mechanism(s) of metal toxicity (as well as oxidative stress responses and metal homeostasis). The evidence discussed in the first part of this paper is consistent with a major role for free radical-generation in metal toxicity. While free radical/ROS-independent mechanisms of toxicity have also been suggested for certain metals in *S. cerevisiae* and other organisms, few of these alternatives have been tested or proven. Specifically, the evidence described here has established lipid peroxidation and protein oxidation as the major causes of Cd and Cr toxicity, respectively. Work is ongoing to identify particular protein species that are targets of the action of Cr and other metals.

Recent development of techniques for studying the individuality of cell behaviour has allowed exploration of metal toxicity effects that vary between single cells. This approach uncovers phenomena that are otherwise masked by conventional sampling techniques, in which metal toxicity data are averaged across large numbers of cells. Moreover, heterogeneity may confer major advantages to cell populations subjected to varying episodes of metal exposure. Gene functions that may be important for culture-averaged metal resistance have been found to fall into three categories regarding their roles in heterogeneity; where their expression has either no effect on the degree of heterogeneity in metal resistance, enhances heterogeneity, or suppresses heterogeneity. It seems likely that the opposing latter two effects may have evolved to provide benefits in environments with differing degrees of instability. One major challenge in the coming years will be to try and extrapolate these findings to the natural environment. Specifically, is phenotypic heterogeneity only important for organismal fitness in settings where there is little genetic diversity, or can phenotypic heterogeneity provide an additional layer of diversity in the natural environment, contributing to the overall fitness and survival of organisms that already exhibit genotypic diversity? In addition to the relevance to human health, advances in this and other areas pertinent to metal toxicity will help to understand the fitness of yeasts and filamentous fungi in industrial (and possibly laboratory) settings and their potential control where proliferation is undesirable.

Acknowledgements

The author thanks the British Mycological Society for the honour of the Berkeley Award. The support of the NIH (R01 GM57945), the BBSRC (BB/C506656/1) and the NERC (NER/B/S/2003/00730) are gratefully acknowledged. Members of the Avery lab past and present, and John Houghton and Anupama Shanmuganathan (Georgia State University), are thanked for the input to the studies referred to.

References

Allen, J., Davey, H. M., Broadhurst, D., Heald, J. K., Rowland, J. J., Oliver, S. G. & Kell, D. B. (2003). High-throughput classification of yeast mutants for functional genomics using metabolic footprinting. *Nature Biotechnology*, **21**, 692–6.

Assmann, S., Sigler, K. & Hofer, M. (1996). Cd^{2+}-induced damage to yeast plasma membrane and its alleviation by Zn^{2+}: studies on *Schizosaccharomyces pombe* cells and reconstituted plasma membrane vesicles. *Archives of Microbiology*, **165**, 279–84.

Avery, S. V. (1995). Caesium accumulation by microorganisms: uptake mechanisms, cation competition, compartmentalization and toxicity. *Journal of Industrial Microbiology*, **14**, 76–84.

Avery, S. V. (2001). Metal toxicity in yeasts and the role of oxidative stress. *Advances in Applied Microbiology*, **49**, 111–42.

Avery, S. V. (2005a). Cell individuality: the bistability of competence development. *Trends in Microbiology*, **13**, 459–62.

Avery, S. V. (2005b). Phenotypic diversity and fungal fitness. *The Mycologist*, **19**, 74–80.

Avery, S. V. & Tobin, J. M. (1993). Mechanism of adsorption of hard and soft metal ions to *Saccharomyces cerevisiae* and influence of hard and soft anions. *Applied and Environmental Microbiology*, **59**, 2851–6.

Avery, A. M. & Avery, S. V. (2001). *Saccharomyces cerevisiae* expresses three phospholipid hydroperoxide glutathione peroxidases. *Journal of Biological Chemistry*, **276**, 33730–5.

Avery, S. V., Howlett, N. G. & Radice, S. (1996). Copper toxicity towards *Saccharomyces cerevisiae*: dependence on plasma-membrane fatty acid composition. *Applied and Environmental Microbiology*, **62**, 3960–6.

Avery, A. M., Willetts, S. A. & Avery, S. V. (2004). Genetic dissection of the phospholipid hydroperoxidase activity of yeast Gpx3 reveals its functional importance. *Journal of Biological Chemistry*, **279**, 46652–8.

Baryla, A., Laborde, C., Montillet, J. L., Triantaphylides, C. & Chagvardieff, P. (2000). Evaluation of lipid peroxidation as a toxicity bioassay for plants exposed to copper. *Environmental Pollution*, **109**, 131–5.

Basu, U., Southron, J. L., Stephens, J. L. & Taylor, G. J. (2004). Reverse genetic analysis of the glutathione metabolic pathway suggests a novel role of *PHGPX* and *URE2* genes in aluminum resistance in *Saccharomyces cerevisiae*. *Molecular Genetics and Genomics*, **271**, 627–37.

Beckman, K. B. & Ames, B. N. (1997). Oxidative decay of DNA. *Journal of Biological Chemistry*, **272**, 19633–66.

Bergman, A. & Siegal, M. L. (2003). Evolutionary capacitance as a general feature of complex gene networks. *Nature*, **424**, 549–52.

Blake, W. J., Kaern, M., Cantor, C. R. & Collins, J. J. (2003). Noise in eukaryotic gene expression. *Nature*, **10**, 633–7.

Brennan, R. J. & Schiestl, R. H. (1996). Cadmium is an inducer of oxidative stress in yeast. *Mutation Research – Fundamental Molecular Mechanisms of Mutagenesis*, **356**, 171–8.

Cabiscol, E., Piulats, E., Echave, P., Herrero, E. & Ros, J. (2000). Oxidative stress promotes specific protein damage in *Saccharomyces cerevisiae*. *Journal of Biological Chemistry*, **275**, 27393–8.

Cakmak, I. (2000). Possible roles of zinc in protecting plant cells from damage by reactive oxygen species. *New Phytologist*, **146**, 185–205.

Casalino, E., Sblano, C. & Landriscina, C. (1997). Enzyme activity alteration by cadmium administration to rats – the possibility of iron involvement in lipid peroxidation. *Archives of Biochemistry and Biophysics*, **346**, 171–9.

Chen, C. Y., Su, Y. J., Wu, P. F. & Shyu, M. M. (2002). Nickel-induced plasma lipid peroxidation and effect of antioxidants in human blood: involvement of hydroxyl radical formation and depletion of alpha-tocopherol. *Journal of Toxicology and Environmental Health, Part A*, **65**, 843–52.

Chiang, K. T., Shinyashiki, M., Switzer, C. H., Valentine, J. S., Gralla, E. B., Thiele, D. J. & Fukuto, J. M. (2000). Effects of nitric oxide on the copper-responsive transcription factor Ace1 in *Saccharomyces cerevisiae*: cytotoxic and cytoprotective actions of nitric oxide. *Archives of Biochemistry and Biophysics*, **377**, 296–303.

Cooke, M. S., Evans, M. D., Dizdaroglu, M. & Lunec, J. (2003). Oxidative DNA damage: mechanisms, mutation, and disease. *FASEB Journal*, **17**, 1195–214.

Costa, W. M. V., Amorim, M. A., Quintanilha, A. & Moradas-Ferreira, P. (2002). Hydrogen peroxide-induced carbonylation of key metabolic enzymes in *Saccharomyces cerevisiae*: the involvement of the oxidative stress response regulators Yap1 and Skn7. *Free Radical Biology and Medicine*, **33**, 1507–15.

Culotta, V. C., Joh, H.-D., Lin, S.-J., Slekar, K. H. & Strain, J. (1995). A physiological role for *Saccharomyces cerevisiae* copper/zinc superoxide dismutase in copper buffering. *Journal of Biological Chemistry*, **270**, 29991–7.

Davey, H. M. & Kell, D. B. (1996). Flow cytometry and cell sorting of heterogeneous microbial populations: the importance of single-cell analyses. *Microbiological Reviews*, **60**, 641–96.

Deutschbauer, A. M., Jaramillo, D. F., Proctor, M., Kumm, J., Hillenmeyer, M. E., Davis, R. W., Nislow, C. & Giaever, G. (2005). Mechanisms of haploinsufficiency revealed by genome-wide profiling in yeast. *Genetics*, **169**, 1915–25.

Delaunay, A., Pflieger, D., Barrault, M. B., Vinh, J. & Toledano, M. B. (2002). A thiol peroxidase is an H_2O_2 receptor and redox-transducer in gene activation. *Cell*, **111**, 471–81.

De Vos, C. H. R., Bookum, W. M. Y., Vooijs, R., Schat, H. & De Kok, L. J. (1993). Effects of copper on fatty acid composition and peroxidation of lipids in the roots of copper tolerant and sensitive *Silene cucabalus*. *Plant Physiology and Biochemistry*, **31**, 151–8.

Dix, T. A. & Aikens, J. (1993). Mechanisms and biological relevance of lipid peroxidation initiation. *Chemical Research in Toxicology*, **6**, 2–18.

Drakulic, T., Temple, M. D., Guido, R., Jarolim, S., Breitenbach, M., Attfield, P. V. & Dawes, I. W. (2005). Involvement of oxidative stress response genes in redox homeostasis, the level of reactive oxygen species, and ageing in *Saccharomyces cerevisiae*. *FEMS Yeast Research*, **5**, 1215–28.

Dunning, J. C., Ma, Y. & Marquis, R. E. (1998). Anaerobic killing of oral streptococci by reduced, transition metal cations. *Applied and Environmental Microbiology*, **64**, 27–33.

Elowitz, M. B., Levine, A. J., Siggia, E. D. & Swain, P. S. (2002). Stochastic gene expression in a single cell. *Science*, **297**, 1183–6.

Fortuniak, A., Zadzinski, R., Bilinski, T. & Bartosz, G. (1996). Glutathione depletion in the yeast *Saccharomyces cerevisiae*. *Biochemistry and Molecular Biology International*, **38**, 901–10.

Fraser, H. B., Hirsh, A. E., Giaever, G., Kumm, J. & Eisen, M. B. (2004). Noise minimization in eukaryotic gene expression. *PLOS Biology*, **2**, 834–8.

Gadd, G. M. (1993). Interactions of fungi with toxic metals. *New Phytologist*, **124**, 25–60.

Gadd, G. M. (2000). Bioremedial potential of microbial mechanisms of metal mobilization and immobilization. *Current Opinion in Biotechnology*, **11**, 271–9.

Gadd, G. M. & Mowll, J. L. (1983). The relationship between cadmium uptake, potassium release and viability in *Saccharomyces cerevisiae*. *FEMS Microbiology Letters*, **16**, 45–8.

Galiazzo, F., Cirilio, M. R., Carri, M. T., Civitareale, P., Marcocci, L., Marmocchi, F. & Rotilio, G. (1991). Activation and induction by copper of Cu/Zn superoxide dismutase in *Saccharomyces cerevisiae* – presence of an inactive proenzyme in anaerobic yeast. *European Journal of Biochemistry*, **196**, 545–9.

Gralla, E. B. (1997). Superoxide dismutase: studies in the yeast *Saccharomyces cerevisiae*. In *Oxidative Stress and the Molecular Biology of Antioxidant Defenses*, ed. J. G. Scandialos. Cold Spring Harbor: Cold Spring Harbor Laboratory Press, pp. 495–525.

Greco, M. A., Hrab, D. I., Magner, W. & Kosman, D. J. (1990). Cu, Zn superoxide-dismutase and copper deprivation and toxicity in *Saccharomyces cerevisiae*. *Journal of Bacteriology*, **172**, 317–25.

Halliwell, B. & Gutteridge, J. M. C. (1990). Role of free radicals and catalytic metal ions in human disease: an overview. *Methods in Enzymology*, **186**, 1–88.

Halliwell, B. & Gutteridge, J. M. C. (1999). *Free Radicals in Biology and Medicine*, 3rd edn. Oxford, UK: Oxford University Press.

Harris, Z. L. & Gitlin, J. D. (1996). Genetic and molecular basis for copper toxicity. *American Journal of Clinical Nutrition*, **63**, S836–S841.

Hepburn, D. D. D., Burney, J. M., Woski, S. A. & Vincent, J. B. (2003). The nutritional supplement chromium picolinate generates oxidative DNA damage and peroxidized lipids in vivo. *Polyhedron*, **22**, 455–63.

Hippeli, S. & Elstner, E. F. (1999). Transition metal ion-catalyzed oxygen activation during pathogenic processes. *FEBS Letters*, **443**, 1–7.
Howlett, N. G. & Avery, S. V. (1997a). Induction of lipid peroxidation during heavy metal stress in *Saccharomyces cerevisiae* and influence of plasma membrane fatty acid unsaturation. *Applied and Environmental Microbiology*, **63**, 2971–6.
Howlett, N. G. & Avery, S. V. (1997b). Relationship between cadmium sensitivity and degree of plasma membrane fatty acid unsaturation in *Saccharomyces cerevisiae*. *Applied Microbiology and Biotechnology*, **48**, 539–45.
Howlett, N. G. & Avery, S. V. (1999). Flow cytometric investigation of heterogeneous copper sensitivity in asynchronously-grown *Saccharomyces cerevisiae*. *FEMS Microbiology Letters*, **176**, 379–86.
Hughes, M. N. & Poole, R. K. (1991). Metal speciation and microbial growth – the hard (and soft) facts. *Journal of General Microbiology*, **137**, 725–34.
Hughes, T. R., Marton, M. J., Jones, A. R., Roberts, C. J., Stoughton, R., Armour, C. D., Bennett, H. A., Coffey, E., Dai, H. Y., He, Y. D. D., Kidd, M. J., King, A. M., Meyer, M. R., Slade, D., Lum, P. Y., Stepaniants, S. B., Shoemaker, D. D., Gachotte, D., Chakraburtty, K., Simon, J., Bard, M. & Friend, S. H. (2000). Functional discovery via a compendium of expression profiles. *Cell*, **102**, 109–26.
Inoue, Y., Matsuda, T., Sugiyama, K.-I., Izawa, S. & Kimura, A. (1999). Genetic analysis of glutathione peroxidase in oxidative stress response of *Saccharomyces cerevisiae*. *Journal of Biological Chemistry*, **274**, 27002–9.
Jin, Y. H., Clark, A. B., Slebos, R. J. C., Al-Refai, H., Taylor, J. A., Kunkel, T. A., Resnick, M. A. & Gordenin, D. A. (2003). Cadmium is a mutagen that acts by inhibiting mismatch repair. *Nature Genetics*, **34**, 326–9.
Jungmann, J., Reins, H. A., Schobert, C. & Jentsch, S. (1993). Resistance to cadmium mediated by ubiquitin-dependent proteolysis. *Nature*, **361**, 369–71.
Kaern, M., Elston, T. C., Blake, W. J. & Collins, J. J. (2005). Stochasticity in gene expression: from theories to phenotypes. *Nature Reviews Genetics*, **6**, 451–64.
Kale, S. P. & Jazwinski, S. M. (1996). Differential response to UV stress and DNA damage during the yeast replicative life span. *Developmental Genetics*, **18**, 154–60.
Karlstrom, A. R. & Levine, R. L. (1991). Copper inhibits the protease from human immunodeficiency virus-1 by both cysteine-dependent and cysteine-independent mechanisms. *Proceedings of the National Academy of Sciences of the USA*, **88**, 5552–6.
Kasprzak, K. S. (2002). Oxidative DNA and protein damage in metal-induced toxicity and carcinogenesis. *Free Radical Biology and Medicine*, **32**, 958–67.
Kryukov, G. V., Kumar, R. A., Koc, A., Sun, Z. H. & Gladyshev, V. N. (2002). Selenoprotein R is a zinc-containing stereo-specific methionine sulfoxide reductase. *Proceedings of the National Academy of Sciences of the USA*, **99**, 4245–50.
Larison, J. R., Likens, G. E., Fitzpatrick, J. W. & Crock, J. G. (2000). Cadmium toxicity among wildlife in the Colorado Rocky Mountains. *Nature*, **406**, 181–3.
Lee, J. K., Kim, J. M., Kim, S. W., Nam, D. H., Yong, C. S. & Huh, K. (1996). Effect of copper-ion damage in superoxide dismutase-deficient *Saccharomyces cerevisiae*. *Archives of Pharmaceutical Research*, **19**, 178–82.
Lehmann, M., Reidel, K., Adler, K. & Kunze, G. (2000). Amperometric measurement of copper ions with a deputy substrate using a novel *Saccharomyces cerevisiae* sensor. *Biosensors and Bioelectronics*, **15**, 211–19.
Leonard, S. S., Bower, J. J. & Shi, X. L. (2004). Metal-induced toxicity, carcinogenesis, mechanisms and cellular responses. *Molecular and Cellular Biochemistry*, **255**, 3–10.
Li, Z. S., Lu, Y. P., Zhen, R. G., Szczypka, M., Thiele, D. J. & Rea, P. A. (1997). A new pathway for vacuolar cadmium sequestration in *Saccharomyces cerevisiae*:

YCF1-catalyzed transport of bis(glutathionato)cadmium. *Proceedings of the National Academy of Sciences USA*, **94**, 42–7.

Lin, C.-M., Crawford, B. F. & Kosman, D. J. (1993). Distribution of Cu^{64} in *Saccharomyces cerevisiae*: cellular locale and metabolism. *Journal of General Microbiology*, **139**, 1605–15.

Liochev, S. I. & Fridovich, I. (1999). Superoxide and iron: partners in crime. *IUBMB Life*, **48**, 157–61.

Llanos, R. M. & Mercer, J. F. B. (2002). The molecular basis of copper homeostasis and copper-related disorders. *DNA and Cell Biology*, **21**, 259–70.

Lloyd, D. R., Phillips, D. H. & Carmichael, P. L. (1997). Generation of putative intrastrand cross-links and strand breaks in DNA by transition metal ion-mediated oxygen radical attack. *Chemical Research in Toxicology*, **10**, 393–400.

Lu, Y., Roe, J. A., Bender, C. J., Peisach, J., Banci, L., Bertini, I., Gralla, E. B. & Valentine, J. S. (1996). New type 2 copper-cysteinate proteins. Copper site histidine-to-cysteine mutants of yeast copper-zinc superoxide dismutase. *Inorganic Chemistry*, **35**, 1692–700.

Lum, P. Y., Armour, C. D., Stepaniants, S. B., Cavet, G., Wolf, M. K., Butler, J. S., Hinshaw, J. C., Garnier, P., Prestwich, G. D., Leonardson, A., Garrett-Engele, P., Rush, C. M., Bard, M., Schimmack, G., Phillips, J. W., Roberts, C. J. & Shoemaker, D. D. (2004). Discovering modes of action for therapeutic compounds using a genome-wide screen of yeast heterozygotes. *Cell*, **116**, 121–37.

Malik, A. (2004). Metal bioremediation through growing cells. *Environment International*, **30**, 261–78.

Mastrolorenzo, A., Scozzafava, A. & Supuran, C. T. (2000). Antifungal activity of silver and zinc complexes of sulfadrug derivatives incorporating arylsulfonylureido moieties. *European Journal of Pharmaceutical Sciences*, **11**, 99–107.

Mukhopadhyay, R., Shi, J. & Rosen, B. P. (2000). Purification and characterization of Acr2p, the *Saccharomyces cerevisiae* arsenate reductase. *Journal of Biological Chemistry*, **275**, 21149–57.

Murakami, K. & Yoshino, M. (1999). Dipicolinic acid as an antioxidant: protection of glutathione reductase from the inactivation by copper. *Biomedical Research – Tokyo*, **20**, 321–6.

Naganuma, A., Miura, N., Kaneko, S., Mishina, T., Hosoya, S., Miyairi, S., Furuchi, T. & Kuge, S. (2000). GFAT as a target molecule of methylmercury toxicity in *Saccharomyces cerevisiae*. *FASEB Journal*, **14**, 968–72.

Nicoletti, G., Domalewska, E. & Borland, R. (1999). Fungitoxicity of oxine and copper oxinate: activity spectrum, development of resistance and synergy. *Mycological Research*, **103**, 1073–84.

Ohsumi, Y., Kitamoto, K. & Anraku, Y. (1988). Changes induced in the permeability barrier of the yeast plasma membrane by cupric ion. *Journal of Bacteriology*, **170**, 2676–82.

Ozbudak, E. M., Thattai, M., Kurtser, I., Grossman, A. D. & van Oudenaarden, A. (2002). Regulation of noise in the expression of a single gene. *Nature Genetics*, **31**, 69–73.

Park, J. I., Grant, C. M., Davies, M. J. & Dawes, I. W. (1998). The cytoplasmic Cu, Zn superoxide dismutase of *Saccharomyces cerevisiae* is required for resistance to freeze-thaw stress. Generation of free radicals during freezing and thawing. *Journal of Biological Chemistry*, **273**, 22921–8.

Pereira, M. D., Herdeiro, R. S., Fernandes, P. N., Eleutherio, E. C. & Panek, A. D. (2003). Targets of oxidative stress in yeast *sod* mutants. *Biochimica et Biophysica Acta*, **1620**, 245–51.

Pourahmad, J. & O'Brien, P. J. (2000). A comparison of hepatocyte cytotoxic mechanisms for Cu^{2+} and Cd^{2+}. *Toxicology*, **143**, 263–73.

Predki, P. F. & Sarkar, B. (1992). Effect of replacement of zinc finger zinc on estrogen-receptor DNA interactions. *Journal of Biological Chemistry*, **267**, 5842–6.

Rae, T. D., Schmidt, P. J., Pufahl, R. A., Culotta, V. C. & O'Halloran T. V. (1999). Undetectable intracellular free copper: the requirement of a copper chaperone for superoxide dismutase. *Science*, **284**, 805–8.

Raser, J. M. & O'Shea, E. K. (2004). Control of stochasticity in eukaryotic gene expression. *Science*, **304**, 1811–14.

Requena, J. R., Groth, D., Legname, G., Stadtman, E. R., Prusiner, S. B. & Levine, R. L. (2001). Copper-catalyzed oxidation of the recombinant SHa(29–231) prion protein. *Proceedings of the National Academy of Sciences of the USA*, **98**, 7170–5.

Rotilio, G., Rossi, L., Demartino, A., Ferreira, A. M. D. S. & Ciriolo, M. R. (1995). Free-radicals, metal-ions and oxidative stress – chemical mechanisms of damage and protection in living systems. *Journal of the Brazilian Chemical Society*, **6**, 221–7.

Rutherford, J. C. & Bird, A. J. (2004). Metal-responsive transcription factors that regulate iron, zinc, and copper homeostasis in eukaryotic cells. *Eukaryotic Cell*, **3**, 1–13.

Sarkar, S., Poonam, Y. & Bhatnagar, D. (1997). Cadmium-induced lipid peroxidation and the antioxidant enzymes in rat tissues – role of vitamin E and selenium. *Trace Elements and Electrolytes*, **14**, 41–5.

Shanmuganathan, A., Avery, S. V., Willetts, S. A. & Houghton, J. E. (2004). Copper-induced oxidative stress in *Saccharomyces cerevisiae* targets enzymes of the glycolytic pathway. *FEBS Letters*, **556**, 253–9.

Shenton, D. & Grant, C. M. (2003). Protein S-thiolation targets glycolysis and protein synthesis in response to oxidative stress in the yeast *Saccharomyces cerevisiae*. *Biochemical Journal*, **374**, 513–19.

Shringarpure, R., Grune, T. & Davies, K. J. A. (2001). Protein oxidation and 20S proteasome-dependent proteolysis in mammalian cells. *Cellular and Molecular Life Sciences*, **58**, 1442–50.

Srinivasan, C., Liba, A., Imlay, J. A., Valentine, J. S. & Gralla, E. B. (2000). Yeast lacking superoxide dismutase(s) show elevated levels of 'free iron' as measured by whole cell electron paramagnetic resonance. *Journal of Biological Chemistry*, **275**, 29187–92.

Stohs, S. J. & Bagchi, D. (1995). Oxidative mechanisms in the toxicity of metal ions. *Free Radical Biology and Medicine*, **18**, 321–36.

Strain, J. & Culotta, V. C. (1996). Copper ions and the regulation of *Saccharomyces cerevisiae* metallothionein genes under aerobic and anaerobic conditions. *Molecular and General Genetics*, **251**, 139–45.

Sumner, E. R. & Avery, S. V. (2002). Phenotypic heterogeneity: differential stress resistance among individual cells of the yeast *Saccharomyces cerevisiae*. *Microbiology*, **148**, 345–51.

Sumner, E. R., Avery, A. M., Houghton, J. E., Robins, R. A. & Avery, S. V. (2003). Cell cycle- and age-dependent activation of Sod1p drives the formation of stress-resistant cell subpopulations within clonal yeast cultures. *Molecular Microbiology*, **50**, 857–70.

Sumner, E. R., Shanmuganathan, A., Sideri, T. C., Willetts, S. A., Houghton, J. E. & Avery S. V. (2005). Oxidative protein damage causes chromium toxicity in yeast. *Microbiology*, **151**, 1939–48.

Szuster-Ciesielska, A., Stachura, A., Slotwinska, M., Kaminska, T., Sniezko, R., Paduch, R., Abramczyk, D., Filar, J. & Kandefer-Szerszen, M. (2000). The inhibitory effect of zinc on cadmium-induced cell apoptosis and reactive oxygen species (ROS) production in cell cultures. *Toxicology*, **145**, 159–71.

Tamai, K. T., Gralla, E. B., Ellerby, L. M., Valentine, J. S. & Thiele, D. J. (1993). Yeast and mammalian metallothioneins functionally substitute for yeast copper-zinc superoxide dismutase. *Proceedings of the National Academy of Sciences of the USA*, **90**, 8013–17.

Thattai, M. & van Oudenaarden, A. (2004). Stochastic gene expression in fluctuating environments. *Genetics*, **167**, 523–30.

Tolker-Nielsen, T., Holmstrom, K., Boe, L. & Molin, S. (1998). Non-genetic population heterogeneity studied by *in situ* polymerase chain reaction. *Molecular Microbiology*, **27**, 1099–105.

Uetz, P., Giot, L., Cagney, G., Mansfield, T. A., Judson, R. S., Knight, J. R., Lockshon, D., Narayan, V., Srinivasan, M., Pochart, P., Qureshi-Emili, A., Li, Y., Godwin, B., Conover, D., Kalbfleisch, T., Vijayadamodar, G., Yang, M. J., Johnston, M., Fields, S. & Rothberg, J. M. (2000). A comprehensive analysis of protein-protein interactions in *Saccharomyces cerevisiae*. *Nature*, **403**, 623–7.

Van Ginkel, G. & Sevanian, A. (1994). Lipid peroxidation-induced membrane structural alterations. *Methods in Enzymology*, **233**, 273–88.

Verma, N. & Singh, M. (2005). Biosensors for heavy metals. *Biometals*, **18**, 121–9.

Viarengo, A., Burlando, B., Ceratto, N. & Panfoli, I. (2000). Antioxidant role of metallothioneins: A comparative review. *Cellular and Molecular Biology*, **46**, 407–17.

Watabe, S., Hasegawa, H., Takimoto, K., Yamamoto, Y. & Takahashi, S. Y. (1995). Possible function of SP-22, a substrate of mitochondrial ATP-dependent protease, as radical scavenger. *Biochemical and Biophysical Research Communications*, **213**, 1010–16.

Wei, J. P. J., Srinivasan, C., Han, H., Valentine, J. S. & Gralla, E. B. (2001). Evidence for a novel role of copper-zinc superoxide dismutase in zinc metabolism. *Journal of Biological Chemistry*, **276**, 44798–803.

Wemmie, J. A., Szczypka, M. S., Thiele, D. J. & Moye-Rowley, W. S. (1994). Cadmium tolerance mediated by the yeast AP-1 protein requires the presence of an ATP-binding cassette transporter-encoding gene, *YCF1*. *Journal of Biological Chemistry*, **269**, 32592–7.

White, C., Sharman, A. K. & Gadd, G. M. (1998). An integrated microbial process for the bioremediation of soil contaminated with toxic metals. *Nature Biotechnology*, **16**, 572–5.

Winzeler, E. A. *et al.* (1999). Functional characterization of the *S. cerevisiae* genome by gene deletion and parallel analysis. *Science*, **285**, 901–6.

11
Bioluminescence-based fungal biosensors

H. J. WEITZ

Introduction

To apply suitable bioremediation techniques, an understanding of the physical, chemical and biological attributes of an environmental matrix is required. Effective bioremediation is based on optimizing these attributes to enhance the biodegradation of target pollutants. Fundamental to these processes is the concept of bioavailability and bioaccessibility of these pollutants at a suitable and relevant scale (Alexander, 2000; Semple *et al.*, 2004). Environmental analyses are still based on chemical approaches that usually require an exhaustive extraction step prior to chromatographic analysis. This extracted fraction is commonly modelled to assess the portion that may cause harm to a particular target receptor. It is widely acknowledged that modelled values may be appropriate for human risk assessment (though inherently conservative) but yield little information for hazard assessment in a wider ecological or environmental context (Alexander, 2000). Many authors have demonstrated that chemical analysis alone does not provide information regarding the bioavailable fraction of compounds nor about their effects on selected biological receptors (Power *et al.*, 1998; Hansen & Sørensen, 2001; Belkin, 2003; Paton & Killham, 2003). Biological assays are able to complement chemical analysis by considering the effects of all pollutants present in a sample, including those not detected by chemical analysis or those unable to be fitted in a model. Bioassays are used for monitoring the progress of bioremediation because they determine the bioavailable fraction of compounds that in part determines the biodegradability of a compound (Hansen & Sørensen, 2001; Paton & Killham, 2003). However, Semple *et al.* (2004) redefined bioavailability and suggested that it is also important to know the bioaccessible fraction for bioremediation. Chemical analysis and biological assays, performed with

Exploitation of Fungi, eds. G. D. Robson, P. van West and G. M. Gadd. Published by Cambridge University Press.

suitable biosensors or bioindicators, thus need to be used in combination for assessing bioremediation practices (Dennison & Turner, 1995; Power *et al.*, 1998; Koehler *et al.*, 2000; D'Souza, 2001; Hansen & Sørensen, 2001; Baeumner, 2003; Belkin, 2003; Paton & Killham, 2003). Whole cell microbial biosensors and, in particular, bioluminescence-based biosensors are being considered in this review, and the development of yeast and filamentous fungal bioluminescence-based biosensors are described and discussed.

Microbial biosensors

A biosensor has been defined as a system composed of a biological sensing component coupled to a transducer that produces a measurable signal when exposed to an analyte (Dennison & Turner, 1995; Ramanathan *et al.*, 1997; Daunert *et al.*, 2000; Belkin, 2003). Biosensing systems encompass a wide array of biological materials including nucleic acids, enzymes, antibodies and microorganisms (Korpan & El'skaya, 1995; Ramanathan *et al.*, 1997; D'Souza, 2001). Biosensors based on whole cells lack the specificity of enzymes or proteins, but they are more cost effective and potentially more stable (Ramanathan *et al.*, 1997; Daunert *et al.*, 2000; D'Souza, 2001; Keane *et al.*, 2002). Whole cell microbial biosensors have advantages in toxicity testing: they are generally routine and cost effective to culture, easy to apply, and they often respond rapidly to the target analytes (in minutes rather than days or months) (Dennison & Turner, 1995; Korpan & El'skaya, 1995; Ramanathan *et al.*, 1997; Keane *et al.*, 2002; Paton & Killham, 2003). Bioassays using whole cell microbial biosensors have been developed for the detection of compounds and elements in the environment. Microbial biosensors are sensitive to a wide range of organic and inorganic pollutants and they can be used for the monitoring of pollutants on sites to be bioremediated. Biosensors have been divided into two main groupings: 'non-specific' (metabolic) biosensors and analyte 'specific' biosensors (Hansen & Sørensen, 2001; Paton & Killham, 2003). Non-specific biosensors respond to a wide range of pollutants and are commonly used for environmental toxicity testing. Progress in the research and development of biosensors and their applications has been the subject of numerous reviews (Dennison & Turner, 1995; Steinberg *et al.*, 1995; Daunert *et al.*, 2000; D'Souza, 2001; Hansen & Sørensen, 2001; Keane *et al.*, 2002; Baeumner, 2003; Belkin, 2003).

Most microbial biosensors that have been described to date are based on bacterial cells. The use of yeasts and filamentous fungi rather than bacteria in biosensors has been reviewed by Baronian (2004). As yeasts and filamentous fungi are eukaryotic, biosensors based on them can sometimes provide information of relevance to other eukaryotes that prokaryotic biosensors cannot. Yeasts and filamentous fungi also occupy a wide range of ecological niches. However, the number of fungal biosensors and bioassays developed is small when compared to the number of bacterial biosensors. Walmsley & Keenan (2000) proposed several advantages of using yeasts rather than bacteria as whole cell biosensors: these include their rapid growth, ease of manipulation and growth on a broad range of substrates. Hrenovic *et al.* (2005) used a yeast toxicity test based on the inhibition of saccharose fermentation by *Saccharomyces cerevisiae* and found that results for the toxicity of wastewater measured by it agreed with other toxicity tests, such as inhibition of bacterial bioluminescence, aerobic bacterial growth inhibition and TTC-dehydrogenase activity inhibition. Baronian (2004) described both wild-type and genetically-modified fungi that have been used as biosensors for toxicity and genotoxicity testing. Signal detection for biosensor application of fungi has been restricted to a relatively small number of transducers. The most commonly used transducer method has been amperometric. Most yeast biosensors refer to biological oxygen demand (BOD) based on, for example, the perturbation of the respiratory activity of yeast cells by a toxic compound (Campanella *et al.*, 1995) or oxygen consumption by yeast cells during their assimilation of compounds (Alkasrawi *et al.*, 1999).

Several reporter genes encoding, for example, green fluorescent protein (GFP), firefly luciferase and bacterial luciferase have been used in whole cell biosensors to provide a phenotype, for example fluorescence or bioluminescence, which is easily measured and quantified through an appropriate analytical device (Gupta *et al.*, 2003). Recent reviews provide an overview of the various reporter genes used, along with the advantages and disadvantages of these reporters (Daunert *et al.*, 2000; Koehler *et al.*, 2000; Keane *et al.*, 2002). The reporter genes used for modifying yeasts include the bacterial (*lux*) and firefly (*luc*) bioluminescence genes as well as the green fluorescence protein (*gfp*) genes. The *gfp* genes have, for example, been used in the development of some yeast biosensors for genotoxicity testing (Billinton *et al.*, 1998). A *gfp*-marked *Aureobasidium pullulans* has been used for testing sensitivity to biocides (Webb *et al.*, 2001; Sabev *et al.*, 2004). The use of GFP as a reporter in yeast and filamentous fungi was reviewed by Lorang *et al.* (2001). The use of bioluminescent and fluorescent reporter genes in microbial biosensors has been compared (Hansen & Sørensen, 2001; Keane *et al.*, 2002; Belkin, 2003). Bioluminescence enables faster and more sensitive detection than fluorescence because the former is a measure of enzyme (luciferase) activity whereas the latter quantifies the presence of a protein (GFP). The advantage of GFP and other fluorescent proteins is their high stability (Belkin, 2003). However, bioluminescence has several advantages, including the ease, sensitivity and accuracy with which light can be recorded and measured (Meighen, 1993; Chatterjee & Meighen, 1995; D'Souza, 2001; Paton & Killham, 2003). Consequently, light output can be used as an easily detectable reporter of metabolic activity in whole cell bioluminescence-based biosensors used for toxicity testing (Keane *et al.*, 2002).

Bioluminescence

Bioluminescence is the emission of visible light by living organisms via a reaction of molecular oxygen (O_2) with a substrate (luciferin) catalyzed by an enzyme (luciferase) (Meighen, 1993; Wilson & Hastings, 1998). Organisms that exhibit bioluminescence are diverse and widely distributed in nature, for example bacteria, dinoflagellates, fungi, insects (such as click beetles and fireflies) and fish (Meighen, 1991; Wilson & Hastings, 1998; Daunert *et al.*, 2000). The enzymes that catalyze the bioluminescence reaction show no homology to each other, and the substrates are also chemically unrelated. The requirement for molecular oxygen is the only common feature of bioluminescence reactions, indicating that the bioluminescent systems in most organisms may have evolved independently (Meighen, 1991, 1993; Wilson & Hastings, 1998). The terms 'luciferase' and 'luciferin' are thus generic terms for the enzymes and substrates involved in the different bioluminescent reactions.

The bioluminescence reaction of the firefly *Photinus pyralis* involves the oxidative decarboxylation of luciferin, a heterocyclic carboxylic acid, with ATP to give oxyluciferin, AMP and CO_2 with the emission of light (560 nm) (Daunert *et al.*, 2000; Koehler *et al.*, 2000) (reaction 1):

$$\text{luciferin} + \text{ATP} + O_2 \xrightarrow{\text{eukaryotic luciferase}} \text{oxyluciferin} + PP_i + CO_2 + \text{light (560 nm)} \qquad (1)$$

The *luc* gene, which encodes firefly luciferase, has been isolated from the North American firefly *P. pyralis*. The substrate for the firefly bioluminescence reaction, luciferin, needs to be added exogenously (Koehler *et al.*, 2000).

The bioluminescence reaction of bacteria involves the oxidation of a long-chain aldehyde (RCHO) and reduced flavin mononucleotide ($FMNH_2$) to give the corresponding long chain carboxylic acid (RCOOH), water and

FMN with the emission of blue-green light (490 nm) (Meighen, 1991, 1993; Chatterjee & Meighen, 1995; Daunert *et al.*, 2000) (reaction 2):

$$FMNH_2 + RCHO + O_2 \xrightarrow{\text{bacterial luciferase}} FMN + RCOOH + H_2O + \text{light (490 nm)} \qquad (2)$$

The substrates $FMNH_2$ and O_2 required by the bacterial bioluminescence reaction are found in sufficient levels in bacteria; the reaction is highly specific for $FMNH_2$ (Meighen, 1991). The reaction also requires a long chain aldehyde that is produced from fatty acyl precursors in the cell by a multienzyme reductase complex encoded by the *lux*CDE genes using ATP and NADPH (Meighen, 1991). Bacterial luciferase is encoded by the *lux*AB genes (Meighen, 1991, 1993; Chatterjee & Meighen, 1995). The *lux* genes coding for the enzymes required for bioluminescence have been cloned and sequenced from a number of bacteria, for example *Vibrio fischeri* and *Vibrio harveyi* (Meighen, 1991, 1993). If only the *lux*AB genes are used, the aldehyde, usually *n*-decanal, needs to be added exogenously as the substrate for the bioluminescence reaction (Meighen, 1991; Hill *et al.*, 1993; Chatterjee & Meighen, 1995; Koehler *et al.*, 2000; Daunert *et al.*, 2000). The natural aldehyde substrate for the reaction is thought to be tetradecanal, but *n*-decanal serves as a suitable alternative substrate, especially as it appears to cross the cell membrane more readily than longer-chain aldehydes (Meighen, 1991; Chatterjee & Meighen, 1995).

Bioluminescence-based bacterial biosensors

Naturally bioluminescent bacteria (e.g. *Photobacterium phosphoreum* and *V. fischeri*) as well as bacteria genetically modified with *lux* and *luc* genes have been used as bioluminescence-based biosensors. Light emission is possible by the transfer of the *luc* or *lux* genes (either *lux*AB or *lux*CDABE) under a suitable promoter into non-bioluminescent bacteria (Meighen, 1991; Chatterjee & Meighen, 1995; Koehler *et al.*, 2000; Daunert *et al.*, 2000). Non-specific biosensors have been developed to provide information on overall metabolic status through the fusion of the *lux* or *luc* genes to constitutive promoters. These bacterial cells express bioluminescence constitutively and they are used for non-specific detection of pollutants by indicating their metabolic status (Daunert *et al.*, 2000; Koehler *et al.*, 2000; Hansen & Sørensen, 2001; Belkin, 2003). Crucial to their environmental applications, it is only metabolically active cells that emit light and any pollutant that impairs cell integrity and metabolism, thus compromising cellular activity and viability, leads to a reduction in light emission. Consequently, biosensor response to toxicity corresponds to a change in light output (Chatterjee & Meighen, 1995; Steinberg *et al.*, 1995; Ulitzur, 1997; Kelly *et al.*, 1999). Several studies have shown non-specific bacterial biosensors, for example *Escherichia coli* HB101 pUCD607 (Rattray *et al.*, 1990) and *Pseudomonas putida* F1 mini-Tn*5* *lux*CDABE (Weitz *et al.*, 2001a), to be sensitive to a range of pollutants (e.g. Sousa *et al.*, 1998; Kelly *et al.*, 1999; Chaudri *et al.*, 1999; Strachan *et al.*, 2002; Bhattacharyya *et al.*, 2005). The use of naturally bioluminescent bacteria has been commercialized and standardized under different trade names (Farré *et al.*, 2004): these bioassays are also accepted by regulatory authorities (Bhattacharyya *et al.*, 2005). Several specific bacterial biosensors have been developed to detect the presence of target analytes, for example organic compounds, metals or metalloids, and environmental stresses through the fusion of *lux* or *luc* genes to inducible promoters and regulatory genes of environmental relevance that are responsive to the analytes or stresses to be detected (Ramanathan *et al.*, 1997; Daunert *et al.*, 2000; Koehler *et al.*, 2000; Hansen & Sørensen, 2001).

Bioluminescence-based bacterial biosensors have been used for measuring toxicity of pollutants, determining the relationship between bioavailability and toxicity and monitoring bioremediation processes (Sousa *et al.*, 1998; Bundy *et al.*, 2001; Mowat & Bundy, 2001; Paton & Killham, 2003; Bhattacharyya *et al.*, 2005; Paton *et al.*, 2006). *Pseudomonas fluorescens* HK44 was the first genetically-engineered microorganism approved for field testing for bioremediation purposes (Ripp *et al.*, 2000). The utilization of a bioluminescence-based biosensor was shown to be an option for *in situ* determination of environmental contaminant bioavailability and biodegradation process monitoring and control. The use of bioluminescence-based bacteria for environmental and other applications has been reviewed by Hill *et al.* (1993), Meighen (1993), Chatterjee & Meighen (1995), Steinberg *et al.* (1995), Ulitzur (1997) and D'Souza (2001). Bhattacharyya *et al.* (2005) discussed the value of bioluminescence-based bacterial biosensors and described their application in hazard assessment and bioremediation programmes.

Bioluminescence-based yeast biosensors

There have been several reports of the expression of firefly luciferase in *S. cerevisiae* (e.g. Aflalo, 1990; Vieites *et al.*, 1994; Hollis *et al.*, 2000). The production of firefly luciferase only requires the transcription of one cistron and therefore can be expressed in eukaryotes under control of a single promoter (Boylan *et al.*, 1989). Vieites *et al.* (1994) assessed the possibility of the *luc* gene as a reporter gene in yeast. The bioassay conditions for the *luc*-marked yeast were optimized with regard to oxygen concentration, temperature and cell concentration in the assay mixture. As the luciferin is an amphipathic molecule with a charged carboxyl group at physiological pH, it cannot pass easily across cell membranes, leading to problems during the exogenous addition of it for bioassays. Vieites *et al.* (1994) observed that there was a significant increase in light emission below pH 3. This was thought to be as a result of an increase in the amount of uncharged luciferin permeable through the cell membrane (Aflalo, 1990). Hollis *et al.* (2000) thus developed a bioassay involving the acidification of the biosensor preparation after exposure to the toxic compound and before measurement of bioluminescence. *S. cerevisiae* LucΔ, which was genetically-modified to express firefly luciferase, was used as a bioluminescence-based biosensor for toxicity testing. The bioluminescent yeast was shown to detect chemicals known to be toxic to eukaryotes but not toxic to prokaryotes (Hollis *et al.*, 2000).

The expression of bacterial luciferase in yeast required the fusion of *lux*A and *lux*B genes resulting in monocistronic *lux*AB genes coding for active fused luciferases. These fused luciferases can be expressed in eukaryotes under control of a single promoter without the necessity of having *lux*A and *lux*B genes under separate promoters as eukaryotes are unable to efficiently translate prokaryotic polycistronic messages (Boylan *et al.*, 1989; Kirchner *et al.*, 1989; Meighen, 1991; Chatterjee & Meighen, 1995). A fused bacterial luciferase from *V. harveyi* was expressed in *S. cerevisiae* (Boylan *et al.*, 1989; Kirchner *et al.*, 1989). Boylan *et al.* (1989) found that the relative level of light emission (light units ml^{-1} cells) *in vivo* in yeast was much lower than in *Escherichia coli* even though *in vitro* luciferase activities (light units mg^{-1} protein) in extracts prepared from the bacterial and yeast cells were similar. The difference in light emission was suggested to be due to less $FMNH_2$ being available in yeast cells than in bacterial cells (Boylan *et al.*, 1989). Hollis *et al.* (2001) found that *n*-decanal, the substrate used in bacterial bioassays with *lux*AB, is toxic to yeast cells. The use of the bacterial *lux*AB genes to produce bioluminescent eukaryotes thus initially met with some problems due to the difficulties in supplying $FMNH_2$ and aldehyde substrate for the bioluminescence reaction.

Unsaturated long chain aldehydes, which are relatively volatile and soluble and also gave high levels of bioluminescence with bacterial luciferase, were tested to determine if they can cross the yeast cell membrane and have low toxicity (Szittner *et al.*, 2003). Szittner *et al.* (2003) showed that the aldehyde Z-9-tetradecenal gave high levels of light in a yeast expressing bacterial luciferase (*lux*AB), *S. cerevisiae* pLG-*lux*AB, without causing toxicity. As Z-9-tetradecenal is commercially available, inexpensive, delivered as a liquid and easily manipulated, it may be the aldehyde of choice for yeast containing *lux*AB. The introduction of the *frp* gene (pRS-frp) encoding for a NADPH-dependent FMN reductase from *V. harveyi* into the yeast already containing the fused *lux*AB genes (Szittner *et al.*, 2003) resulted in light levels at least 100 times brighter than those previously measured by Boylan *et al.* (1989). This suggested that the limiting factor in generating $FMNH_2$ for the bacterial bioluminescence reaction in yeast was the level of FMN reductase and not the levels of NAD(P)H or FMN (Szittner *et al.*, 2003). The effect of three fungicides on the bioluminescent yeast was tested and the results indicated that bioluminescence showed the toxicity of compounds on the yeast biosensor (Szittner *et al.*, 2003).

The expression of fused *lux*AB genes was low in a variety of eukaryotic cells, and an aldehyde had to be added to the yeast cells as a substrate for the bioluminescence reaction. Another disadvantage was that these constructs failed to fold properly at higher temperatures and thus had low bioluminescence levels (Escher *et al.*, 1989; Gupta *et al.*, 2003). Gupta *et al.* (2003) demonstrated the functional expression of the *lux*A, B, C, D and E genes from *Photorhabdus luminescens* in *S. cerevisiae*. Rather than using fused *lux*AB genes, Gupta *et al.* (2003) used an alternative approach using yeast expression vectors for the bi-directional expression of the individual *lux*A, B, C, D and E genes. The *lux*A and *lux*B genes were expressed on a plasmid (pBEVY-U). Luminescence resulting from constitutive expression of *lux*A and *lux*B genes with exogenously added *n*-decanal was shown to be approximately 20 times greater that that reported for constitutive expression from fused *lux*AB in *S. cerevisiae* (Kirchner *et al.*, 1989). Another plasmid (pBEVY-L) was used for the bi-directional expression of the *lux*C, D and E genes; the *lux*D gene was fused to a yeast internal ribosome entry site (IRES), which allowed translation of multiple genes from a single promoter in eukaryotes, to ensure high bicistronic expression (Gupta *et al.*, 2003; Sanseverino *et al.*, 2005). For bioluminescence, the fatty acyl precursors needed to be readily available in yeast and the *lux*CDE genes had to be readily expressed in eukaryotes (Gupta *et al.*, 2003). The aldehyde substrate for the bioluminescent reaction was synthesized in the yeast cells and the levels did not seem to inhibit growth. Gupta *et al.* (2003) suggested that the bioluminescence levels were low as a result of $FMNH_2$ pools being limited in the yeast cells rather than *n*-decanal having toxic effects on the cells. When the *frp* gene encoding the NADPH-dependent FMN oxidoreductase from

V. harveyi was cloned and co-expressed with the *lux* genes, bioluminescence levels increased up to five times compared to yeast cells without the *frp* gene (Gupta *et al.*, 2003). The resulting yeast strain, *S. cerevisiae* BLYEV, produced bioluminescence without the necessity to exogenously add the substrate for the reaction and was bioluminescent at levels comparable to bacterial bioluminescence-based biosensors containing the *lux*CDABE genes (Gupta *et al.*, 2003; Sanseverino *et al.*, 2005).

Bioluminescence-based filamentous fungal biosensors

The significance of filamentous fungi in key ecological processes, such as nutrient cycling and decomposition, makes them an important taxonomic group for environmental detection and monitoring of pollutants. Despite the ecological importance of filamentous fungi, there are few fungal techniques for toxicity testing and growth measurements are generally used. Fungal growth is commonly determined by measurement of colony diameter (for example, Chiu *et al.*, 1998; Blaudez *et al.*, 2000), absorbance (for example, Hoiland & Dybdahl, 1993; Hoiland, 1995) and biomass (for example, Alleman *et al.*, 1992; Chiu *et al.*, 1998). However, such toxicity bioassays can be labour intensive, insensitive, poor to replicate and time-consuming (Webb *et al.*, 2001). The recent development of a bioluminescence-based, filamentous fungal bioassay for toxicity testing using the naturally bioluminescent fungi *Armillaria mellea* (Vahl : Fries) Karsten (ATCC 1113) and *Mycena citricolor* (Berkley et Curtis) Saccardo (ATCC 34884) is described by Weitz *et al.* (2002).

Naturally bioluminescent fungi have been found both in temperate and tropical climates and there are currently thought to be more than 40 species of bioluminescent fungi within nine genera, all of which are members of the division Basidiomycota (Wassink, 1978; Herring, 1994). Examples of luminescent fungi include *A. mellea*, *M. citricolor*, *Omphalotus olearius* and *Panellus stipticus*. They are commonly wood-degrading fungi, for example *Armillaria* species (e.g. *A. gallica*, *A. mellea*, *A. ostoyae* and *A. tabescens*) and litter-degrading fungi, for example *Mycena* species. Bioluminescent fungi emit a greenish light (520–530 nm) and the light emission is continuous and occurs only in living cells (O'Kane *et al.*, 1990). No correlation of fungal bioluminescence with cell structure has been found. Bioluminescence may occur in both mycelia and fruiting bodies, as in *P. stipticus* and *O. olearius*, or only in mycelia and young rhizomorphs, as in *A. mellea* (Wassink, 1978). The biochemistry of the fungal bioluminescence system is not fully characterized. However, the preparation of bioluminescent, cell-free extracts has allowed the characterization of the in vitro requirements of fungal bioluminescence. Fungal bioluminescence has been shown to involve a 'luciferin-luciferase' reaction requiring O_2 and either NADH or NADPH in *A. mellea* (Airth & Foerster, 1962; Kamzolkina *et al.*, 1984) and *M. citricolor* (Kuwabara & Wassink, 1966). The luciferase(s) and luciferin(s) in fungal bioluminescence have not yet been identified. Conditions that affect growth, such as pH, light and temperature, have been found to influence fungal bioluminescence (Weitz *et al.*, 2001b). This link between growth conditions and bioluminescence levels suggests a link between metabolic activity and fungal bioluminescence. *A. mellea* and *M. citricolor* grew as bioluminescent, globular mycelia in shaken cultures and their growth curves also point to a relationship between growth and bioluminescence (Weitz *et al.*, 2002), which has also been suggested by O'Kane *et al.* (1986) and Shimomura (1992). As fungal bioluminescence can be linked to metabolic activity, it has the potential to provide a rapid, real-time indication of the metabolic status of these filamentous fungi. Bioluminescent, globular mycelia of *A. mellea* and *M. citricolor* have been used in a bioluminescence-based fungal bioassay as described by Weitz *et al.* (2002).

Comparative performance of bioluminescence-based biosensors

Application in aqueous standards

Bioassays were carried out to test the fungal bioluminescence response to pentachlorophenol (PCP), Cu and Zn. The EC_{50} values for the fungal biosensors were compared with the bioluminescence-based bacterial biosensors *P. putida* F1 Tn*5* *lux*CDABE and *P. fluorescens* 10586 pUCD607, and the bioluminescence-based yeast biosensor *S. cerevisiae* LucΔ (Table 11.1). *A. mellea* and *M. citricolor* were significantly less sensitive to Zn than *P. putida* F1 Tn*5* *lux*CDABE. Palmer *et al.* (1998) reported the EC_{50} values for PCP for *P. fluorescens* 10586 pUDC607 and *Rhizobium leguminosarum* bv. *trifolii* F6 pUCD607 to be 0.81 and 1.08 mg l^{-1}, respectively, which are higher than the EC_{50} value for *M. citricolor* (0.30 mg l^{-1}). Therefore, the EC_{50} values for *A. mellea* and *M. citricolor* were found to be similar to the EC_{50} values for previously characterized bioluminescence-based bacterial biosensors (Paton *et al.*, 1997; Weitz *et al.*, 2001a) for PCP and Cu, but not for Zn. The EC_{50} values for *A. mellea* and *M. citricolor* were found to be similar to the EC_{50} values for Cu (3.2 mg l^{-1}) and Zn (>100 mg l^{-1}) for the yeast biosensor *S. cerevisiae* LucΔ (Hollis, 1999; Hollis *et al.*, 2000). The insensitivity of the filamentous fungi to Zn compared to the bacterial biosensors was also observed for *S. cerevisiae* LucΔ. The sensitivity of *A. mellea* and

Table 11.1 *Comparison of EC_{50} values for fungal biosensors* A. mellea *and* M. citricolor, *yeast biosensor* S. cerevisiae *Luc*Δ *and bacterial biosensors* P. putida *F1 Tn5* lux*CDABE and* P. fluorescens *10586 pUCD607. (Adapted from Weitz* et al., *2002)*

	EC_{50} value (mg l^{-1})		
	PCP	Cu	Zn
A. mellea	8.08	0.08	32.1
M. citricolor	0.36	2.25	>90
S. cerevisiae LucΔ[1]	nd	3.2	>100
P. putida F1 Tn*5 lux*CDABE[2]	nd	0.17	0.04
P. fluorescens 10586 pUCD607	0.81	0.34	0.10

Notes:
nd – not determined
[1] Hollis (1999)
[2] Weitz *et al.* (2001a)

Table 11.2 *Comparison of EC_{50} values for fungal biosensors* A. mellea *and* M. citricolor *and bacterial biosensors* E. coli *HB101 pUCD607 and* P. fluorescens *10586 pUCD607 after exposure to a range of potentially toxic elements (mg l^{-1}). (Chadwick, Weitz & Paton, unpublished data)*

PTE	*A. mellea*	*M. citricolor*	*E. coli* HB101 pUCD607	*P. fluorescens* 10586 pUCD607
Cd	>100	5.04	0.12	0.21
Cr	>100	>100	1.23	2.21
Cu	0.08	2.25	0.21	0.34
Pb	>100	>100	1.15	2.24
Hg	1.72	1.74	0.64	0.75
Sn	>100	>100	2.44	3.32
Zn	32.1	>90	0.10	0.10

M. citricolor to a range of potentially toxic elements (PTEs) was also tested and compared to the sensitivity of bioluminescence-based bacterial biosensors (Table 11.2).

The bioluminescence-based, filamentous fungal bioassay provides a rapid and sensitive method to assess toxicity. The results suggest that the fungal biosensor is biologically relevant and as sensitive as existing bioluminescence-based biosensors. The different bacterial and fungal responses to Zn and some other PTEs highlight the importance of using both prokaryotic and eukaryotic biosensors in toxicity testing (Hollis *et al.*,

2000; Weitz *et al.*, 2002). The filamentous fungal biosensors are also important as a measurement of toxicity to multicellular fungi, thus complementing the single cell yeast biosensors. The bioluminescence-based, filamentous fungal bioassay thus complements the existing range of bacterial and yeast bioluminescent-based biosensors available.

Environmental applications

There have been few applications of a range of biological tests for assessing contamination in soils. Tandy *et al.* (2005) compared a range of standard biological tests and reported that bioluminescence-based bacterial biosensors (although lacking sensitivity to Cu) predicted the relatively longer term changes associated with microbial biomass and other standard measurements. Paton *et al.* (2006) used bioluminescence-based bacterial and fungal biosensors in combination with basal respiration, nematode numbers and general heterotrophic measurements for an assessment of contaminated forest soils. The results posed a challenge to interpret but in some cases the fungi offered a level of sensitivity that the bacteria did not exhibit while for other samples the bacteria were more sensitive. The importance in using a number of relevant biological tests when assessing ecotoxicological responses to contaminated soils was shown.

In the study reported here, five different soils were amended with Cu and Zn at environmentally-relevant concentrations. These soils were selected to reflect different metal binding capacities so that those soils with a high capacity would bind the metals while those with a low capacity would allow the metals to be readily extractable. The soils were incubated with the metal amendments and the soil solution was then extracted. Dose-response curves were generated to enable a comparison of the different biosensors to the soil extracts. Campbell *et al.* (2001) published the response of a range of bacterial biosensors to these metals and found that there was a lack of sensitivity to Cu when compared to Zn. The intention of the experiment was to reflect metal concentrations associated with sewage sludge amendments in soils. The data presented enables a direct comparison of the individual soils in their dose-response curves (Tables 11.3 and 11.4). The fungi were less sensitive to Zn than the bacteria, and this observation has been previously reported. However, the bacterial biosensors were much less sensitive to Cu than *A. mellea*, which was most responsive to the metal doses tested. These metal doses were at ecologically relevant values and, as a consequence, this particular fungal biosensor may have an important role in toxicity monitoring.

Table 11.3 *The EC_{50} values of a range of fungal and bacterial biosensors after exposure to Zn amended soils representing different metal binding capacities. (Chadwick, Weitz, Dawson & Paton, unpublished data)*

	EC_{50} (mg kg^{-1}) for Zn			
	A. mellea	*M. citricolor*	*E. coli* HB101 pUCD607	*P. fluorescens* 10586 pUCD607
Weak binding (Culbin)	640	615	35	46
Medium binding (Boghall)	720	740	212	292
Strong binding (Nethermains)	>800	>800	614	712
Very strong binding (St. Fergus)	>800	>800	>800	>800

Table 11.4 *The EC_{50} values of a range of fungal and bacterial biosensors after exposure to Cu amended soils representing different metal binding capacities. (Chadwick, Weitz, Dawson & Paton, unpublished data)*

	EC_{50} (mg kg^{-1}) for Cu			
	A. mellea	*M. citricolor*	*E. coli* HB101 pUCD607	*P. fluorescens* 10586 pUCD607
Weak binding (Culbin)	88	>400	50	78
Medium binding (Boghall)	72	>400	>400	>400
Strong binding (Nethermains)	>400	>400	>400	>400
Very strong binding (St. Fergus)	>400	>400	>400	>400

The results for the environmental applications showed that the level of sensitivity to the soil extracts was significantly different to aqueous standards. Hence, if an ecologically relevant sample can be extracted, the response of the biosensor can be predicted and indeed validated. There has been a great deal of research into the development and application of suitable extraction techniques. In general, the bioassays must complement the extraction techniques such that they have a level of sensitivity to interrogate the environmental matrix under investigation. To this end, the fungal biosensors offer a promising realization of this potential.

Conclusions

The consideration of both biological and chemical data can often result in significant advances in hazard assessment and act as a suitable baseline for risk assessments to be made for site investigation and bioremediation. Chemical analysis and biological assays using bioluminescence-based biosensors should be used in combination for assessing bioremediation. Their combined use allows identification of the bioremediation potential of contaminated sites as well as the monitoring of bioremediation processes. If biosensor application is coupled to chemical extraction or sample manipulation techniques, then the bioluminescence-based biosensors may be further exploited to assess constraints to bioremediation and, importantly, help identify optimal bioremediation strategies (Sousa *et al.*, 1998; Paton & Killham, 2003; Bhattacharyya *et al.*, 2005). The bioluminescence-based bioassays used should include a range of microbial biosensors. Both bioluminescence-based bacterial and fungal biosensors are available for looking at environmental toxicity, although the yeast and filamentous fungal bioassays still need to be used in bioremediation studies.

Acknowledgements

The work on the bioluminescence-based filamentous fungal bioassay was funded through the University of Aberdeen as part of the University of Aberdeen/Macaulay Institute Soil Health Initiative. Katherine L. Chadwick and Julian J. C. Dawson are acknowledged for the provision of the data for Tables 11.2, 11.3 and 11.4. The author would like to thank Graeme I. Paton for his help with the manuscript preparation.

References

Aflalo, C. (1990). Targeting of cloned firefly luciferase to yeast mitochondria. *Biochemistry*, **29**, 4758–66.

Airth, R. L. & Foerster, G. E. (1962). The isolation of catalytic components required for cell-free fungal bioluminescence. *Archives of Biochemistry and Biophysics*, **97**, 567–73.

Alexander, M. (2000). Aging, bioavailability, and overestimation of risk from environmental pollutants. *Environmental Science and Technology*, **34**, 4259–65.

Alkasrawi, M., Nandakumar, R., Margesin, R., Schinner, F. & Mattiasson, B. (1999). A microbial biosensor based on *Yarrowia lipolytica* for the off-line determination of middle-chain alkanes. *Biosensors and Bioelectronics*, **14**, 723–7.

Alleman, B. C., Logan, B. E. & Gilbertson, R. L. (1992). Toxicity of pentachlorophenol to six species of white rot fungi as a function of chemical dose. *Applied and Environmental Microbiology*, **58**, 4048–50.

Baeumner, A. J. (2003). Biosensors for environmental pollutants and food contaminants. *Analytical and Bioanalytical Chemistry*, **377**, 434–45.

Baronian, K. H. R. (2004). The use of yeast and moulds as sensing elements in biosensors. *Biosensors and Bioelectronics*, **19**, 953–62.

Belkin, S. (2003). Microbial whole-cell sensing systems of environmental pollutants. *Current Opinion in Microbiology*, **6**, 206–12.

Bhattacharyya, J., Read, D., Amos, S., Dooley, S., Killham, K. & Paton, G. I. (2005). Biosensor-based diagnostics of contaminated groundwater: assessment and remediation strategy. *Environmental Pollution*, **134**, 485–92.

Billinton, N., Barker, M. G., Michel, C. E., Knight, A. W., Heyer, W.-D., Goddard, N. J., Fielden, P. R. & Walmsley, R. M. (1998). Development of a green fluorescent protein reporter for a yeast genotoxicity biosensor. *Biosensors and Bioelectronics*, **13**, 831–8.

Blaudez, D., Jacob, C., Turnau, K., Colpaert, J. V., Ahonen-Jonnarth, U., Finlay, R., Botton, B. & Chalot, M. (2000). Differential responses of ectomycorrhizal fungi to heavy metals *in vitro*. *Mycological Research*, **104**, 1366–71.

Boylan, M., Pelletier, J. & Meighen, E. A. (1989). Fused bacterial luciferase subunits catalyze light emission in eukaryotes and prokaryotes. *Journal of Biological Chemistry*, **264**, 1915–18.

Bundy, J. G., Campbell, C. D. & Paton, G. I. (2001). Comparison of response of six different luminescent bacterial bioassays to bioremediation of five contrasting oils. *Journal of Environmental Monitoring*, **3**, 404–10.

Campanella, L., Favero, G. & Tomassetti, M. (1995). Immobilised yeast cells biosensor for total toxicity testing. *The Science of the Total Environment*, **171**, 227–34.

Campbell, C. D., Paton, G. I., Towers, W., Paterson, E., Dawson, J. C. C., Cameron, C. M., Coull, M. C. & Christie, P. (2001). A biological classification scheme to assess the sensitivity of Scottish and Northern Ireland soils to heavy metals. *SNIFFER Report No SR (00) 08*.

Chatterjee, J. & Meighen, E. A. (1995). Biotechnological applications of bacterial bioluminescence (*lux*) genes. *Photochemistry and Photobiology*, **62**, 641–50.

Chaudri, A. M., Knight, B. P., Barbosa-Jefferson, V. L., Preston, S., Paton, G. I., Killham, K., Coad, N., Nicholson, F. A., Chambers, B. J. & McGrath, S. P. (1999). Determination of acute Zn toxicity in pore water from soils previously treated with sewage sludge using bioluminescence assays. *Environmental Science and Technology*, **33**, 1880–5.

Chiu, S. W., Ching, M. L., Fong, K. L. & Moore, D. (1998). Spent oyster mushroom substrate performs better than mushroom mycelia in removing the biocide pentachlorophenol. *Mycological Research*, **102**, 1553–62.

Daunert, S., Barrett, G., Feliciano, J. S., Shetty, R. S., Shrestha, S. & Smith-Spencer, W. (2000). Genetically engineered whole-cell sensing systems: coupling biological recognition with reporter genes. *Chemical Reviews*, **100**, 2705–38.

Dennison, M. J. & Turner, A. P. F. (1995). Biosensors for environmental monitoring. *Biotechnology Advances*, **13**, 1–12.

D'Souza, S. F. (2001). Microbial biosensors. *Biosensors and Bioelectronics*, **16**, 337–53.

Escher, A., O'Kane, D. J., Lee, J. & Szalay, A. A. (1989). Bacterial luciferase $\alpha\beta$ fusion protein is fully active as a monomer and highly sensitive *in vivo* to elevated temperature. *Proceedings of the National Academy of Sciences of the United States of America*, **86**, 6528–32.

Farré, M., Arranz, F., Ribó, J. & Barceló, D. (2004). Interlaboratory study of the bioluminescence inhibition tests for rapid wastewater toxicity assessment. *Talanta*, **62**, 549–58.

Gupta, R. K., Patterson, S. S., Ripp, S., Simpson, M. L. & Sayler, G. S. (2003). Expression of the *Photorhabdus luminescens lux* genes (*luxA, B, C, D*, and *E*) in *Saccharomyces cerevisiae. FEMS Yeast Research*, **4**, 305–13.

Hansen, L. H. & Sørensen, S. J. (2001). The use of whole-cell biosensors to detect and quantify compounds or conditions affecting biological systems. *Microbial Ecology*, **42**, 483–93.

Herring, P. J. (1994). Luminous fungi. *Mycologist*, **8**, 181–3.

Hill, P. J., Rees, C. E. D., Winson, M. K. & Stewart, G. S. A. B. (1993). The application of *lux* genes. *Biotechnology and Applied Biochemistry*, **17**, 3–14.

Hoiland, K. (1995). Reaction of some decomposer basidiomycetes to toxic elements. *Nordic Journal of Botany*, **15**, 305–18.

Hoiland, K. & Dybdahl, H. G. (1993). A micro-well method for estimating fungal response to metal ions. Response to aluminium by some saprophytic basidiomycetes. *Nordic Journal of Botany*, **13**, 691–6.

Hollis, R. P. (1999). Construction and application of a luminescent eukaryotic biosensor. Ph.D. thesis, University of Aberdeen.

Hollis, R. P., Killham, K. & Glover, L. A. (2000). Design and application of a biosensor for monitoring toxicity of compounds to eukaryotes. *Applied and Environmental Microbiology*, **66**, 1676–9.

Hollis, R. P., Lagido, C., Pettitt, J., Porter, A. J. R., Killham, K., Paton, G. I. & Glover, L. A. (2001). Toxicity of the bacterial luciferase substrate, *n*-decyl aldehyde, to *Saccharomyes cerevisiae* and *Caenorhabditis elegans. FEBS Letters*, **506**, 140–2.

Hrenovic, J., Stilinovic, B. & Dvoracek, L. (2005). Use of prokaryotic and eukaryotic biotests to assess toxicity of wastewater from pharmaceutical sources. *Acta Chimica Slovenica*, **52**, 119–25.

Kamzolkina, O. V., Bekker, Z. E. & Egorov, N. S. (1984). Extraction of the luciferin-luciferase system from the fungus *Armillaria mellea. Biologicheskiye Nauki*, **1**, 73–7.

Keane, A., Phoenix, P., Ghoshal, S. & Lau P. C. K. (2002). Exposing culprit organic pollutants: a review. *Journal of Microbiological Methods*, **49**, 103–19.

Kelly, C. J., Lajoie, C. A., Layton, A. C. & Sayler, G. S. (1999). Bioluminescent reporter bacterium for toxicity monitoring in biological wastewater treatment systems. *Water Environment Research*, **71**, 31–5.

Kirchner, G., Roberts, J. L., Gustafson, G. D. & Ingolia, T. D. (1989). Active bacterial luciferase from a fused gene: expression of a *Vibrio harveyi lux*AB translational fusion in bacteria, yeast and plant cells. *Gene*, **81**, 349–54.

Koehler, S., Belkin, S. & Schmid, R. D. (2000). Reporter gene bioassays in environmental analysis. *Fresenius Journal of Analytical Chemistry*, **366**, 769–79.

Korpan, Y. I. & El'skaya, A. V. (1995). Microbial sensors: achievements, problems, and prospects. *Biochemistry (Moscow)*, **60**, 1517–24.

Kuwabara, S. & Wassink, E. C. (1966). Purification and properties of the active substance of fungal luminescence. In *Bioluminescence in Progress*, eds. F. H. Johnson & Y. Haneda. Princeton: Princeton University Press, pp. 233–45.

Lorang, J. M., Tuori, R. P., Martinez, J. P., Sawyer, T. L., Redman, R. S., Rollins, J. A., Wolpert, T. J., Johnson, K. B., Rodriguez, R. J., Dickman, M. B. & Cuiffeti, L. M. (2001). Green fluorescent protein is lighting up fungal biology. *Applied and Environmental Microbiology*, **67**, 1987–94.

Meighen, E. A. (1991). Molecular biology of bacterial bioluminescence. *Microbiological Reviews*, **55**, 123–42.

Meighen, E. A. (1993). Bacterial bioluminescence: organization, regulation, and application of the *lux* genes. *FASEB Journal*, **7**, 1016–22.

Mowat, F. S. & Bundy, K. J. (2001). Correlation of field-measured toxicity with chemical concentration and pollutant availability. *Environment International*, **27**, 479–89.

O'Kane, D. J., Lingle, W. L., Porter, D. & Wampler, J. E. (1986). Development and localization of bioluminescence in the fruiting bodies of the mushroom *Panellus stypticus*. *Photochemistry and Photobiology*, **43**, 100s.

O'Kane, D. J., Lingle, W. L., Porter, D. & Wampler, J. E. (1990). Spectral analysis of bioluminescence of *Panellus stypticus*. *Mycologia*, **82**, 607–16.

Palmer, G., McFadzean, R., Killham, K., Sinclair, A. & Paton, G. I. (1998). Use of *lux*-based biosensors for rapid diagnosis of pollutants in arable soils. *Chemosphere*, **36**, 2683–97.

Paton, G. I., Rattray, E. A. S., Campbell, C. D., Cresser, M. S., Glover, L. A., Meeussen, J. C. L. & Killham, K. (1997). Use of genetically modified microbial biosensors for soil ecotoxicity testing. In *Biological Indicators of Soil Health and Sustainable Productivity*, eds. C. Pankhurst, B. Doube & V. Gupta. Wallingford: CAB International, pp. 397–418.

Paton, G. I. & Killham, K. (2003). Intelligent site assessment – a role for ecotoxicology. In *Bioremediation: A Critical Review*, eds. I. M. Head, I. Singleton & M. G. Milner. Wymondham: Horizon Scientific Press, pp. 157–83.

Paton, G. I., Viventsova, R. E., Kumpene, J., Wilson, M. J., Weitz, H. J. & Dawson, J. J. C. (2006). An ecotoxicity assessment of contaminated forest soils from the Kola Peninsula. *Science of the Total Environment*, **355**, 106–17.

Power, M., van der Meer, J. R., Tchelet, R., Egli, T. & Eggen, R. (1998). Molecular-based methods can contribute to assessments of toxicological risks and bioremediation strategies. *Journal of Microbiological Methods*, **32**, 107–19.

Ramanathan, S., Ensor, M. & Daunert, S. (1997). Bacterial biosensors for monitoring toxic metals. *Trends in Biotechnology*, **15**, 500–6.

Rattray, E. A. S., Prosser, J. I., Killham, K. & Glover, L. A. (1990). Luminescence-based non-extractive technique for *in situ* detection of *Escherichia coli* in soil. *Applied and Environmental Microbiology*, **56**, 3368–74.

Ripp, S., Nivens, D. E., Ahn, Y., Werner, C., Jarrell IV, J., Easter J. P., Cox, C. D., Burlage, R. S. & Sayler, G. S. (2000). Controlled field release of a bioluminescent genetically engineered microorganism for bioremediation process monitoring and control. *Environmental Science and Technology*, **34**, 846–53.

Sabev, H. A., Handley, P. S. & Robson, G. D. (2004). *In situ* quantification of biocide efficacy using GFP transformed *Aureobasidium pullulans*. *Journal of Applied Microbiology*, **97**, 1132–9.

Sanseverino, J., Gupta, R. K., Layton, A. C., Patterson, S. S., Rip, S. A., Saidak, L., Simpson, M. L., Schultz, T. W. & Sayler, G. S. (2005). Use of *Saccharomyces cerevisiae* BLYES expressing bacterial bioluminescence for rapid, sensitive detection of estrogenic compounds. *Applied and Environmental Microbiology*, **71**, 4455–60.

Semple, K. T., Doick, K. J., Jones, K. C., Burauel, P., Craven, A. & Harms, H. (2004). Defining bioavailability and bioaccessibility of contaminated soil and sediment is complicated. *Environmental Science and Technology*, **38**, 209A–232A.

Shimomura, O. (1992). Role of superoxide dismutase in regulating the light emission of luminescent fungi. *Journal of Experimental Botany*, **43**, 1519–25.

Sousa, S., Duffy, C., Weitz, H., Glover, L. A., Baer, E., Henkler, R. & Killham, K. (1998). Use of a *lux*-modified bacterial biosensor to identify constraints to bioremediation of BTEX-contaminated sites. *Environmental Toxicology and Chemistry*, **17**, 1039–45.

Steinberg, S. M., Poziomek, E. J., Engelmann, W. H. & Rogers, K. R. (1995). A review of environmental applications of bioluminescence measurements. *Chemosphere*, **30**, 2155–97.

Strachan, G., Capel, S., Maciel, H., Porter, A. J. R. & Paton, G. I. (2002). Application of cellular and immunological biosensor techniques to assess herbicide toxicity in soils. *European Journal of Soil Science*, **53**, 37–44.

Szittner, R., Jansen, G., Thomas, D. Y. & Meighen, E. (2003). Bright stable luminescent yeast using bacterial luciferase as a sensor. *Biochemical and Biophysical Research Communications*, **309**, 66–70.

Tandy, S., Barbosa, V., Tye, A., Preston, S., Paton, G., Zhang, H. & McGrath, S. (2005). Comparison of different microbial bioassays to assess metal-contaminated soils. *Environmental Toxicology and Chemistry*, **24**, 530–6.

Ulitzur, S. (1997). Established technologies and new approaches in applying luminous for analytical purposes. *Journal of Bioluminescence and Chemiluminescence*, **12**, 179–92.

Vieites, J. M., Navarro-Garcia, F., Perez-Diaz, R., Pla, J. & Nombela, C. (1994). Expression and *in vivo* determination of firefly luciferase as gene reporter in *Saccharomyces cerevisiae*. *Yeast*, **10**, 1321–7.

Walmsley, R. M. & Keenan, P. (2000). The eukaryote alternative: advantage of using yeasts in place of bacteria in microbial biosensor development. *Biotechnology and Bioprocess Engineering*, **5**, 387–94.

Wassink, E. C. (1978). Luminescence in fungi. In *Bioluminescence in Action*, ed. P. J. Herring. London: Academic Press, pp. 171–95.

Webb, J. S, Barratt, S. R., Sabev, H., Nixon, M., Eastwood, I. M., Greenhalgh, M., Handley, P. S. & Robson, G. D. (2001). Green fluorescent protein as a novel indicator of antimicrobial susceptibility in *Aureobasidium pullulans*. *Applied and Environmental Microbiology*, **67**, 5614–20.

Weitz, H. J., Ritchie, J. M., Bailey, D. A., Horsburgh, A. M., Killham, K. & Glover, L. A. (2001a). Construction of a modified mini-Tn*5* *lux*CDABE transposon for the development of bacterial biosensors for ecotoxicity testing. *FEMS Microbiology Letters*, **197**, 159–65.

Weitz, H. J., Ballard, A. L., Campbell, C. D. & Killham, K. (2001b). The effect of culture conditions on the mycelial growth and luminescence of naturally bioluminescent fungi. *FEMS Microbiology Letters*, **202**, 165–70.

Weitz, H. J., Campbell, C. D. & Killham, K. (2002). Development of a novel, bioluminescence-based, fungal bioassay for toxicity testing. *Environmental Microbiology*, **4**, 422–9.

Wilson, T. & Hastings, J. W. (1998). Bioluminescence. *Annual Review of Cell and Developmental Biology*, **14**, 197–230.

12
White-rot fungi and xenobiotics

P. J. HARVEY AND C. E. SCHEER

Introduction

The word 'xenobiotic' comes from the Greek word 'xenos', which means 'foreign', and describes foreign compounds that are in direct contact with a living environment. Man-made xenobiotics have been dispersed directly into the environment for many years, dumped as waste products, applied as agrochemicals, or as a result of major accidents, or indirectly, in the form of emissions from incineration processes. Xenobiotic structures are not readily recognized by existing degradative biological systems and have accumulated in the environment, and although substantial progress has been made in reducing chronic industrial derived pollution there is a growing bank of contaminated derelict industrial land – so called 'brownfield sites' – in towns and cities all over the country. In order that these sites may be repurposed for housing or for building up new commercial areas, powerful and cost-effective decontamination strategies are needed.

The design of a decontamination strategy for a given site depends on the nature and concentration of contaminants, the site characteristics (especially water movement), and the extent of contamination. Directed bioremediation, an activity in which micro- and phyto biological processes are used to degrade or transform contaminants into less toxic or non-toxic forms holds considerable potential as a strategy for *in situ* decontamination. It is generally cost-effective and less disruptive to soil and the natural landscape than *ex situ* techniques.

However, successful implementation of *in situ* bioremediation requires knowledge of the biochemistry, metabolism, structure and function of the plants and microbial populations concerned. Over the last decade enormous strides have been taken to gain such knowledge as evidenced by the rise in national and international scientific programmes established in the

Exploitation of Fungi, eds. G. D. Robson, P. van West and G. M. Gadd. Published by Cambridge University Press.

area (DTI-LINK Bioremediation programme in the UK; COST actions e.g. COST 859 Phytotechnologies www.gre.ac.uk/cost859 in Europe). As a result, the number of practical methods and the number and type of contaminants to which bioremediation can be applied has expanded enormously.

In the case of heavy metal contamination of soil, a number of practical demonstration sites now exist showing the successful application of land reclamation strategies based on the use of plants (see Mench *et al.*, 2003). Heavy metals share many similar physico–chemical properties with essential plant nutrients and trace elements, allowing them to be sequestered by plants or stabilized by plant root exudates.

The biological decontamination of dispersed organic xenobiotic substances from soil poses a more complex challenge, not least because the more recalcitrant of these have been designed for longevity (synthetic dyes), or for their ability to act as metabolic poisons (wood preservatives, insecticides, mustard gas). However, heterotrophic microbes exhibit a remarkably wide range of different metabolic pathways elaborated as a result of evolutionary pressure for their survival, laying the foundations for novel biotechnological solutions for xenobiotic decontamination. *Shewanella oneidensis*, for example, oxidizes organic carbon in air, but under anaerobic conditions will use oxidized metals as well as thiosulphate and elemental sulphur as the terminal electron acceptor instead of oxygen, to produce insoluble metal sulphides (Bencharit & Ward, 2005). The *Shewanella*-Federation (www.shewanella.org/home.sjsp), a multi-investigator and cross-institutional consortium, is now driving exploitation of this understanding to implement techniques for the removal of dilute metal pollutants, particularly radionuclides, in both contained-storage and natural sites. More recently, *Pseudomonas putida*, well-known for its ability to metabolize low molecular weight polyaromatic hydrocarbons as a sole carbon source, was reported to degrade the detoxification products of mustard gas (Ermakova *et al.*, 2004), triggering renewed interest in bioremediation strategies to detoxify the Russian chemical weapons stockpile.

Two approaches to *in situ* bioremediation are recognized: biostimulation and bioaugmentation. In biostimulation, a limiting nutrient is delivered to the soil to stimulate the growth and activity of indigenous microbial populations. In bioaugmentation, specific microorganisms possibly supplemented with nutrients are introduced to the environment to augment the resident population.

It is in the context of designing a bioaugmentation strategy for degrading xenobiotics that white-rot fungi have attracted interest. White-rot fungi are a functional group of basidiomycete fungi that derive their name from their ability to attack and degrade the highly recalcitrant natural plant protectant, lignin, giving rise to a bleached appearance of wood following lignin digestion. Tien and Kirk (1983) isolated the first ligninolytic enzyme, lignin peroxidase (LiP), responsible for degrading lignin in the extracellular 'soup' of the white-rot fungus *Phanerochaete chrysosporium*, opening up the possibility of developing a biological approach for wood pulp bleaching. These ligninolytic enzymes now provide the basis for using white-rot fungi in bioaugmentation strategies to clean quickly hydrocarbon-contaminated sites, and interest in their bioremediation potential is evidenced by the fact that, recently, *P. chrysosporium* became the first hyphal fungus to have its genome sequenced (Martinez *et al.*, 2004).

In this chapter, knowledge concerning ligninolytic enzyme systems and their induction are reviewed together with progress made in introducing white-rot fungi into 'real world' soil for xenobiotic degradation.

Oxidation of xenobiotics by ligninolytic enzyme-redox mediator systems

The capacity for white-rot fungi to attack and degrade polymeric xenobiotics derives from their ability to attack and degrade polymeric lignin in order to gain access to plant cell wall polysaccharides. This centres on three main extracellular ligninolytic redox enzymes: laccase, lignin peroxidase (LiP) and Mn-peroxidase (MnP). Lignin bears only passing resemblance to organic xenobiotics – it is a hydrophobic, highly branched and optically-inactive amorphous material comprised of randomly arranged monomers of coniferyl, sinapyl and p-coumaryl alcohols (Adler, 1977).

However, each ligninolytic enzyme has been shown to attack both lignin and a broad range of xenobiotic compounds non-specifically, using redox mediators (see Fig. 12.1 and Table 12.1).

Fungal redox mediator systems confer a number of advantages over bacterial enzyme systems deployed for xenobiotics degradation. Bacterial

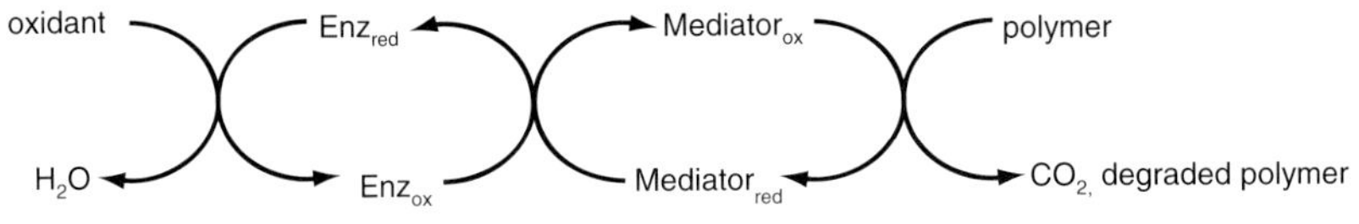

Figure 12.1 Scheme showing operation of a redox mediator between a ligninolytic enzyme (MnP, LiP or laccase) and polymeric electron donor.

Table 12.1 *Ligninolytic enzyme–redox mediator systems.*

Enzyme	Redox mediator	Reference
Lignin peroxidase	Veratryl alcohol, a natural fungal secondary metabolite	Harvey *et al.*, 1985; Candeias & Harvey, 1995; Bietti *et al.*, 1998
Mn-peroxidase	Mn-chelates (oxalate, succinate)	Gold & Alic, 1993
laccase	Synthetic: ABTS (2,2′-azinobis (3-ethylbenzthiazoline-6-sulfonate), 1-hydroxybenzotriazole, 3,4-hydroxyanthranilic acid Natural: phenol, aniline, 4-hydroxy-benzoic acid, 4-hydroxybenzyl alcohol, methionine, cysteine and reduced glutathione	Johannes *et al.*, 1996; Li *et al.*, 1999; Johannes & Majcherczyk, 2000; Mougin *et al.*, 2002a

systems are largely based on intracellular, highly specific, NAD(P)H-dependent and membrane-associated cytochrome P450 enzymes and dioxygenases (Cerniglia & Sutherland, 2001). Redox mediators are small and non-stereospecific as regards reductant, making them ideally suited to attack xenobiotics. They extend the capacity of the enzyme system to oxidize compounds beyond the reach of the parent enzyme alone. For example, the normal standard redox potential of PAHs ranges between 1.3 and 1.8 V/NHE (Majcherczyk *et al.*, 1998). This places them beyond the reach of laccase, whose redox potential of the oxidized:reduced species is only around 0.8 V/NHE (Xu, 1996). PAHs are, nonetheless, oxidized by the laccase-ABTS couple in high concentrations of ABTS, in a redox chain involving oxidation of ABTS to the relatively stable intermediate $ABTS^{+\cdot}$, thence $ABTS^{2+}$ (E_0 $ABTS^{+\cdot}/ABTS^{2+} = 1.1$ V/NHE). High concentrations of ABTS lower the redox potential of the electron donor in the redox couples according to the Nernst equation. Oxidation of the PAHs by the final oxidant $ABTS^{2+}$ is then driven in a forward direction by the irreversible formation of diones. Redox mediators operate externally to the fungal cell wall, precluding the need for specificity in membrane transporters to ferry xenobiotics inside the fungal cell for metabolism, and do not require NAD(P)H for functionality, unlike their bacterial counterparts.

Different fungi appear to be able to achieve essentially the same effect with different combinations of LiP, MnP and laccase, and in *Phanerochaete sordida* MnP is the only ligninolytic enzyme seen (Ruttimannjohnson *et al.*, 1994). LiP and MnP are both heme (protoporphyrin IX)-containing

Table 12.2 *Comparison between lignin peroxidase (LiP), manganese peroxidase (MnP) and laccase in the oxidation of methoxybenzene congeners (from Kersten* et al., *1990).*

Enzyme		LiP	MnP	Laccase
Oxidant		H_2O_2	H_2O_2	O_2
Methoxybenzene	***E1/2*** **(V)**			
1,3,5-trimethoxybenzene	1.49	+	–	–
1,2-dimethoxybenzene	1.45	+	–	–
1,2,3-trimethoxybenzene	1.42	+	–	–
1,2,4-trimethoxybenzene	1.12	+	+	–
1,2,3,5-tetramethoxybenzene	1.09	+	+	–
1,2,4,5-tetramethoxybenzene	0.81	+	+	+

enzymes that use H_2O_2 as the oxidant, while laccase is a blue multi-copper protein that uses O_2. The main difference between the enzymes in terms of their functionality, resides in the redox potential that can be sustained by the oxidized enzyme intermediate and the relative stability of the corresponding redox mediator in its oxidized form, which impacts on the range of substrates that can be oxidized (see Table 12.2).

Lignin peroxidase (LiP)

LiP is able to oxidize polymeric lignin and non-phenolic electron-rich aromatic compounds with redox potentials up to 1.4 V/NHE, the redox potential of the veratryl alcohol radical cation ($VA^{+\cdot}$)/VA couple (Candeias & Harvey, 1995) and substrates with ionization potentials of $\leq$7.55 eV. Compounds with a higher standard redox potential than the $VA^{+\cdot}$/VA couple can also be oxidized by the Lip-VA system, provided $VA^{+\cdot}$ can be generated in a sufficiently high (local) concentration. Such is the case for the oxidation of the polymeric dye Poly R (Candeias & Harvey, 1995) and isoeugenol (ten Have *et al.*, 1999). However, neither the aliphatic ethers, polyethylene glycol (PEG) and methyl tert-butyl ether (MTBE), which are significant pollutants of soil and groundwater (Kayshoemake & Watwood, 1996), nor BTEX compounds are oxidized by the LiP system. Aliphatic ethers do not have an extended pi(∗) molecular orbital system to allow for delocalization of the cation/free radical and a decrease in the energy required for intermediate formation, consequently, the energy required to generate a cation radical intermediate is beyond the capacity of the system. BTEX compounds have higher ionization potentials than that required for LiP activity (9.24 eV for benzene; 8.82 eV for toluene; 8.56 eV for *o*-xylene and 8.44 eV for *p*-xylene).

Electroenzymatic oxidation of veratryl alcohol by lignin peroxidase, and the subsequent oxidation of lignin and PAH by the resulting radical has been demonstrated (Field *et al.*, 1995, 1996; Mori, 2002; Lee & Moon, 2003). The role of veratryl alcohol and oxalate in lignin degradation has been reviewed by Zapanta & Tien (1997).

Manganese peroxidase (MnP)

MnP oxidizes Mn^{II} to Mn^{III}, which has a high standard redox potential of 1.5 V/NHE in its unchelated form, but chelation with organic acids reduces this to the order of 1.12 V/SCE (1.362 V/NHE) (Popp & Kirk, 1991). If the concentration of chelated Mn^{III} could be maintained sufficiently high, a broad range of organic substrates might be oxidized by the MnP oxidation system. In practice, however, the rate of generation of Mn^{III} by MnP seems to be a limiting factor in restricting the substrate range of MnP (Wariishi *et al.*, 1989; Tuor *et al.*, 1992). Nevertheless, fluorine, which cannot be degraded by LiP, is oxidized by MnP. This reaction takes place in a cometabolism with fatty acids (Bogan *et al.*, 1996a). *Phanerochaete laevis* HHB-1625 degrades PAH using MnP (Bogan & Lamar, 1996; Bogan *et al.*, 1996b), and crude extract of MnP from *Nematoloma frowardii* can degrade PAH (Sack & Gunther, 1993; Sack & Fritsche, 1997; Sack *et al.*, 1997a, b, c). Recently, work has been published on MnP from *Panus tigrinus* (Lisov *et al.*, 2003) and *Pleurotus ostreatus* (Banci *et al.*, 2003).

Laccase

Laccase possesses a highly specific binding pocket for oxygen, but the binding pocket for reducing substrates appears to be shallow and relatively non-stereospecific (Gianfreda *et al.*, 1999; Bertrand *et al.*, 2002). Dec and Bollag (1990, 1994) studied the use of differently substituted phenols as substrates of laccase. The efficiency of oxidation decreased with increasing molecular weight of the substituent and its changing position on the aromatic ring, in the order, $4 > 3 > 2$. However, the governing feature of whether a compound will or will not be oxidized is dictated by the redox potential differences between the reducing substrate and type I copper in the active site of the protein. Laccase in the oxidized form is able to oxidize a broad range of substrates with redox potentials less than 1 V/NHE (Reinhammer, 1972; Kersten *et al.*, 1990; Xu, 1996; Gianfreda *et al.*, 1999), but the substrate range can be extended with redox mediators. Mougin *et al.* (2002a) showed that up to 14 polycyclic aromatic hydrocarbons (PAHs) including three- and five-ring compounds could be transformed in a commercial mixture by the two-electron oxidized intermediate $ABTS^{2+}$ generated by laccase (see also Collins *et al.*, 1996; Rama *et al.*, 1998; Valderrama *et al.*, 2003). Since then, a variety of naturally occurring compounds have been shown to act as a mediator for laccase in PAH oxidation (Johannes & Majcherczyk, 2000). These observations have paved the way for devising bioremediation strategies in which xenobiotic substrates are incorporated into humic material and in which phenols and aromatic compounds are transformed into water-insoluble polymers.

Expression of ligninolytic enzyme–redox mediator systems by white-rot fungi

A number of factors militate against development of bioremediation strategies for treating xenobiotics based on ligninolytic enzyme systems per se:

- Costs associated with the supply of redox mediators.
- The instability of ligninolytic enzymes in hostile environments. These enzymes, rich reserves of nitrogen, are readily degraded by proteases (Dosoretz *et al.*, 1990a); they may also be unstable in the organic solvents used to solubilize hydrophobic contaminants, and in the case of peroxidases, are rapidly deactivated by high levels of hydrogen peroxide (see Torres *et al.*, 2003).
- Interfering soil contaminants. For example, Mougin *et al.* (2002a) found that PAHs extracted from industrial soil probably contained cyanides and metals able to reduce $ABTS^{2+}$ generated by the laccase-ABTS couple.

Instead, bioremediation research with white-rot fungi has concentrated on approaches aimed at creating a favourable environment for growth and expression of ligninolytic enzyme systems by white-rot fungi in soil.

Expression of lignin peroxidase and manganese peroxidase

In *P. chrysosporium*, the expression of LiP and MnP is associated with oxidant stress (Rothschild *et al.*, 1999; Zacchi *et al.*, 2000a, b, c; Belinky *et al.*, 2003) and is not induced by PAHs (Sack & Gunther, 1993) or lignin. In laboratory cultures, oxidant stress leading to the induction of LiP and MnP activities is typically achieved by exposure of cultures to a pure oxygen atmosphere under conditions of nutrient (either nitrogen or carbon) limitation (Kirk *et al.*, 1978; Bar-Lev & Kirk, 1981; Leisola *et al.*, 1985; Dosoretz *et al.*, 1990b; Zacchi *et al.*, 2000b, c; Belinky *et al.*, 2003). These conditions do not lend themselves to bioremediation strategies in the field. However, from their studies using cellulose as the carbon source, Zacchi *et al.* (2000a) showed that oxidant stress and LiP synthesis could be induced by any metabolic condition which increased the availability of oxygen relative to the level of reducing equivalents extracted from carbon metabolism, i.e. the ratio of [O_2] : [reducing equivalents such as NADH, NAD(P)H]. For example, in liquid cultures with cellulose as the carbon source, the rate of supply of glucose for metabolism and the supply of reducing equivalents was regulated by the hydrolytic action of cellulases. By keeping the supply of reducing equivalents low, LiP was synthesized in air without requiring the cultures to be exposed to a pure oxygen atmosphere (Zacchi *et al.*, 2000a, see Fig. 12.2a).

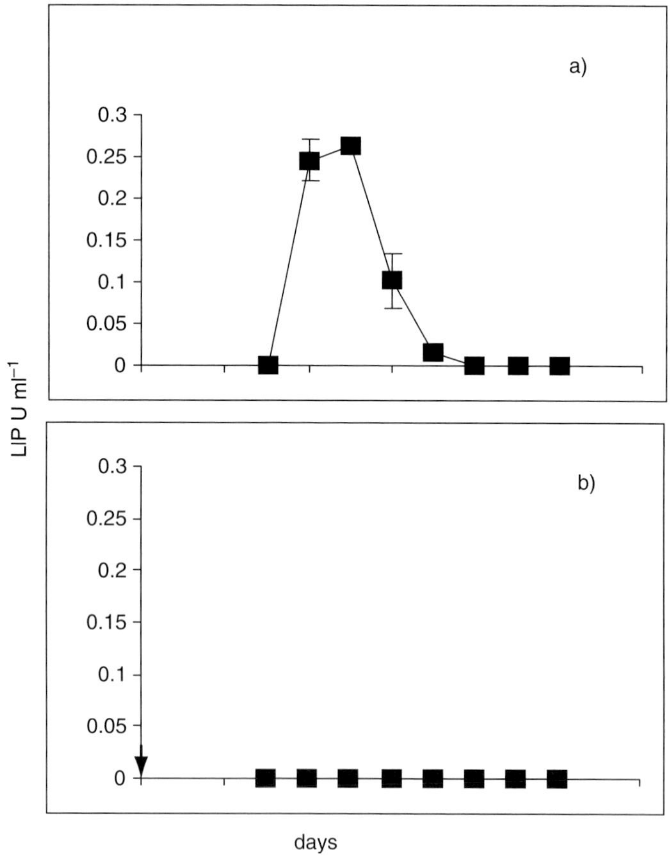

Figure 12.2 LiP production by air-agitated, cellulose-maintained (0.4% w/v) submerged cultures of *Phanerochaete chrysosporium*. (a) Cellulose alone, (b) 0.4% glucose supplement added at the outset of the experiment (adapted from Zacchi *et al.*, 2000a).

In contrast, cells grown on abundant glucose (Fig. 12.2b) switched to an anaerobic mode of metabolism and their mitochondrial function became impaired (Zacchi *et al.*, 2000c). On exposure to oxygen these cells were no longer able to supply sufficient levels of reducing equivalents by aerobic respiration. The cells became hyperoxidant and synthesized LiP because of an inability to reduce oxygen to water using mitochondrial cytochrome oxidase. Oxidant stress can also be achieved by Mn^{2+} deficiency, where the inhibition of manganese-containing superoxide dismutase leads to an increase in oxygen radicals and synthesis of the LiP enzyme system (Rothschild *et al.*, 1999).

Expression of laccase

The situation is different for laccase, which appears to be induced in response to the presence of oxidizable xenobiotics. For example, laccase is induced in high titres on exposure of cultures of *T. versicolor* to 2,5-xylidine (Fahraeus & Reinhammar, 1967). *T. versicolor* also increases production of laccase in response to fluorene, and even more so in response to its metabolite, fluorenone. However, it should be noted that no such increase was seen with benzo[a]pyrene or benzo[a]pyrene quinone (Mougin *et al.*, 2002b).

Metabolic pathways associated with xenobiotic metabolism

Extracellular degradation of aromatic xenobiotics by ligninolytic enzyme systems initiates biochemical pathways associated with processing degraded lignin, suggesting they serve as a source of reducing equivalents and carbon skeletons for metabolism and growth. In general, aromatic pollutants are converted into benzoquinones, hydroquinones and other intermediates (Valli *et al.*, 1992a; Joshi & Gold, 1993) and follow the same metabolic pathway used for vanillate metabolism (Tai *et al.*, 1990), which entails aromatic ring-cleavage of 1,2,4-trihydroxybenzene with the corresponding dioxygenase (Rieble *et al.*, 1994; Fig. 12.3). Details of the metabolism of major groups of xenobiotics are given below.

Chlorophenols, pentachlorophenol

Pentachlorophenol (PCP) is a respiratory poison listed as a high priority pollutant by the US Environmental Protection Agency (www.epa.gov). PCP (Valli & Gold, 1991; Reddy *et al.*, 1998) and a range of persistent chlorinated pollutants (Bumpus *et al.*, 1985) are oxidized by white-rot fungi using the LiP and MnP enzyme systems, and dechlorinated by a reductive process. In the case of PCP, for example, oxidation yields either tetrachlorobenzoquinone directly, or tetrachlorodihydroxybenzene (Reddy & Gold, 2000). Subsequent reductive dechlorination reactions are catalyzed by a glutathione S-transferase (GST) and a glutathione-conjugate reductase (GCR) working in tandem (Reddy & Gold, 2001) to form a hydroquinone (see Bumpus *et al.*, 1985; Lamar *et al.*, 1990b; Chung & Aust, 1995; Ruttimannjohnson & Lamar, 1996; Fig. 12.4).

Polychlorinated dioxin and furans (polyCDDs, polyCDFs)

Polychlorinated dibenzo-p-dioxins (polyCDDs) and dibenzofurans (polyCDFs) are well-known environmental by-products of industrial

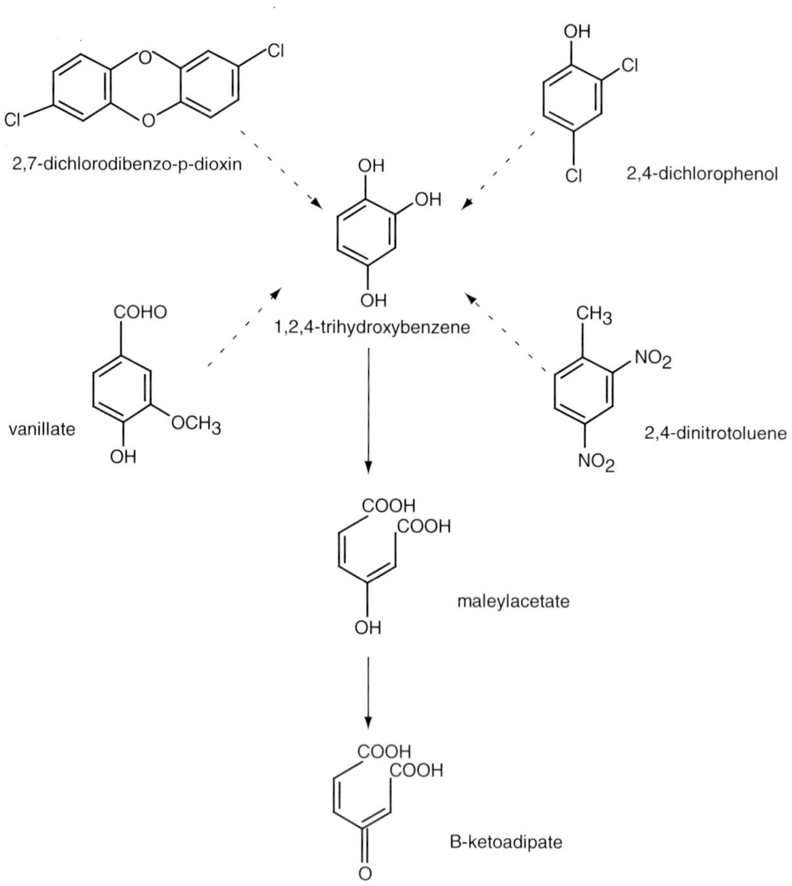

Figure 12.3 Intracellular metabolic pathways associated with the metabolism of xenobiotics and vanillate in a white-rot fungus (adapted from Rieble *et al.*, 1994).

production processes and extremely harmful carcinogenic substances. Bumpus *et al.* (1985) were the first to report successful degradation experiments of tetra- to octa-CDDs and CDFs by *P. chrysosporium*, the metabolic pathway of which has since been elucidated by Valli *et al.* (1992b). Bumpus *et al.* (1985) postulated a C– O– C bond cleavage caused by one-electron oxidation catalyzed by LiP. Degradation resulted in the formation of benzoquinones. However, in studies with the white-rot fungus *Phlebia lindtneri*, Mori (2002) confirmed a hydroxylation pathway, more common for mammals.

Organochlorine pesticides

Most organophosphate insecticides contain strong covalent P–S bonds, which are relatively resistant to enzymatic hydrolysis. However, Amitai *et al.* (1998) demonstrated degradation of phosphothiolates using laccase isolated from *Pleurotus ostreatus*, although the exact mechanism of this process is unknown.

BTEX (benzene, toluene, ethylbenzene and xylene)

Highly carcinogenic and neurotoxic, BTEX compounds are widely distributed in the environment as a result of the manufacture of solvents, pesticides, plastics and various other products (Shen, 1998). A considerable amount of petroleum and gasoline also contaminates soil and water during, e.g. leakage from storage tanks, accidental spills and inadequate waste disposal practices. *P. chrysosporium* can degrade BTEX compounds (Yadav & Reddy, 1993) and in vitro experiments using liquid cultures showed that almost 99.5% of ethylbenzene, 53.3% m- and p-xylene, 44.8% o-xylene, 41.4% toluene and 18.1% benzene

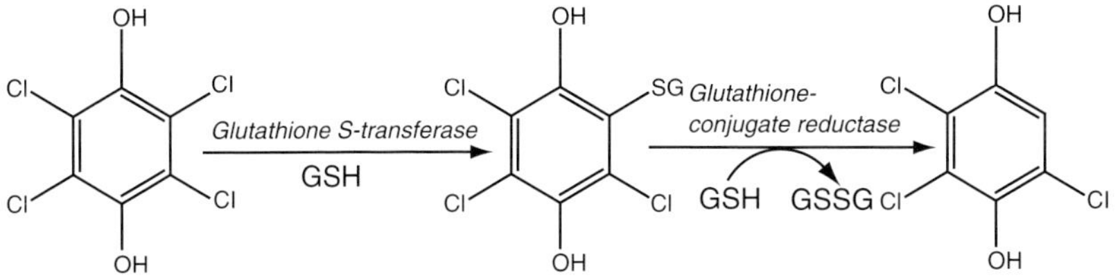

Figure 12.4 Scheme for the two-step reductive dehalogenation of tetrachlorohydroquinone (TCHQ) to trichlorohydroquinone (TrCHQ) via S-glutathionyltrichloro-1,4-hydroquinone (GS-TrCHQ) (from Reddy & Gold, 2001).

disappeared from the medium after five days. However, the mechanism is unknown.

Olive mill wastes

Olive mill waste-water (OMW) is a liquid fluid with a water quality of 100 to 200 g l^{-1} chemical oxygen demand (COD) and with organic fractions containing sugars, tannins, lipids, alcohols, phenols and polyphenols. Polyphenolic compounds are similar in structure to lignin giving the sludge its characteristic brownish-black colour. They are also responsible for the toxicity and antibacterial properties of OMW, hampering the application of conventional biological technologies based on phytoremediation and bacterial degradation. A very effective decolourization of OMW in liquid cultures containing *P. chrysosporium* and *P. flavido-alba* has been observed (Sayadi & Ellouz, 1995) and *Pleurotus sajor caju, Lentinus tigrinus, Pycnoporus coccineus, Coriolopsis polyzona* are reported to be particularly effective with higher polluted COD levels of 75 to 100 g l^{-1} (Jaouani *et al.*, 2003).

Explosive wastes

The production, storage, use and destruction of explosives has contributed to contamination of numerous sites over the world. TNT is the most widespread explosive and is mutagenic in bacterial and mammalian assay systems and carcinogenic in animal studies (www.epa.gov, see Lewis *et al.*, 2004). *P. chrysosporium* degrades nitroaromatics to CO_2 and water using the ligninolytic enzyme system (Aust, 1990; Valli *et al.*, 1992a), which involves cycles of intracellular reduction and extracellular oxidation because nitrotoluenes are not substrates for the subsequent obligatory extracellular ligninolytic peroxidatic oxidations. Intracellular reduction of the aromatic nitro group to an aromatic amine by a membrane-associated aromatic multi-component NAD(P)H-dependent nitroreductase, which is also sensitive to the presence of oxygen, is followed by extracellular oxidation. A manganese-dependent peroxidase isolated from *Phlebia radiata* oxidized the first TNT metabolic product 4-NO-2,6-DNT to 4-OHA-2,6-DNT in a process that could be accelerated by adding GSH oxidized to GSSG by Mn^{3+} (Van Aken *et al.*, 1999). Initial reduction of nitroaromatics is not uncommon amongst microbes, but none have been found to have the capacity of *P. chrysosporium* to completely degrade nitroaromatics.

Dyes

Textile, paper and leather industries, the major producers and users of dyes, produce effluents that are usually resistant to biological treatment. Gold *et al.* (1988) reported the decolourization of several polymeric dyes, Poly B-411, Poly R-481 and Poly Y-606, by *P. chrysosporium*. Poly R-478 has become an important tool for the screening of ligninolytic white-rot fungi for extracellular ligninolytic enzymes because it remains extracellular and has a high extinction coefficient allowing rapid quantitative spectrophotometric assays. Abadulla *et al.* (2000) observed a high laccase activity when they used *Trametes hirsuta* to decolourize textile dyes (triarylmethane, indigoid, azo and anthraquinonic dyes). Enzyme inhibitors, i.e. salt, surfactants, sodium azide and chelators, are common in textile waste-water, but by immobilizing laccases on alumina pellets, agarose or sepharose carriers, the toxicity of dyes could be reduced by up to 80% (Abadulla *et al.*, 2000). In studies with stationary and shaken cultures of the white-rot fungus, *Irpex lacteus*, a decolourization potential of textile industry effluents was measured up to 93%.

Polyaromatic hydrocarbons

One of the most significant groups of xenobiotic pollutants found in soils are polyaromatic hydrocarbons (PAHs), a group of hydrocarbons and substituted hydrocarbons consisting of two or more aromatic and five-member rings in linear, angular or cluster arrangements (Fig. 12.5).

PAHs are of particular interest because they are distributed widely as a result of their production during the incomplete combustion of fossil fuels, they are persistent in soil, and many are toxic or are metabolized in humans to form carcinogenic products. PAHs comprise two or more fused benzene rings. They are largely insoluble in water (0.1 μg to 4.5 mg/l) but highly lipophilic with octanol-water partition coefficients (log K_{ow}) values ranging from three to seven. Lipophilicity and water insolubility tend to increase as a function of size along with a decrease in vapour pressure. PAHs partition preferentially into the humic fractions of soils rather than the aqueous phases, and also sorb strongly to clay in soil (Hwang *et al.*, 2003; Hwang & Cutright, 2003a, 2003b). Those with log K_{ow} values of four or less partition readily across membranes into the lipophilic compartments of living cells.

PAH biodegradation depends on the ability to introduce oxygen into the rings, which has the effect of both increasing PAH solubility and chemical reactivity (Sutherland, 1992; Wilson & Jones, 1993; Meulenberg *et al.*,

1997). The two- and three-ringed structures are much more biodegradable than four, five and six-membered rings, whilst angularly-arranged ring structures (phenanthrene, benzo[α]pyrene, pyrene) tend to be more stable than linear arrangements (anthracene, benz[α]anthracene).

The pace of research into the potential for fungal degradation of PAHs is reflected in the number of times the field has been reviewed (Higson, 1991; Cerniglia, 1997; Canet *et al.*, 1999, 2001; Chung *et al.*, 2000; Juhasz & Naidu, 2000; Harvey & Thurston, 2001; Clemente *et al.*, 2001; Mougin, 2002). Both bacteria and fungi are able to degrade PAHs (see Fig. 12.6), and microbial communities in soil previously exposed to PAHs are able

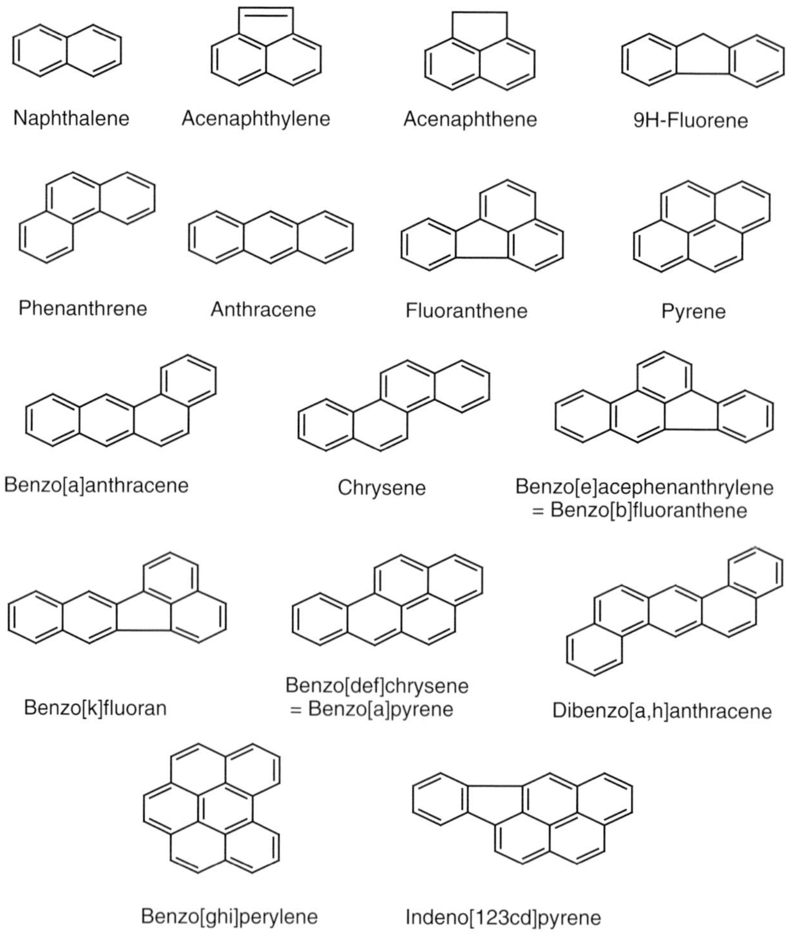

Figure 12.5 Structures of some relevant PAHs.

to degrade fresh contamination more readily (Rathbone *et al.*, 1998). However, degradation of PAHs by bacteria and non-ligninolytic fungi is largely restricted to those structures able to penetrate the cell wall for oxidation by dioxygenases or cytochrome P450-type monooxygenases, typically no greater than four fused rings. White-rot fungi, in contrast, oxidize PAHs using mainly the extracellular oxidative enzyme systems based on LIP, MnP and/or laccase (Bumpus *et al.*, 1985; Haemmerli *et al.*, 1986; Hammel *et al.*, 1986; Hammel, 1995; Bogan & Lamar, 1995, 1996; Bogan *et al.*, 1999, 1996a; Collins *et al.*, 1996). In some white-rot fungi such as *P. ostreatus*, which does not produce LiP, the initial attack on two- and three-ringed PAHs may be by cytochrome P450 (Bezalel *et al.*, 1996a, b, 1997), and a benzo[a]pyrene hydroxylase has been reported from *Pleurotus pulmonaris* (Maspahy *et al.*, 1999).

Many of the reactions associated with degradation may also result in oxygen activation, creating oxygen radicals that perpetuate the oxidative

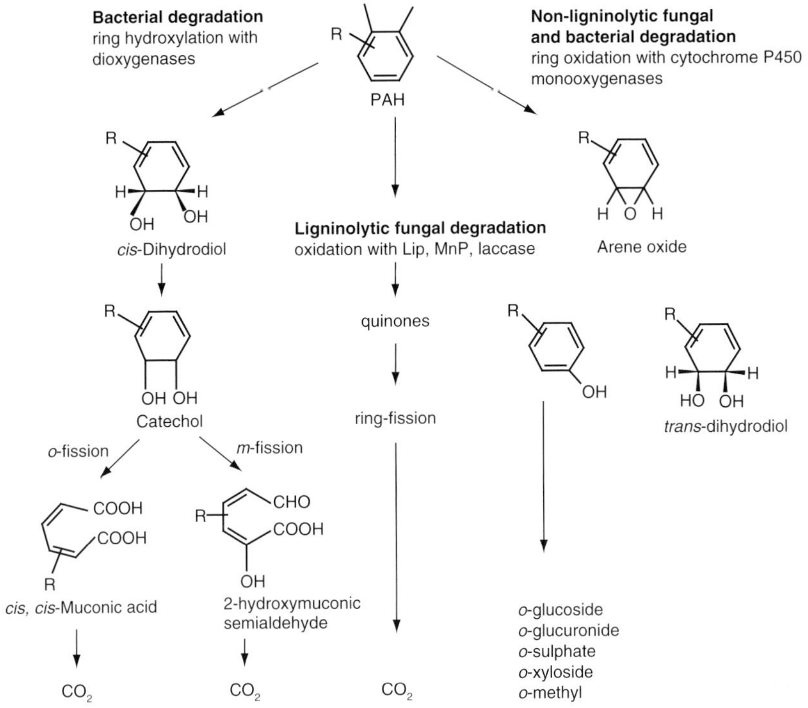

Figure 12.6 Metabolic pathways for the degradation of PAHs by bacteria, ligninolytic and non-ligninolytic fungi (after Cerniglia & Sutherland, 2001).

attack (Schoemaker *et al.*, 1985). The interest in white-rot fungi lies in their ability to oxidize the higher molecular weight, large polyaromatic structures that are not degraded by bacteria in order to increase their water-solubility and bioavailability for degradation by bacteria and fungi in plant rhizospheric communities (Meulenberg *et al.*, 1997). Indeed, when added to soils contaminated with PAH, it was observed that three- and four-ring PAH degradation was not accelerated by addition of fungi, but that of five- to seven-ring was (Gramss *et al.*, 1999a). The degradation kinetics of PAH by white-rot fungi have been investigated by Cutright & Lee (1994b).

Phanerochaete chrysosporium is reported to degrade major compounds found in creosotes (PAH and phenols) (Kennes & Lema, 1994). *Irpex lacteus* can mineralize pyrene (Hwang & Song, 2000) and recently a *Bjerkandera* sp. has proved to be particularly effective at PAH degradation (Field *et al.*, 1996). *Pleurotus* sp. Florida can degrade solid phase PAH in a mixture of sand and straw (Wolter *et al.*, 1997). Phenanthrene degradation in oil-contaminated soils was enhanced by *P. chrysosporium* (Brodkorb & Legge, 1992; May *et al.*, 1997), as was anthracene (Vyas *et al.*, 1994), petroleum hydrocarbons and crude oil (Yateem *et al.*, 1998; Sack *et al.*, 1997a; Pickard *et al.*, 1999; Song, 1999; Zadrazil *et al.*, 1996; Isikhuemhen *et al.*, 1996, 2003; Fasidi *et al.*, 1996; Ilori *et al.*, 1997). Table 12.3 lists white-rot fungi known to degrade PAH.

Fate of breakdown products

Ideally, white-rot fungi would completely degrade the contaminant. However, a low level of degradation is not necessarily indicative of poor detoxification (Kotterman *et al.*, 1998b; Reddy, 1995). The end products may be less toxic than the parent compound. PAH degradation by white-rot fungi produces a variety of metabolites (Haemmerli *et al.*, 1986; Kennes & Lema, 1994; Liao & Tseng, 1996; Wischmann & Steinhart, 1997; Andersson & Henrysson, 1996; Liao *et al.*, 1997), many of which are less toxic than the PAHs themselves (Lau *et al.*, 2003; Bhatt *et al.*, 2002). End-products may be further metabolized. White-rot fungi increase the bioavailability of PAHs, allowing bacteria to mineralize the oxidized contaminant, and it has been suggested that sequential treatment of PAH by white-rot fungi and bacteria may be effective for degradation (Brodkorb & Legge, 1992). Finally, the end-products may be immobilized. *P. chrysosporium* does not degrade the pesticide atrazine, but does immobilize it in soil (Lamar *et al.*, 1990b; Hickey *et al.*, 1994). White-rot fungi enhance binding and immobilization of PCP to humic substances to varying degrees (*P. ostreatus* > *I. lacteus* > *T. versicolor* > *Bjerkandera adusta*)

Table 12.3 *White-rot fungi known to degrade PAHs.*

Phanerochaete chrysosporium, *P. sordida*	Lamar & Dietrich, 1990; Lamar *et al.*, 1990b
Phanerochaete laevis HHB-1625	Bogan & Lamar, 1996; Bogan *et al.*, 1996b
Pleurotus ostreatus	Wolter *et al.*, 1997; Eggen & Majcherczyk, 1998; Martens *et al.*, 1999; Eggen, 1999; Gramss *et al.*, 1999b; Márquez-Rocha *et al.*, 2000; Hestbjerg *et al.*, 2003
Cunninghamella echinulata var. *elegans*	Cutright & Lee, 1994a, b, 1995; Cutright, 1995a, b; Cutright *et al.*, 1996; Liebeg & Cutright, 1999; Ramirez & Cutright, 2000; Hwang & Cutright, 2001, 2002a, b; Mendez-Sanchez *et al.*, 2003; Hwang *et al.*, 2003; Hwang & Cutright, 2003a, b
Chrysosporium lignorum	Morgan *et al.*, 1993
Trametes versicolor	Dehorter & Blondeau, 1992; Song, 1997, 1999
Trametes trogii	Segura *et al.*, 1993; Andersson & Henrysson, 1996; Novotny *et al.*, 1999; Levin *et al.*, 2003
Kuehneromyces mutabilis	Sack & Fritsche, 1997
Agrocybe aegerita	
Staganophora gigaspora	Segura *et al.*, 1993
Pleurotus pulmonaris	
Bjerkandera adusta	Schützendübel *et al.*, 1999; Eggen & Majcherczyk, 1998; Kotterman *et al.*, 1998a
Ramaria sp.	Andersson & Henrysson, 1996
Agaricales sp.	
Pleurotus sajor-caja	
Dichomitus squalens	in der Wiesche *et al.*, 1996
Irpex lacteus	Song, 1997, 1999; Hwang & Song, 2000; Novotny *et al.*, 2000
Nematoloma frowardii	Sack *et al.*, 1997b, c
Marasmiellus troyanus	Wunch *et al.*, 1999

(Ruttimannjohnson & Lamar, 1997). Humification of PAHs by *P. ostreatus* in soil is dependent on a prior degree of association of the PAH with humic substances (Ruttimannjohnson & Lamar, 1997; Bogan *et al.*, 1999). This may be due to polymerization reactions (Lamar & Dietrich, 1990; Lamar *et al.*, 1990a, b; Ruttimannjohnson & Lamar, 1996).

Effect of the soil environment on white-rot fungal physiology

From the preceding sections, the advantages of using white-rot fungi in a bioaugmentation strategy can be summarized.

- White-rot fungi can attack mixed pollutants.
- White-rot fungi do not need a minimum concentration of pollutant to induce the production of degradative enzymes.

- White-rot fungi can degrade highly oxidized substrates.
- White-rot fungi can degrade contaminants extracellularly, thus reducing their toxicity prior to uptake by other organisms (Barr & Aust, 1994).

However, soil is not a natural environment for white-rot fungi. Lessons learned from wood degradation studies suggest that, at the very least, the following conditions should be satisfied for successful soil colonization.

- The presence of oxygen. Lignin degradation and expression of the ligninolytic enzyme system is an aerobic process.
- A moisture level of c. 20%. Boyle *et al.* (1998) found that white-rot fungi grew best in soil at 18% moisture.
- An energy source. Lignin metabolism does not result in a net gain in energy for growth. In soil, good results for fungal colonization were obtained using straw in a 4:1, soil to straw ratio (Morgan *et al.*, 1993), or when soil organic matter was present to provide a carbon and energy source (Gramss *et al.*, 1999a). In steam-sterilized soils, Boyle (1995) found that fungal growth was limited by carbon and nitrogen, but nutrient supplements in unsterile soil frequently benefited indigenous fungi more than the inoculated fungi, unless the former were inhibited with benomyl at 115 ppm.
- An acidic environment. The optimum pH value for culture of *P. chrysosporium* is around pH 4.5 (Kirk *et al.*, 1978).

Most approaches for delivering wood-rot fungi to soil have centred on techniques developed by mushroom growers, namely the production of fungal spawn on lignocellulose waste such as corn cobs, straw, wood chips and sawdust (Morgan *et al.*, 1993). More recently, this approach has been extended to deliver pelleted solid substrates coated with a sodium alginate suspension of fungal spores. When incubated until overgrown with fungal mycelium, the pellets were introduced to soil, and the fungi survived and proliferated (Lestan & Lamar, 1996). Schmidt *et al.* (2005) noted a strong correlation between the amount of fungal inoculum used in soil augmentation and white-rot fungal colonization of soil, with younger (<7 d) fungal inoculum faring better than old (>10 d), and pine sawdust proving a better carrier, especially when supplemented with corn, than poplar sawdust.

Nevertheless, results in soil have been variable for several reasons. Soil particles may inhibit fungal growth. Tucker *et al.* (1995) reported that the effect was greater in silt than sand, and greatest in clay, while Mougin found that laccase from *Pycnoporus cinnabarinus* plus a mediator oxidized some PAHs in a complex mixture to quinones, but the reactions were inhibited by soil from a manufactured gas plant due to sorption to clay and organic material, and reduction of mediator by metals (Mougin *et al.*, 2002a). Contaminated soil may contain inhibitors of metabolism or of ligninolytic enzymes. Mercury (50–100 ppm) and cadmium (100–500 ppm) have been reported to reduce the colonization of soil by *P. ostreatus* and inhibit ligninolytic enzymes (Baldrian *et al.*, 2000). Cyanide (2 mM) on the other hand is metabolized by *P. chrysosporium*, but binds to LiP with a Kd of 10 μM and inhibits veratryl alcohol oxidation ($I_{50} = 2$–3 μM; Shah *et al.*, 1991). Competition from native microbiota may inhibit successful colonization in soil (see below). Consequently, any application at the field scale must be preceded by microcosm experiments.

Interaction with native soil microbiota

White-rot fungi have been reported to both enhance degradation of PAHs in non-sterile soil by soil microbiota (Brodkorb & Legge, 1992; Sack & Fritsche, 1997; Andersson *et al.*, 2003), as well as to have no effect (Abbondanzi *et al.*, 2003). Treatment of soil with a *Bjerkandera* sp. followed by further degradation by indigenous microflora was effective in mineralizing benzo[a]pyrene (Kotterman *et al.*, 1998b). *Pleurotus* sp. (Bhutan) strongly inhibited soil bacteria, while growth of *Dichomitus squalens* was inhibited (Lang *et al.*, 2000). Some fungal/bacterial co-cultures were found to enhance PAH degradation (Boonchan *et al.*, 2000). Other co-cultures had a negative effect on degradation rates. Where synergism occurred, fungi and bacteria could be sourced from different sites meaning that co-evolution was obviously not a requirement (Chavez-Gomez *et al.*, 2003). There are also reports that fungal degradation of xenobiotics cannot be seen in non-autoclaved soil (Andersson *et al.*, 2000, 2003). However, Radtke *et al.* (1994) showed that antagonism by fluorescent pseudomonads towards the growth of *P. chrysosporium* was variable according to the pH value and carbon and nitrogen sources used in the growth medium, with no antagonism being evident under conditions which were optimal for fungal growth, e.g. pH 4. Indeed, under conditions where the effect of bioaugmentation with *P. chrysosporium* was reported as synergistic with soil microbiota, the pH value of the soil was lowered to between pH 5 and 5.5 with periodic additions of HCl at three-day intervals to encourage fungal growth (Brodkorb & Legge, 1992).

Practical approaches – field trials

Pollutant bioavailability is a major restricting factor in the bioremediation of organic pollutants. Bioavailability depends on the soil structure (content of humic substances, pH value, water content and porosity) and on properties of the pollutant. Humic substances represent the main cause for the restrictions placed on bioavailability. These compounds, which represent as much as 60–80% of soil organic matter, are polydispersed polymers of aromatic and aliphatic units that are resistant to degradation by covalent binding of their reactive sites to metal ions and clay minerals. They possess a high content of oxygen-containing functional groups including –COOH, phenolic and/or enolic –OH, alcoholic –OH and the –C=O double bonds of quinones, hydroquinones and α,β-unsaturated ketones (Stevenson, 1994). They are able to bind phenolics, anilines and PAHs in a process known as ageing (Alexander, 1995). Often the sites in which these pollutants are entrapped have effective diameters of less than 1 μm, rendering them inaccessible to bacteria. Humic substances may also serve as carriers of these pollutants, increasing their solubility and motility in soil and water.

There have been few field-scale applications of white-rot fungi to remediate contaminated soil. White-rot fungi were used at the field scale to treat soil contaminated with PCP and creosote with some success (Davis *et al.*, 1993; Lamar *et al.*, 1993). In these studies it was found that a significant role was played by polymerization, which resulted in reduced bioavailability. Degradation of PCP was not a significant route (Lamar & Dietrich, 1990; Lamar *et al.*, 1990a, b; Holroyd & Caunt, 1997). Benzo[a]pyrene was not degraded by *P. ostreatus*, but degradation was improved by the addition of a non-ionic surfactant to enhance desorption of pollutants from the soil matrix (Márquez-Rocha *et al.*, 2000). It has also been suggested that improvements using white-rot fungi in the field might be achieved by increasing fungal biomass (Reddy, 1995).

Monitoring of field trials is intrinsically more prone to error than laboratory-scale investigations, due to the greater heterogeneity of natural soils and pollutant distribution. While it is possible to mitigate against this by mixing soil across a site, the increased aeration caused by the initial homogenization of contaminated soil may have an effect far greater than the treatment subsequently applied (Joner *et al.*, 2003).

Conclusions

From intensive research in the 1980s to develop biotechnological approaches to break down lignin in the pulp and paper industry, to current requirements for decontaminating xenobiotic contamination, the ligninolytic enzyme system of white-rot fungi emerges as a highly oxidizing, non-specific, biological system that could solve a number of problems associated with organic pollutant decontamination. However, incorporation of the technology into a cost-effective bioaugmentation treatment programme is likely to depend on use of the whole organism rather than individual enzymes, and it is here that major effort will be required to understand on a case-by-case basis, the prevailing environmental conditions (abiotic and biotic) so that these can be manipulated to accommodate the growth and expression of the necessary fungal enzymes in a soil environment – and on the scale often required. Good progress has been made in devising strategies to introduce fungi into soil using carrier matrixes such as alginate pellets, but there is also a need to develop cost-effective approaches for large-scale implementation.

References

Abadulla, E., Tzanov, T., Costa, S., Robra, K. H., Cavaco-Paulo, A. & Gubitz, G. M. (2000). Decolorization and detoxification of textile dyes with a laccase from *Trametes hirsuta*. *Applied and Environmental Microbiology*, **66**, 3357–62.

Abbondanzi, F., Campisi, T., Gussoni, A., Iacondini, A., Malaspina, F., Mingozzi, L., Raccagni, M. & Visani, M. (2003). Bioremediation of PAHs in soils: bioaugmentation test in pilot plant. In *Proceedings of the Second European Bioremediation Conference*, ed. N. Kalogerakis. Crete, Greece: Technical University Chania, pp. 103–6.

Adler, E. (1977). Lignin chemistry – past, present and future. *Wood Science and Technology*, **11**, 169–218.

Alexander, M. (1995). How toxic are toxic chemicals in soil? *Environmental Science and Technology*, **29**, 2713–17.

Amitai, G. Adani, R., Sod-Moriah, G., Rabinovitz, I., Vincze, A., Leader, H., Chefetz, B., Leibovitz-Persky, L., Friesem, D. & Hadar, Y. (1998). Oxidative biodegradation of phosphorothiolates by fungal laccase. *FEBS Letters*, **438**, 195–200.

Andersson, B. E. & Henrysson, T. (1996). Accumulation and degradation of dead-end metabolites during treatment of soil contaminated with polycyclic aromatic hydrocarbons with five strains of white-rot fungi. *Applied Microbiology and Biotechnology*, **46**, 647–52.

Andersson, B. E., Lundstedt, S., Tornberg, K., Schnurer, Y., Oberg, L. G. & Mattiasson, B. (2003). Incomplete degradation of polycyclic aromatic hydrocarbons in soil inoculated with wood-rotting fungi and their effect on the indigenous soil bacteria. *Environmental Toxicology and Chemistry*, **22**, 1238–43.

Andersson, B. E., Welinder, L., Olsson, P. A., Olsson, S. & Henrysson, T. (2000). Growth of inoculated white-rot fungi and their interactions with the bacterial community in soil contaminated with polycyclic aromatic hydrocarbons, as measured by phospholipid fatty acids. *Bioresource Technology*, **73**, 29–36.

Aust, S. D. (1990). Degradation of environmental pollutants by *Phanerochaete chrysosporium*. *Microbial Ecology*, **20**, 197–209.

Baldrian, P., in der Wiesche, C. I., Gabriel, J., Nerud, F. & Zadrazil, F. (2000). Influence of cadmium and mercury on activities of ligninolytic enzymes and degradation of polycyclic aromatic hydrocarbons by *Pleurotus ostreatus* in soil. *Applied and Environmental Microbiology*, **66**, 2471–8.

Banci, L., Camarero, S., Martinez, A. T., Martinez, M. J., Perez-Boada, M., Pierattelli, R. & Ruiz-Duenas, F. J. (2003). NMR study of manganese(II) binding by a new versatile peroxidase from the white-rot fungus *Pleurotus eryngii*. *Journal of Biological Inorganic Chemistry*, **8**, 751–60.

Bar-Lev, S. S. & Kirk, T. K. (1981). Effects of molecular oxygen on lignin degradation. *Biochemical and Biophysical Research Communications*, **99**, 373–8.

Barr, D. P. & Aust, S. D. (1994). Mechanisms white rot fungi use to degrade pollutants. *Environmental Science and Technology*, **28**, 78–87.

Belinky, P. A., Flikshtein, N., Lechenko, S., Gepstein, S. & Dosoretz, C. G. (2003). Reactive oxygen species and the induction of lignin peroxidase in *Phanerochaete chrysosporium*. *Applied and Environmental Microbiology*, **69**, 6500–6.

Bencharit, S. & Ward, M. J. (2005). Chemotactic responses to metals and anaerobic electron acceptors in *Shewanella oneidensis* MR-1. *Journal of Bacteriology*, **187**, 5049–53.

Bertrand, T., Jolivalt, C., Caminade, E., Joly, N., Mougin, C. & Briozzo, P. (2002). Purification and preliminary crystallographic study of *Trametes versicolor* laccase in its native form. *Acta Crystallographica Section D-Biological Crystallography*, **58**, 319–21.

Bezalel, L., Hadar, Y. & Cerniglia, C. E. (1997). Enzymatic mechanisms involved in phenanthrene degradation by the white rot fungus *Pleurotus ostreatus*. *Applied and Environmental Microbiology*, **63**, 2495–501.

Bezalel, L., Hadar, Y., Fu, P. P., Freeman, J. P. & Cerniglia, C. E. (1996a). Metabolism of phenanthrene by the white rot fungus *Pleurotus ostreatus*. *Applied and Environmental Microbiology*, **62**, 2547–53.

Bezalel, L., Hadar, Y., Fu, P. P., Freeman. J. P. & Cerniglia, C. E. (1996b). Initial oxidation products in the metabolism of pyrene, anthracene, fluorene, and dibenzothiophene by the white rot fungus *Pleurotus ostreatus*. *Applied and Environmental Microbiology*, **62**, 2554–9.

Bhatt, M., Cajthaml, T. & Sasek, V. (2002). Mycoremediation of PAH-contaminated soil. *Folia Microbiologica*, **47**, 255–8.

Bietti, M., Baciocchi, E. & Steenken, S. (1998). Lifetime, reduction potential and base-induced fragmentation of the veratryl alcohol radical cation in aqueous solution. Pulse radiolysis studies on a ligninase "mediator". *Journal of Physical Chemistry A*, **102**, 7337–42.

Bogan, B. W. & Lamar, R. T. (1995). One-electron oxidation in the degradation of creosote polycyclic aromatic-hydrocarbons by *Phanerochaete chrysosporium*. *Applied and Environmental Microbiology*, **61**, 2631–5.

Bogan, B. W. & Lamar, R. T. (1996). Polycyclic aromatic hydrocarbon-degrading capabilities of *Phanerochaete laevis* HHB-1625 and its extracellular ligninolytic enzymes. *Applied and Environmental Microbiology*, **62**, 1597–603.

Bogan, B. W., Lamar, R. T., Burgos, W. D. & Tien, M. (1999). Extent of humification of anthracene, fluoranthene, and benzo[alpha]pyrene by *Pleurotus ostreatus* during growth in PAH-contaminated soils. *Letters in Applied Microbiology*, **28**, 250–4.

Bogan, B. W., Lamar, R. T. & Hammel, K. E. (1996a). Fluorene oxidation *in vivo* by *Phanerochaete chrysosporium* and *in vitro* during manganese peroxidase-dependent lipid peroxidation. *Applied and Environmental Microbiology*, **62**, 1788–92.

Bogan, B. W., Schoenike, B., Lamar, R. T. & Cullen, D. (1996b). Manganese peroxidase mRNA and enzyme activity levels during bioremediation of polycyclic aromatic hydrocarbon-contaminated soil with *Phanerochaete chrysosporium*. *Applied and Environmental Microbiology*, **62**, 2381–6.

Boonchan, S., Britz, M. L. & Stanley, G. A. (2000). Degradation and mineralization of high-molecular-weight polycyclic aromatic hydrocarbons by defined fungal-bacterial co-cultures. *Applied and Environmental Microbiology*, **66**, 1007–19.

Boyle, C. D. (1995). Development of a practical method for inducing white rot fungi to grow into and degrade organopollutants in soil. *Canadian Journal of Microbiology*, **41**, 345–53.

Boyle, D., Wiesner, C. & Richardson, A. (1998). Factors affecting the degradation of polyaromatic hydrocarbons in soil by white-rot fungi. *Soil Biology and Biochemistry*, **30**, 873–82.

Brodkorb, T. S. & Legge, R. L. (1992). Enhanced biodegradation of phenanthrene in oil tar-contaminated soils supplemented with *Phanerochaete chrysosporium*. *Applied and Environmental Microbiology*, **58**, 3117–21.

Bumpus, J. A., Tien, M., Wright, D. & Aust, S. D. (1985). Oxidation of persistent environmental pollutants by a white rot fungus. *Science*, **228**, 1434–6.

Candeias, L. P. & Harvey, P. J. (1995). Lifetime and reactivity of the veratryl alcohol radical cation. Implications for lignin peroxidase catalysis. *Journal of Biological Chemistry*, **270**, 16745–8.

Canet, R., Birnstingl, J. G., Malcolm, D. G., Lopez-Real, J. M. & Beck, A. J. (2001). Biodegradation of polycyclic aromatic hydrocarbons (PAHs) by native microflora and combinations of white-rot fungi in a coal-tar contaminated soil. *Bioresource Technology*, **76**, 113–17.

Canet, R., Lopez-Real, J. M. & Beck, A. J. (1999). Overview of polycyclic aromatic hydrocarbon biodegradation by white-rot fungi. *Land Contamination and Reclamation*, **7**, 191–7.

Cerniglia, C. E. (1997). Fungal metabolism of polycyclic aromatic hydrocarbons: past, present and future applications in bioremediation. *Journal of Industrial Microbiology and Biotechnology*, **19**, 324–33.

Cerniglia, C. E. & Sutherland, J. B. (2001). Bioremediation of polycyclic aromatic hydrocarbons by ligninolytic and non-ligninolytic fungi. In *Fungi in Bioremediation*, ed. G. M. Gadd. Cambridge: Cambridge University Press, pp. 136–87.

Chavez-Gomez, B., Quintero, R., Esparza-Garcia, F., Mesta-Howard, A. M., Zavala Diaz de la Serna, F. J., Hernandez-Rodriguez, C. H., Gillen, T., Poggi-Varaldo, H. M., Barrera-Cortes, J. & Rodriguez-Vazquez, R. (2003). Removal of phenanthrene from soil by co-cultures of bacteria and fungi pregrown on sugarcane bagasse pith. *Bioresource Technology*, **89**, 177–83.

Chung, N. & Aust, S. (1995). Degradation of pentachlorophenol in soil by *Phanerochaete chrysosporium*. *Journal of Hazardous Materials*, **41**, 177–83.

Chung, N. H., Lee, I. S., Song, H. S. & Bang, W. G. (2000). Mechanisms used by white-rot fungus to degrade lignin and toxic chemicals. *Journal of Microbiology and Biotechnology*, **10**, 737–52.

Clemente, A. R., Anazawa, T. A. & Durrant, L. R. (2001). Biodegradation of polycyclic aromatic hydrocarbons by soil fungi. *Brazilian Journal of Microbiology*, **32**, 255–61.

Collins, P. J., Kotterman, M. J. J., Field, J. A. & Dobson, A. D. W. (1996). Oxidation of anthracene and benzo[a]pyrene by laccases from *Trametes versicolor*. *Applied and Environmental Microbiology*, **62**, 4563–7.

Cutright, T. J. (1995a). Polycyclic aromatic hydrocarbon biodegradation and kinetics using *Cunninghamella echinulata* var. *elegans*. *International Biodeterioration and Biodegradation*, **35**, 397–408.

Cutright, T. J. (1995b). A feasible approach to the bioremediation of contaminated soil – from lab-scale to field-test. *Fresenius Environmental Bulletin*, **4**, 67–73.

Cutright, T. J. & Lee, S. G. (1994a). Quantitative and qualitative-analysis of PAH contaminated soil. *Fresenius Environmental Bulletin*, **3**, 42–8.

Cutright, T. J. & Lee, S. G. (1994b). Bioremediation kinetics for PAH contaminated soils. *Fresenius Environmental Bulletin*, **3**, 597–603.

Cutright, T. J. & Lee, S. G. (1995). *In-situ* soil remediation – bacteria or fungi. *Energy Sources*, **17**, 413–19.

Cutright, T. J., Midha, C. & Lee, S. (1996). Preliminary statistical analysis of PAH-contaminated soils. *Energy Sources*, **18**, 51–6.

Davis, M. W., Glaser, J. A., Evans, J. W. and Lamar, R. T. (1993). Field-evaluation of the lignin-degrading fungus *Phanerochaete sordida* to treat creosote-contaminated soil. *Environmental Science and Technology*, **27**, 2572–6.

Dec, J. & Bollag, J.-M. (1990). Detoxification of substituted phenols by oxidoreductive enzymes through polymerisation reactions. *Archives of Environmental Contamination and Toxicology*, **19**, 543–50.

Dec, J. & Bollag, J.-M. (1994). Use of plant material for the decontamination of water polluted with phenols. *Biotechnology and Bioengineering*, **44**, 1132–9.

Dehorter, B. & Blondeau, R. (1992). Extracellular enzyme activities during humic acid degradation by the white rot fungi *Phanerochaete chrysosporium* and *Trametes versicolor*. *FEMS Microbiology Letters*, **94**, 209–16.

Dosoretz, C. G., Chen, H. C. & Grethlein, H. E. (1990a). Effect of environmental conditions on extracellular protease activity in lignolytic cultures of *Phanerochaete chrysosporium*. *Applied and Environmental Microbiology*, **56**, 395–400.

Dosoretz, C. G., Chen U.-C. & Grethlein H. E. (1990b). Effect of oxygenation conditions on submerged cultures of *Phanerochaete chrysosporium*. *Applied Microbiology and Biotechnology*, **34**, 131–7.

Eggen, T. (1999). Application of fungal substrate from commercial mushroom production – *Pleurotus ostreatus* – for bioremediation of creosote contaminated soil. *International Biodeterioration and Biodegradation*, **44**, 117–26.

Eggen, T. & Majcherczyk, A. (1998). Removal of polycyclic aromatic hydrocarbons (PAH) in contaminated soil by white rot fungus *Plerotus ostreatus*. *International Biodeterioration and Biodegradation*, **41**, 111–17.

Ermakova, I. T., Safrina, N. S., Starovoitov, I. I., Liubun, E. V., Shcherbakov, A. A., Makarov, O. E., Petrova, A. A. & Shpilkov, P. A. (2004). Microbial degradation of mustard gas reaction masses: isolation and selection of degradative microorganisms, analysis of organic components of reaction masses and their biodegradation. *Mikrobiologiia*, **73**, 358–63.

Fahraeus, G. & Reinhammar, B. (1967). Large scale production and purification of laccase from cultures of the fungus *Polyporus versicolor* and some properties of laccase A. *Acta Chemica Scandinavica*, **21**, 2367–78.

Fasidi, I. O., Isikhuemhen, O. S. & Zadrazil, F. (1996). Bioreactors for solid state fermentation of lignocellulosics. *Journal of Scientific and Industrial Research*, **55**, 450–6.

Field, J. A., Boelsma, F., Baten, H. & Rulkens, W. H. (1995). Oxidation of anthracene in water/solvent mixtures by the white-rot fungus, *Bjerkandera* sp. Strain BOS55. *Applied Microbiology and Biotechnology*, **44**, 234–40.

Field, J. A., Baten, H., Boelsma, F. & Rulkens, W. H. (1996). Biological elimination of polycyclic aromatic hydrocarbons in solvent extracts of polluted soil by the white rot fungus, *Bjerkandera* sp. Strain BOS55. *Environmental Technology*, **17**, 317–23.

Gianfreda, L., Xu, F. & Bollag, J. M. (1999). Laccases: a useful group of oxidoreductive enzymes. *Bioremediation Journal*, **3**, 1–25.

Gold, M. H. & Alic, M. (1993). Molecular biology of the lignin-degrading basidiomycete *Phanerochaete chrysosporium*. *Microbiological Reviews*, **57**, 605–22.

Gold, M. H., Glenn, J. K. & Alic, M. (1988). Use of polymeric dyes in lignin biodegradation assays. *Methods in Enzymology*, **161**, 74–8.

Gramss, G., Voigt, K. D. & Kirsche, B. (1999a). Degradation of polycyclic aromatic hydrocarbons with three to seven aromatic rings by higher fungi in sterile and unsterile soils. *Biodegradation*, **10**, 51–62.

Gramss, G., Ziegenhagen, D. & Sorge, S. (1999b). Degradation of soil humic extract by wood- and soil-associated fungi, bacteria and commercial enzymes. *Microbial Ecology*, **37**, 140–51.

Haemmerli, S. D., Leisola, M. S. A., Sangard, D. & Fiechter, A. (1986). Oxidation of benzo(a)pyrene by extracellular ligninases of *Phanerochaete chrysosporium*: veratryl alcohol and stability of ligninase. *Journal of Biological Chemistry*, **261**, 6900–2.

Hammel, K. E. (1995). Mechanisms for polycyclic aromatic hydrocarbon degradation by ligninolytic fungi. *Environmental Health Perspectives*, **103** (Suppl 5), 41–3.

Hammel, K. E., Kalyanaraman, B. & Kirk, T. K. (1986). Oxidation of polycyclic aromatic hydrocarbons and dibenzo[p]-dioxins by *Phanerochaete chrysosporium* ligninase. *Journal of Biological Chemistry*, **261**, 16948–52.

Harvey, P. J., Schoemaker, H. E., Bowen, R. M. & Palmer, J. M. (1985). Single-electron transfer processes and the reaction mechanism of enzymic degradation of lignin. *FEBS Letters*, **183**, 13–16.

Harvey, P. J. & Thurston, C. F. (2001). The biochemistry of ligninolytic fungi. In *Fungi in Bioremediation*, ed. G. M. Gadd. Cambridge: Cambridge University Press, pp. 27–51.

Hestbjerg, H., Willumsen, P. A., Christensen, M., Andersen, O. & Jacobsen, C. S. (2003). Bioaugmentation of tar-contaminated soils under field conditions using *Pleurotus ostreatus* refuse from commercial mushroom production. *Environmental Toxicology and Chemistry*, **22**, 692–8.

Hickey, W. J., Fuster, D. J. & Lamar, R. T. (1994). Transformation of atrazine in soil by *Phanerochaete chrysosporium*. *Soil Biology and Biochemistry*, **26**, 1665–71.

Higson, F. K. (1991). Degradation of xenobiotics by white rot fungi. *Reviews of Environmental Contamination and Toxicology*, **122**, 111–52.

Holroyd, M. L. & Caunt, P. (1997). Field-scale use of white-rot fungi for soil remediation in Finland. In *Bioremediation: Principles and Practice*, eds. S. K. Sikdar and R. L. Irvine, *Vol. III Bioremediation Technologies*. Lancaster: Technomic Publishing Co., pp. 245–57.

Hwang, S. & Cutright, T. J. (2001). Bioavailability of phenanthrene and pyrene aged for 0 and 200-d in a silty-sand soil. *Abstracts of Papers of the American Chemical Society*, **222**, 122-IEC.

Hwang, S. & Cutright, T. J. (2002a). Statistical impact of the extent of desorption, compound aging, and bacteria inoculation on polycyclic aromatic hydrocarbon biodegradation. *Polycyclic Aromatic Compounds*, **22**, 1057–74.

Hwang, S. & Cutright, T. J. (2002b). Biodegradability of aged pyrene and phenanthrene in a natural soil. *Chemosphere*, **47**, 891–9.

Hwang, S. & Cutright, T. J. (2003a). Effect of expandable clays and cometabolism on PAH biodegradability. *Environmental Science and Pollution Research*, **10**, 277–80.

Hwang, S. & Cutright, T. J. (2003b). Statistical implications of pyrene and phenanthrene sorptive phenomena: effects of sorbent and solute properties. *Archives of Environmental Contamination and Toxicology*, **44**, 152–9.

Hwang, S., Ramirez, N., Cutright, T. J. & Ju, L. K. (2003). The role of soil properties in pyrene sorption and desorption. *Water Air and Soil Pollution*, **143**, 65–80.

Hwang, S. S. & Song, H. G. (2000). Biodegradation of pyrene by the white rot fungus, *Irpex lacteus*. *Journal of Microbiology and Biotechnology*, **10**, 344–8.

Ilori, M. O., Fasida, I. O. & Isikhuemhen, O. S. (1997). Mushroom research and commercial cultivation in Nigeria. *Food Reviews International*, **13**, 489–96.

in der Wiesche, C., Martens, R. & Zadrazil, F. (1996). Two-step degradation of pyrene by white-rot fungi and soil microorganisms. *Applied Microbiology and Biotechnology*, **46**, 653–9.

Isikhuemhen, O. S., Anoliefo, G. O. & Oghale, O. I. (2003). Bioremediation of crude oil polluted soil by the white rot fungus, *Pleurotus tuberregium* (Fr.) Sing. *Environmental Science and Pollution Research*, **10**, 108–12.

Isikhuemhen, O. S., Zadrazil, F. & Fasidi, I. O. (1996). Cultivation of white rot fungi in solid state fermentation. *Journal of Scientific and Industrial Research*, **55**, 388–93.

Jaouani, A., Sayadi, S., Vanthournhout, M. & Penninckx, M. J. (2003). Potent fungi for decolourisation of olive oil mill wastewaters. *Enzyme and Microbial Technology*, **33**, 802–9.

Johannes, C. & Majcherczyk, A. (2000). Natural mediators in the oxidation of polycyclic aromatic hydrocarbons by laccase mediator systems. *Applied and Environmental Microbiology*, **66**, 524–8.

Johannes, C., Majcherczyk, A. & Huttermann, A. (1996). Degradation of anthracene by laccase of *Trametes versicolor* in the presence of different mediator compounds. *Applied Microbiology and Biotechnology*, **46**, 313–17.

Joner, E. J., Hirmann, D., Szolar, O. H. J., Todovoric, D., Leyval, C. & Loibner, A. P. (2003). Priming effects on PAH degradation and ecotoxicity during a phytoremediation experiment. *Environmental Pollution*, **128**, 429–35.

Joshi, D. K. & Gold, M. H. (1993). Degradation of 2,4,5-trichlorophenol by the lignin-degrading basidiomycete *Phanerochaete chrysosporium. Applied and Environmental Microbiology*, **59**, 1779–85.

Juhasz, A. L. & Naidu, R. (2000). Bioremediation of high molecular weight polycyclic aromatic hydrocarbons: a review of the microbial degradation of benzo[a]pyrene. *International Biodeterioration and Biodegradation*, **45**, 57–88.

Kayshoemake, J. L. & Watwood, M. E. (1996). Limitations of the lignin peroxidase system of the white rot fungus, *Phanerochaete chrysosporium. Applied Microbiology and Biotechnology*, **46**, 438–42.

Kennes, C. & Lema, J. M. (1994). Degradation of major compounds of creosotes (PAH and phenols) by *Phanerochaete chrysosporium. Biotechnology Letters*, **16**, 759–64.

Kersten, P. J., Kalyanaraman, B., Hammel, K. E., Reinhammar, B. & Kirk, T. K. (1990). Comparison of lignin peroxidase, horseradish peroxidase and laccase in the oxidation of methoxybenzenes. *Biochemical Journal*, **268**, 475–80.

Kirk, T. K., Schultz, E., Connors, W. J., Lorenz, L. F. & Zeikus, J. G. (1978). Influence of culture parameters on lignin metabolism by *Phanerochaete chrysosporium. Archives of Microbiology*, **117**, 277–85.

Kotterman, M. J. J., Rietberg, H. J., Hage, A. & Field, J. A. (1998a). Polycyclic aromatic hydrocarbon oxidation by the white-rot fungus *Bjerkandera* sp. Strain BOS55 in the presence of nonionic surfactants. *Biotechnology and Bioengineering*, **57**, 220–7.

Kotterman, M. J. J., Vis, E. H. & Field, J. A. (1998b). Successive mineralization and detoxification of benzo[a]pyrene by the white rot fungus *Bjerkandera* sp. Strain BOS55 and indigenous microflora. *Applied and Environmental Microbiology*, **64**, 2853–8.

Lamar, R. T. & Dietrich, D. M. (1990). *In situ* depletion of pentachlorophenol from contaminated soil by *Phanerochaete spp. Applied and Environmental Microbiology*, **56**, 3093–100.

Lamar, R. T., Evans, J. W. & Glaser, J. A. (1993). Solid-phase treatment of a pentachlorophenol-contaminated soil using lignin-degrading fungi. *Environmental Science and Technology*, **27**, 2566–71.

Lamar, R. T., Larsen, M. J. & Kirk, T. K. (1990a). Sensitivity to and degradation of pentachlorophenol by *Phanerochaete spp. Applied and Environmental Microbiology*, **56**, 3519–26.

Lamar, T., Glaser, J. A. & Kirk, T. K. (1990b). Fate of pentachlorophenol (PCP) in sterile soils inoculated with the white-rot basidiomycete *Phanerochaete chrysosporium*: mineralization, volatilization and depletion of PCP. *Soil Biology and Biochemistry*, **22**, 433–40.

Lang, E., Kleeberg, I. & Zadrazil, F. (2000). Extractable organic carbon and counts of bacteria near the lignocellulose-soil interface during the interaction of soil microbiota and white rot fungi. *Bioresource Technology*, **75**, 57–65.

Lau, K. L., Tsang, Y. Y. & Chiu, S. W. (2003). Use of spent mushroom compost to bioremediate PAH-contaminated samples. *Chemosphere*, **52**, 1539–46.

Lee, K. & Moon, S. H. (2003). Electroenzymatic oxidation of veratryl alcohol by lignin peroxidase. *Journal of Biotechnology*, **102**, 261–8.

Leisola, M. S. A., Thanei-Wyss, U. & Fiechter, A. (1985). Strategies for production of high ligninase activities by *Phanerochaete chrysosporium*. *Journal of Biotechnology*, **3**, 97–107.

Lestan, D. & Lamar, R. T. (1996). Development of fungal inocula for bioaugmentation of contaminated soils. *Applied and Environmental Microbiology*, **62**, 2045–52.

Levin, L., Viale, A. & Forchiassin, A. (2003). Degradation of organic pollutants by the white rot basidiomycete *Trametes trogii*. *International Biodeterioration and Biodegradation*, **52**, 1–5.

Lewis, T. A., Newcombe, D. A. & Crawford, R. L. (2004). Bioremediation of soils contaminated with explosives. *Journal of Environmental Management*, **70**, 291–307.

Li, K., Xu, F. & Eriksson, K. E. (1999). Comparison of fungal laccases and redox mediators in oxidation of a nonphenolic lignin model compound. *Applied and Environmental Microbiology*, **65**, 2654–60.

Liao, W. L. & Tseng, D. H. (1996). Biotreatment of naphthalene by PAH-acclimated pure culture with white-rot fungus *Phanerochaete chrysosporium*. *Water Science and Technology*, **34**, 73–9.

Liao, W. L., Tseng, D. H., Tsai, Y. C. & Chang, S. C. (1997). Microbial removal of polycyclic aromatic hydrocarbons by *Phanerochaete chrysosporium*. *Water Science and Technology*, **35**, 255–64.

Liebeg, E. W. & Cutright, T. J. (1999). The investigation of enhanced bioremediation through the addition of macro and micro nutrients in a PAH contaminated soil. *International Biodeterioration and Biodegradation*, **44**, 55–64.

Lisov, A. V., Leontievsky, A. A. & Golovleva, L. A. (2003). Hybrid Mn-peroxidase from the ligninolytic fungus *Panus tigrinus* 8/18. Isolation, substrate specificity, and catalytic cycle. *Biochemistry – Moscow*, **68**, 1027–35.

Majcherczyk, A., Johannes, C. & Huttermann, A. (1998). Oxidation of polycyclic aromatic hydrocarbons (PAH) by laccase of *Trametes versicolor*. *Enzyme Microbial Technology*, **22**, 335–41.

Martens, R., Wolter, M., Bahadir, M. & Zadrazil, F. (1999). Mineralization of ^{14}C-labelled highly-condensed polycyclic aromatic hydrocarbons in soils by *Pleurotus* sp. Florida. *Soil Biology & Biochemistry*, **31**, 1893–9.

Martinez, D., Larrondo, L. F., Putnam, N., Gelpke, M. D., Huang, K., Chapman, J., Helfenbein, K. G., Ramaiya, P., Detter, J. C., Larimer, F., Coutinho, P. M., Henrissat, B., Berka, R., Cullen, D. & Rokhsar, D. (2004). Genome sequence of the lignocellulose degrading fungus *Phanerochaete chrysosporium* strain RP78. *Natur Biotechnology*, **22**, 695–700.

Márquez-Rocha, F. J., Hernandez-Rodriguez, V. Z. & Vazquez-Duhalt, R. (2000). Biodegradation of soil-adsorbed polycyclic aromatic hydrocarbons by the white rot fungus *Pleurotus ostreatus*. *Biotechnology Letters*, **22**, 469–72.

Maspahy, S., Lamb, D. C. & Kelly, S. L. (1999). Purification and characterization of a benzo[a]pyrene hydroxylase from *Pleurotus pulmonarius*. *Biochemical and Biophysical Research Communications*, **266**, 326–9.

May, R., Schröder, P. & Sandermann Jr, H. (1997). *Ex situ* process for treating PAH-contaminated soil with *Phanerochaete chrysosporium*. *Environmental Science and Technology*, **31**, 2626–33.

Mench, M., Bussiere, S., Boisson, J., Castaing, E., Vangronsveld, J., Ruttens, A., De Koe, T., Bleeker, P., Assuncao, A. & Manceau, A. (2003). Progress in remediation and revegetation of the barren Jales gold mine spoil after *in situ* treatments. *Plant and Soil*, **249**, 187–202.

Mendez-Sanchez, N., Cutright, T. J. & Qiao, P. Z. (2003). Simultaneous evaluation of composite biodeterioration and changes in the physicochemical and biological water characteristics. *International Biodeterioration and Biodegradation*, **52**, 187–96.

Meulenberg, R., Rijnaarts, H. H. M., Doddema, H. J. & Field, J. A. (1997). Partially oxidised polycyclic aromatic hydrocarbons show an increased bioavailability and biodegradability. *FEMS Microbiology Letters*, **152**, 45–9.

Morgan, P., Lee, S. A., Lewis, S. T., Sheppard, A. N. & Watkinson, R. J. (1993). Growth and biodegradation by white-rot fungi inoculated into soil. *Soil Biology and Biochemistry*, **25**, 279–87.

Mori, T. K. R. (2002). Oxidation of chlorinated dibenzo-p-dioxin and dibenzofuran by white-rot fungus, *Phlebia lindtneri*. *FEMS Microbiology Letters*, **216**, 223–7.

Mougin, C. (2002). Bioremediation and phytoremediation of industrial PAH-polluted soils. *Polycyclic Aromatic Compounds*, **22**, 1011–43.

Mougin, C., Jolivalt, C., Malosse, C. & Chaplain, V. (2002a). Interference of soil contaminants with laccase activity during the transformation of complex mixtures of polycyclic aromatic hydrocarbons in liquid media. *Polycyclic Aromatic Compounds*, **22**, 673–88.

Mougin, C., Kollmann, A. & Jolivalt, C. (2002b). Enhanced production of laccase in the fungus *Trametes versicolor* by the addition of xenobiotics. *Biotechnology Letters*, **24**, 139–42.

Novotný, C., Erbanová, P., Sasek, V., Kubátová, A., Cajthaml, T., Lang, E., Krahl, J. & Zadrazil, F. (1999). Extracellular oxidative enzyme production and PAH removal in soil by exploratory mycelium of white rot fungi. *Biodegradation*, **10**, 156–68.

Novotný, C., Erbanova, P., Cajthaml, T., Rothschild, N., Dosoretz, C. & Sasek, V. (2000). *Irpex lacteus*, a white rot fungus applicable to water and soil bioremediation. *Applied Microbiology and Biotechnology*, **54**, 850–3.

Pickard, M. A., Roman, R., Tinoco, R. & Vazquez-Duhalt, R. (1999). Polycyclic aromatic hydrocarbon metabolism by white rot fungi and oxidation by *Coriolopsis gallica* UAMH 8260 laccase. *Applied and Environmental Microbiology*, **65**, 3805–9.

Popp, J. L. & Kirk, T. K. (1991). Oxidation of methoxybenzenes by manganese peroxidase and by Mn^{3+}. *Archives of Biochemistry and Biophysics*, **288**, 145–8.

Radtke, C., Cook, W. S. & Anderson, A. (1994). Factors affecting antagonism of the growth of *Phanerochaete chrysosporium* by bacteria isolated from soil. *Applied Microbiology and Biotechnology*, **41**, 274–80.

Rama, R., Mougin, C., Boyer, F. D., Kollmann, A., Malosse, C. & Sigoillot, J. C. (1998). Biotransformation of benzo[a]pyrene in bench scale reactor using laccase of *Pycnoporus cinnabarinus*. *Biotechnology Letters*, **20**, 1101–4.

Ramirez, N. & Cutright, T. J. (2000). Sequestration of pyrene by clay minerals in a natural soil. *Abstracts of Papers of the American Chemical Society*, **220**, 51–ENVR.

Rathbone, K., Fuchs, J., Anderson, K., Karthikeyan, R. & Nurhidayat, N. (1998). Effects of PAHs on microbial activity and diversity in freshly contaminated and weathered soils. In *Proceedings of the 1998 Conference on Hazardous Waste Research*. Manhattan, KA: Great Plains/Rocky Mountain Hazardous Substance Research Center, pp. 383–402.

Reddy, C. A. (1995). The potential for white-rot fungi in the treatment of pollutants. *Current Opinion in Biotechnology*, **6**, 320–8.

Reddy, G. V., Gelpke, M. D. & Gold, M. H. (1998). Degradation of 2,4,6-trichlorophenol by *Phanerochaete chrysosporium*: involvement of reductive dechlorination. *Journal of Bacteriology*, **180**, 5159–64.

Reddy, G. V. & Gold, M. H. (2000). Degradation of pentachlorophenol by *Phanerochaete chrysosporium*: intermediates and reactions involved. *Microbiology*, **146**, 405–13.

Reddy, G. V. & Gold, M. H. (2001). Purification and characterization of glutathione conjugate reductase: a component of the tetrachlorohydroquinone reductive dehalogenase system from *Phanerochaete chrysosporium*. *Archives of Biochemistry and Biophysics*, **391**, 271–7.

Reinhammer, B. (1972). Oxidation-reduction potentials of the electron acceptors I laccases and stellacyanin. *Biochimica et Biophysica Acta*, **275**, 245–59.

Rieble, S., Joshi, D. K. & Gold, M. H. (1994). Purification and characterization of a 1,2,4-trihydroxybenzene 1,2-dioxygenase from the basidiomycete *Phanerochaete chrysosporium*. *Journal of Bacteriology*, **176**, 4838–44.

Rothschild, N., Levkowitz, A., Hadar, Y. & Dosoretz, C. G. (1999). Manganese deficiency can replace high oxygen levels needed for lignin peroxidase formation by *Phanerochaete chrysosporium*. *Applied and Environmental Microbiology*, **65**, 483–8.

Ruttimannjohnson, C., Cullen, D. & Lamar, R. T. (1994). Manganese peroxidases of the white-rot fungus *Phanerochaete sordida*. *Applied and Environmental Microbiology*, **60**, 599–605.

Ruttimannjohnson, C. & Lamar, R. T. (1996). Polymerization of pentachlorophenol and ferulic acid by fungal extracellular lignin-degrading enzymes. *Applied and Environmental Microbiology*, **62**, 3890–3.

Ruttimannjohnson, C. & Lamar, R. T. (1997). Binding of pentachlorophenol to humic substances in soil by the action of white rot fungi. *Soil Biology and Biochemistry*, **29**, 1143–8.

Sack, U. & Fritsche, W. (1997). Enhancement of pyrene mineralization in soil by wood-decaying fungi. *FEMS Microbiology Ecology*, **22**, 77–83.

Sack, U. & Gunther, T. (1993). Metabolism of PAH by fungi and correlation with extracellular enzymatic-activities. *Journal of Basic Microbiology*, **33**, 269–77.

Sack, U., Heinze, T. M., Deck, J., Cerniglia, C. E., Martens, R., Zadrazil, F. & Fritsche, W. (1997a). Comparison of phenanthrene and pyrene degradation by different wood-decaying fungi. *Applied and Environmental Microbiology*, **63**, 3919–25.

Sack, U., Hofrichter, M. & Fritsche, W. (1997b). Degradation of phenanthrene and pyrene by *Nematoloma frowardii*. *Journal of Basic Microbiology*, **37**, 287–93.

Sack, U., Hofrichter, M. & Fritsche, W. (1997c). Degradation of polycyclic aromatic hydrocarbons by manganese peroxidase of *Nematoloma frowardii*. *FEMS Microbiology Letters*, **152**, 227–34.

Sayadi, S. & Ellouz, R. (1995). Roles of lignin peroxidase and manganese peroxidase from *Phanerochaete chrysosporium* in the decolorization of olive mill wastewaters. *Applied and Environmental Microbiology*, **61**, 1098–103.

Schmidt, K. R., Chand, S., Gostomski, P. A., Boyd-Wilson, K. S., Ford, C. & Walter, M. (2005). Fungal inoculum properties and its effect on growth and enzyme activity of *Trametes versicolor* in soil. *Biotechnology Progress*, **21**, 377–85.

Schoemaker, H. E., Harvey, P. J., Bowen, R. M. & Palmer, J. M. (1985). On the mechanism of enzymatic lignin breakdown. *FEBS Letters*, **183**, 7–12.

Schützendübel, A., Majcherczyck, A., Johannes, C. & Hüttermann, A. (1999). Degradation of fluorene, anthracene, phenanthrene, fluoranthene, and pyrene lacks connection to the production of extracellular enzymes by *Pleurotus ostreatus* and *Bjerkandera adusta*. *International Biodeterioration and Biodegradation*, **43**, 93–100.

Segura, F., Durany, G., Lozano, R. & Huguet, V. (1993). Detoxification pre-treatment of black liquor derived from non-wood feedstock with white-rot fungi. *Environmental Technology*, **14**, 681–7.

Shah, M. M., Grover, T. A. & Aust, S. D. (1991). Metabolism of cyanide by *Phanerochaete chrysosporium*. *Archives of Biochemistry and Biophysics*, **290**, 173–8.

Shen, Y. (1998). *In vitro* cytotoxicity of BTEX metabolites in HeLa cells. *Archives of Environmental Contamination and Toxicology*, **34**, 229–34.

Song, H. G. (1997). Biodegradation of aromatic hydrocarbons by several white-rot fungi. *Journal of Microbiology*, **35**, 66–71.

Song, H. G. (1999). Comparison of pyrene biodegradation by white rot fungi. *World Journal of Microbiology and Biotechnology*, **15**, 669–72.

Stevenson, F. (1994). *Humus Chemistry*, 2nd edn. New York: John Wiley and Sons.

Sutherland, J. B. (1992). Detoxification of polycyclic aromatic hydrocarbons by fungi. *Journal of Industrial Microbiology*, **9**, 53–62.

Tai, D., Terazawa, M., Chen, C.-L. & Chang, H. (1990). Lignin biodegradation products from birch wood decayed by *Phanerochaete chrysosporium*. Part 2. The constituents of ether-soluble low-molecular-weight fraction. *Holzforschung*, **44**, 257–62.

ten Have, R., De Thouars, R. G., Swarts, H. J. & Field, J. A. (1999). Veratryl alcohol-mediated oxidation of isoeugenyl acetate by lignin peroxidase. *European Journal of Biochemistry*, **265**, 1008–14.

Tien, M. & Kirk, T. K. (1983). Lignin-degrading enzyme from the hymenomycete *Phanerochaete chrysosporium*. *Science*, **221**, 661–3.

Torres, E., Bustos-Jaimes, I. & Le Borgne, S. (2003). Potential use of oxidative enzymes for the detoxification of organic pollutants. *Applied Catalysis B – Environmental*, **46**, 1–15.

Tucker, B., Radtke, C., Kwon, S. I. & Anderson, A. J. (1995). Suppression of bioremediation by *Phanerochaete chrysosporium* by soil factors. *Journal of Hazardous Materials*, **41**, 251–65.

Tuor, U., Wariishi, H., Schoemaker, H. & Gold, M. H. (1992). Oxidation of phenolic arylglycerol -aryl ether lignin model compounds by manganese peroxidase from *Phanerochaete chrysosporium*: oxidative cleavage of and -carbonyl model compound. *Biochemistry*, **31**, 4986–95.

Valderrama, B., Oliver, P., Medrano-Soto, A. & Vazquez-Duhalt, R. (2003). Evolutionary and structural diversity of fungal laccases. *Antonie Van Leeuwenhoek*, **84**, 289–99.

Valli, K. & Gold, M. H. (1991). Degradation of 2,4-dichlorophenol by the lignin-degrading fungus *Phanerochaete chrysosporium*. *Journal of Bacteriology*, **173**, 345–52.

Valli, K., Brock, B. J., Joshi, D. K. & Gold, M. H. (1992a). Degradation of 2,4-dinitrotoluene by the lignin-degrading fungus *Phanerochaete chrysosporium*. *Applied and Environmental Microbiology*, **58**, 221–8.

Valli, K., Wariishi, H. & Gold, M. H. (1992b). Degradation of 2,7-dichlorodibenzo-p-dioxin by the lignin-degrading basidiomycete *Phanerochaete chrysosporium*. *Journal of Bacteriology*, **174**, 2131–7.

Van Aken, B., Hofrichter, M., Scheibner, K., Hatakka, A. I., Naveau, H. & Agathos, S. N. (1999). Transformation and mineralization of 2,4,6-trinitrotoluene (TNT) by manganese peroxidase from the white-rot basidiomycete *Phlebia radiata*. *Biodegradation*, **10**, 83–91.

Vyas, B. R. M., Bakowski, S., Sasek, V. & Matucha, M. (1994). Degradation of anthracene by selected white-rot fungi. *FEMS Microbiology Ecology*, **14**, 65–70.

Wariishi, H., Valli, K. & Gold, M. H. (1989). Oxidative cleavage of a phenolic diarylpropane lignin model dimer by manganese peroxidase from *Phanerochaete chrysosporium*. *Biochemistry*, **28**, 6017–23.

Wilson, S. C. & Jones, K. (1993). Bioremediation of soil contaminated with polynuclear aromatic hydrocarbons (PAHs): a review. *Environmental Pollution*, **81**, 229–49.

Wischmann, H. & Steinhart, H. (1997). The formation of PAH oxidation products in soils and soil/compost mixtures. *Chemosphere*, **35**, 1681–98.

Wolter, M., Zadrazil, F., Martens, R. & Bahadir, M. (1997). Degradation of eight highly condensed polycyclic aromatic hydrocarbons by *Pleurotus* sp. Florida in solid wheat straw substrate. *Applied Microbiology and Biotechnology*, **48**, 398–404.

Wunch, K. G., Alworth, W. L. & Bennett, J. W. (1999). Mineralization of benzo[a]pyrene by *Marasmiellus troyanus*, a mushroom isolated from a toxic waste site. *Microbiological Research*, **154**, 75–9.

Xu, F. (1996). Oxidation of phenols, anilines, and benzenethiols by fungal laccases: correlation between activity and redox potentials as well as halide inhibition. *Biochemistry*, **35**, 7608–14.

Yadav, J. S. & Reddy, C. A. (1993). Degradation of benzene, toluene, ethylbenzene, and xylenes (BTEX) by the lignin-degrading basidiomycete *Phanerochaete chrysosporium*. *Applied and Environmental Microbiology*, **59**, 756–62.

Yateem, A., Balba, M. T., Al-Awadhi, N. & El-Nawawy, A. S. (1998). White rot fungi and their role in remediating oil-contaminated soil. *Environment International*, **24**, 181–7.

Zacchi, L., Burla, G., Zuolong, D. & Harvey, P. J. (2000a). Metabolism of cellulose by *Phanerochaete chrysosporium* in continuously agitated culture is associated with enhanced production of lignin peroxidase. *Journal of Biotechnology*, **78**, 185–92.

Zacchi, L., Morris, I. & Harvey, P. J. (2000b). Disordered ultrastructure in lignin-peroxidase-secreting hyphae of the white-rot fungus *Phanerochaete chrysosporium*. *Microbiology*, **146**, 759–65.

Zacchi, L., Palmer, J. M. & Harvey, P. J. (2000c). Respiratory pathways and oxygen toxicity in *Phanerochaete chrysosporium*. *FEMS Microbiology Letters*, **183**, 153–7.

Zadrazil, F., Kamra, D. N., Isikhuemhen, O. S., Schuchardt, F. & Flachowsky, G. (1996). Bioconversion of lignocellulose into ruminant feed with white rot fungi – review of work done at the FAL, Braunschweig. *Journal of Applied Animal Research*, **10**, 105–24.

Zapanta, L. S. & Tien, M. (1997). The roles of veratryl alcohol and oxalate in fungal lignin degradation. *Journal of Biotechnology*, **53**, 93–102.

13
Metal and mineral transformations: a mycoremediation perspective

M. FOMINA AND G. M. GADD

Introduction

In the periodic table, about 75% of the elements are classified as metals. While many metals are essential, e.g. K, Na, Ca, Mn, Mg, Fe, Cu, Zn, Co and Ni, many others have no apparent essential functions, e.g. Rb, Cs, Al, Cd, Ag, Au, Sr, U, Hg, Pb (Gadd, 1993). However, all can interact with fungi and most of them can exhibit toxicity above a certain threshold concentration, which will vary depending on the organism, metal species and environmental factors (Gadd & Griffiths, 1978; Gadd, 1993).

The redistribution of toxic metals in the biosphere as a result of human activity has become an important process in the biogeochemical cycling of these metals. The main source of metal pollution is industrialization including operational and former mining sites, foundries and smelters, untreated sewage sludge, and as diffuse sources, metal piping, traffic, combustion by-products from coal-burning power stations, and other atmospheric pollution. Arsenic, cadmium, chromium, copper, nickel, lead and mercury are the most common metal pollutants (Knox *et al.*, 2000) with a selection of metal radionuclides also entering the environment from a number of sources (Gadd, 2005).

Unlike degradable organic contaminants, metals are not decomposed in the environment. They can, however, be buried (or otherwise contained), removed and recycled, or transformed into less toxic forms (Knox *et al.*, 2000), these processes employing physical, chemical and biological methods singly or in combination. At present, there is a critical need for the further development of cost-effective efficient remediation technologies that reduce potential risks from toxic metals and radionuclides.

Exploitation of Fungi, eds. G. D. Robson, P. van West and G. M. Gadd. Published by Cambridge University Press.

Bioremediation

A variety of physico-chemical remediation techniques have been applied to clean up metal-contaminated soils, ground waters and other matrices. Chemical remediation technologies use mainly the two opposite approaches of (1) removal of soluble forms of the metal contaminant(s) from matrices and (2) immobilization of metal(s) into chemically stable forms within the contaminated matrices. An example of the first approach is *in situ* soil washing (or flushing or injection/recirculation) based on injection or infiltration of a solution into a zone of contaminated soil/groundwater, followed by extraction of the groundwater and flushing solution mixed with the contaminants. The second approach includes a variety of *in situ* stabilization/solidification techniques, which alter soil contaminant chemistry without changing the contaminant concentration in the matrix. Here, reactive additives such as lime, cement, phosphates, fly-ash, kiln dust, silicates and clays are mechanically blended with the soil to reduce pollutant mobility (Cotter-Howells & Caporn, 1996; Chen *et al.*, 1997; Conca, 1997; Garcia-Sanchez *et al.*, 1999; Knox *et al.*, 2000; Ryan *et al.*, 2001; Brown *et al.*, 2004; Hu *et al.*, 2004; Lima *et al.*, 2004; Tandy *et al.*, 2004; Castaldi *et al.*, 2005; Coleman *et al.*, 2005; Garrido *et al.*, 2005; Terzano *et al.*, 2005; www.marcor.com/, www.gwrtac.org/). These techniques can be rapid and cost-effective, but cause considerable disturbance to the soil ecosystem.

Many studies have demonstrated that both prokaryotic and eukaryotic microorganisms have the ability to remove metals from contaminated water or waste streams, and the use of microorganisms (as well as plants) for environmental protection has generated growing attention with some processes in commercial operation (Raskin *et al.*, 1997). Such technology that uses biological agents and processes to remove or detoxify contaminants from polluted soils or water is called bioremediation, and this general definition can encompass a variety of *in situ* and *ex situ* biological systems.

Toxic metal phytoremediation consists of several plant-based biotechnologies each having a different mechanism of action for the remediation of metal-polluted soil, sediment or water (Raskin *et al.*, 1997; Watanabe, 1997; Prasad & Freitas, 2003; Wenzel *et al.*, 2004; Krämer, 2005). These include extraction and accumulation of contaminants in harvestable plant tissues such as shoots and roots (Raskin *et al.*, 1997; Blaylock & Huang, 2000; Ghosh & Singh, 2005), volatilization of Se, Hg and As (Prasad & Freitas, 2003), and immobilization by adsorption and precipitation (Panfili *et al.*, 2005). The ultimate goal of any soil remediation technique is to return the polluted environment to close to its original state. Revegetation and the establishment of functional microbial communities in the rhizosphere, may be important final steps in any remediation process (Khan, 2005). Phytoremediation approaches may have an advantage in that they do not require significant engineering work and can be cheaper than conventional treatment approaches such as excavation, reburial, incineration, soil washing and *in situ* stabilization/solidification (Watanabe, 1997; Prasad & Freitas, 2003). It may also be less disturbing for the soil ecosystem than physico-chemical remediation (Prasad & Freitas, 2003). However, phytoremediation can be very slow and there is a risk of further transfer of toxic metals accumulated by plants to the food chain as well as air pollution by any volatile metal species. Phytoremediation can be used with chemical amendments to stimulate plant growth (e.g. mineral fertilizers, iron oxyhydroxides, limestone, zeolite and chelates) (Chen *et al.*, 2004*a*; Wu *et al.*, 2004; Lesage *et al.*, 2005; Meers *et al.*, 2005; Sersen *et al.*, 2005).

Metal and mineral transformations by fungi

Being a vital part of the soil microflora, usually with the largest bulk biomass, fungi are well suited for the transformation of metals and minerals (Gadd, 2004, 2005; Gadd *et al.*, 2005). Fungi employ a variety of metabolic and morphological strategies, excrete protons and chelating agents, and are often resistant to environmental factors such as metal toxicity, UV and desiccation (Sterflinger, 2000; Burford *et al.*, 2003; Gadd, 2004; Fomina *et al.*, 2005*a*). Another important feature of fungi is their ability to form mutualistic symbiotic associations with plants, algae and cyanobacteria (Gorbushina *et al.*, 1993; Smith & Read, 1997; Sterflinger, 2000). Fungal communities in soils are very diverse, with mycorrhizal fungi being a particularly important group involved in cycling of elements (Paris *et al.*, 1995; Olsson & Wallander, 1998; Whitelaw *et al.*, 1999; Hoffland *et al.*, 2002). A mycorrhiza is defined as a symbiotic relationship between a fungus and plant roots (Smith & Read, 1997). Depending on the ability of the symbiotic fungi to colonize host root cells intra- or extra-cellularly, the two main types of mycorrhizas are endomycorrhizas (ericoid and arbuscular mycorrhizas) and ectomycorrhizas. The main functions of mycorrhizal mycobionts are assumed to be phosphorus and trace metal acquisition, with an additional contribution to plant adaptation to extreme environments (Smith & Read, 1997; Meharg & Cairney, 2000). Since nearly all land plants (>95%) depend on symbiotic mycorrhizal fungi, fungal abilities in transforming toxic metals may be relevant to, or are already integral components of, some phytoremediation/revegetation approaches.

Fungi are involved in both metal solubilization and immobilization (Gadd, 2001, 2004; Gadd *et al.*, 2005) (Fig. 13.1). They can dissolve minerals releasing mobile metal species via so-called 'heterotrophic leaching' with two main dissolution mechanisms being proton-promoted (or acidification, protonolysis) and ligand-promoted (or complexolysis) (Burgstaller & Schinner, 1993; Burford *et al.*, 2003; Gadd, 2004). Metal mobilization can also occur via redox reactions, e.g. Hg(II) → Hg(0), Mn(IV) → Mn(II), and biomethylation of metalloids resulting in volatile products, e.g. methylated selenium compounds (Gadd, 1993, 2000, 2001). Fungi can immobilize metals by biosorption, transport and intracellular localization, redox immobilization, e.g. Cr(VI) → Cr(III), and precipitation of elemental gold, silver, tellurium and selenium (Gadd, 2001). The environmental significance of fungal mineral solubilization processes includes the mobilization of metals, phosphate and other nutrients, which in turn have potential for leaching of primary or waste ores, and industrial wastes, e.g. fly ash, galvanic sludge, electronic soap (Whitelaw *et al.*, 1999; Brandl, 2001; Gadd, 2001; Hoffland *et al.*, 2002). The environmental and biotechnological significance of metal immobilization by fungi

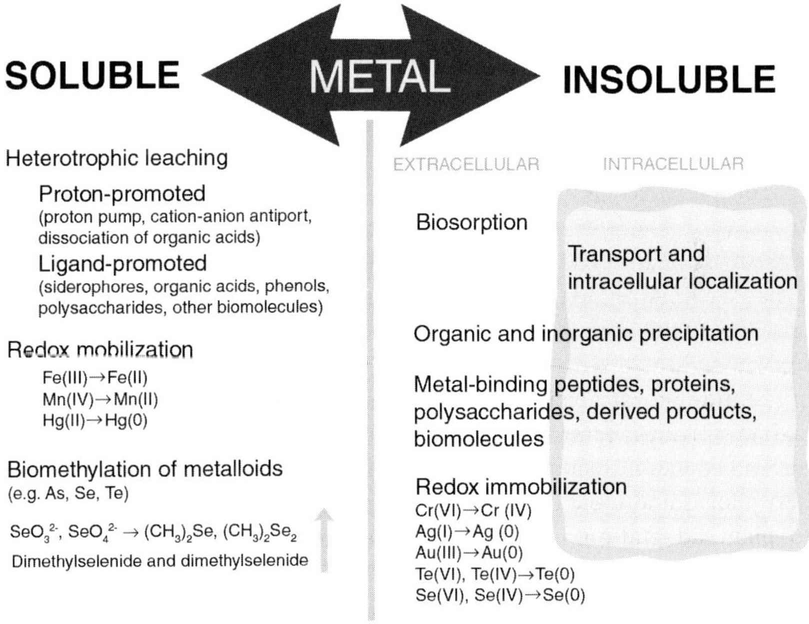

Figure 13.1 Some important processes of metal solubilization and immobilization mediated by fungi.

relates to the formation of new/secondary minerals and the application of fungi to remove metals from effluents, and other means of toxicant stabilization/immobilization (Sayer *et al.*, 1999; Verrecchia, 2000; Gadd, 2001; Tobin, 2001; Burford *et al.*, 2003; Fomina *et al.*, 2005b).

Solubilization

Fungi can attack rocks and minerals by means of biomechanical and biochemical processes (Banfield *et al.*, 1999; Kumar & Kumar, 1999; Sterflinger, 2000; Burford *et al.*, 2003; Money, 2004). Biochemical actions towards minerals and rocks are believed to be more important (Kumar & Kumar, 1999). Dissolution via protonolysis, complexolysis and redoxolysis is assisted by metal accumulation and precipitation by the fungi, functioning as a sink for metals mobilized from minerals and rocks (Burgstaller & Schinner, 1993).

Numerous studies of the interactions between mycorrhizas and toxic metals have shown that mycorrhizal associations enhance plant growth on metal-contaminated soils (Brown & Wilkins, 1985; Jones & Hutchinson, 1988; Jentschke & Goldbold, 2000; Meharg & Cairney, 2000; Van Tichelen *et al.*, 2001; Adriaensen *et al.*, 2003; Meharg, 2003; Colpaert *et al.*, 2004). The most efficient amelioration of metal toxicity to the host was observed when the mycobiont was resistant to the toxic metal (Brown & Wilkins, 1985; Jones & Hutchinson, 1988; Colpaert & Van Assche, 1992; Hartley-Whitaker *et al.*, 2000; Jentschke & Godbold, 2000; Van Tichelen *et al.*, 2001; Adriaensen *et al.*, 2003). Ectomycorrhizal mycobionts can 'filter' toxic metals in the hyphal sheath or Hartig net by sorption, restrict metal mobility due to hydrophobicity of the fungal sheath, chelate metals with released metabolites and adsorb metal species on the external mycelium (Turnau *et al.*, 1996; Jentschke & Godbold, 2000; Christie *et al.*, 2004; Krupa & Kozdroj, 2004). Clearly, the metal and mineral transforming activities of mycorrhizal fungi in soil may alter the physico-chemical environment of the mycorrhizosphere (Leyval & Joner, 2001). In terms of the phosphorus acquisition role of mycorrhizas, the solubilization of toxic metal phosphates in the mycorrhizosphere would be of special interest. Our studies have shown that many ericoid- and ectomycorrhizal fungi were able to solubilize toxic metal phosphates and other toxic metal minerals (Martino *et al.*, 2003; Fomina *et al.*, 2004, 2005c). For the majority of tested mycorrhizal fungi, the main mechanism was acidification (or protonolysis), with a clear correlation between pH and solubilization (Fomina *et al.*, 2004, 2005c). In Petri dish microcosms, mycorrhizal fungi can produce clear solubilization zones in agar amended with copper, cadmium and zinc phosphates and cuprite (Fomina *et al.*, 2005c) (Fig. 13.2). However, solubilization zones were not observed with lead-containing minerals, and there was no correlation between solubility product values and ease of solubilization observed for the tested minerals (Fomina *et al.*, 2005c) (Fig. 13.2). Solubilization of toxic metal minerals could, however, be related to the toxicity of metal species and compounds and the ability of test fungi to tolerate metal toxicity. Many fungi survive and even flourish in metal-contaminated locations employing a variety of biochemical and morphological strategies that underpin their success. Two main groups of biochemical strategies either decrease or prevent cellular metal entry or detoxify metal species intracellularly (Gadd, 1993; Perotto & Martino, 2001; Meharg, 2003) (Fig. 13.3). The former avoidance mechanisms include suppression of influx, enhanced efflux, release of metal-complexing agents, precipitation or metal binding onto cell surfaces and, in some cases, volatilization. Intracellular detoxification mechanisms

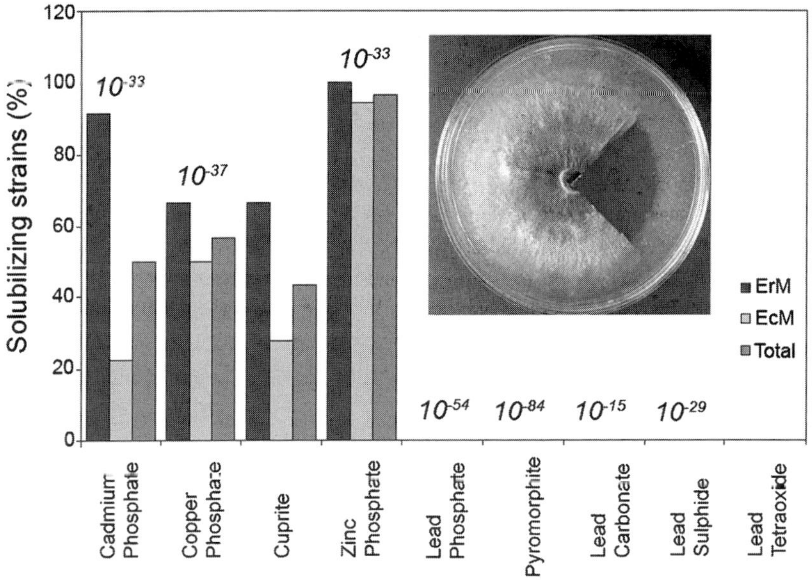

Figure 13.2 Solubilization of toxic metal minerals by ericoid mycorrhizal (ErM) and ectomycorrhizal (EcM) fungi grown on agar. Italicized numbers denote solubility products (the equilibrium constant for the dissociation of the compound in a solvent into its constituent ions) for the tested minerals. Solubility products were not available for cuprite and lead tetraoxide. The inset shows a clear solubilization zone in zinc phosphate-amended agar beneath a colony of *Suillus luteus* 22 (adapted from Fomina *et al.*, 2005c).

include intracellular binding (e.g. metallothionein) and sequestration within vacuoles (Gadd, 1993; Perotto & Martino, 2001; Meharg, 2003).

Our studies of mycorrhizal fungi in axenic culture showed that, in general, the solubilization of toxic metal minerals was related to metal tolerance. Zinc phosphate solubilization correlated with zinc tolerance for both ericoid- and ecto-mycorrhizal fungi. The relationship between toxic metal mineral solubilization and metal tolerance was confirmed by principal component analysis, where copper-tolerant isolates of the ericoid mycorrhizal fungus *Hymescyphus ericae* (from the Devon Consol copper mine area) demonstrated a much higher ability to solubilize cadmium, copper and zinc phosphates than isolates from non-polluted areas (Fig. 13.4). Zinc-tolerant isolates of ectomycorrhizal fungi (*Paxillus involutus, Suillus bovinus, Suillus luteus*) from a zinc smelter location (Lommel, Belgium) demonstrated a higher ability to dissolve zinc and cadmium phosphates than isolates from non-polluted soils (Fomina *et al.*, 2004, 2005c). In mesocosm experiments with ectomycorrhizal associations of Scots pine with *Paxillus involutus* strains, zinc phosphate dissolution, and zinc accumulation by roots and whole plants, depended on the strain

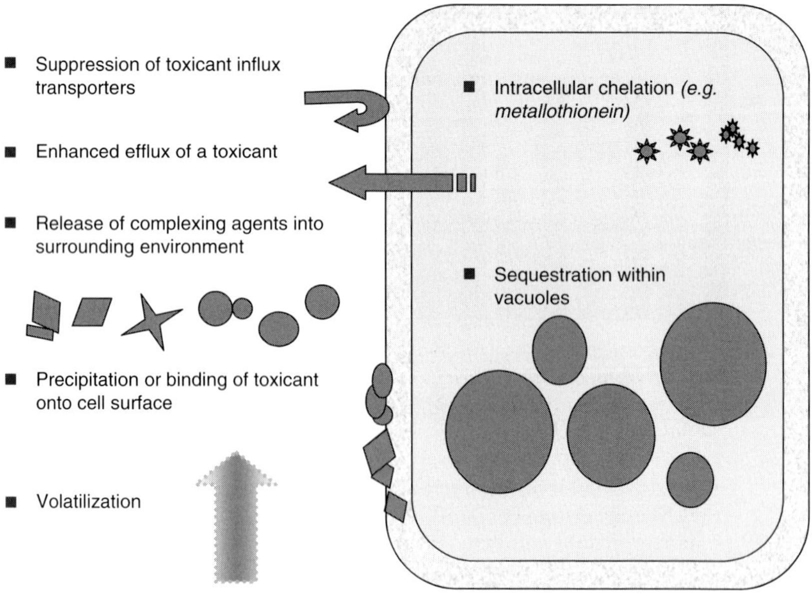

Figure 13.3 Biochemical strategies involved in detoxification of metals, which include (left) mechanisms that restrict metal entry into the cell, and (right) intracellular detoxification or compartmentation.

of the mycobiont, its zinc tolerance, and the phosphorus status of the matrix (Fomina *et al.*, 2006a). Under phosphorus-rich conditions, an ectomycorrhizal association with a zinc-tolerant fungal strain isolated from zinc-polluted soil showed the least zinc mobilization from zinc phosphate and the least zinc accumulation by the whole plant, whereas the highest zinc mobilization and accumulation was observed for non-mycorrhizal plants. In contrast, under phosphorus deficiency, the ectomycorrhizal association with the zinc-tolerant fungus demonstrated the highest zinc mobilization from zinc phosphate, and the highest zinc accumulation by the plant. In both phosphorus-deplete and -replete conditions, ectomycorrhizal mycobionts efficiently assisted phosphorus acquisition and much more phosphorus was accumulated in ectomycorrhizal roots than in

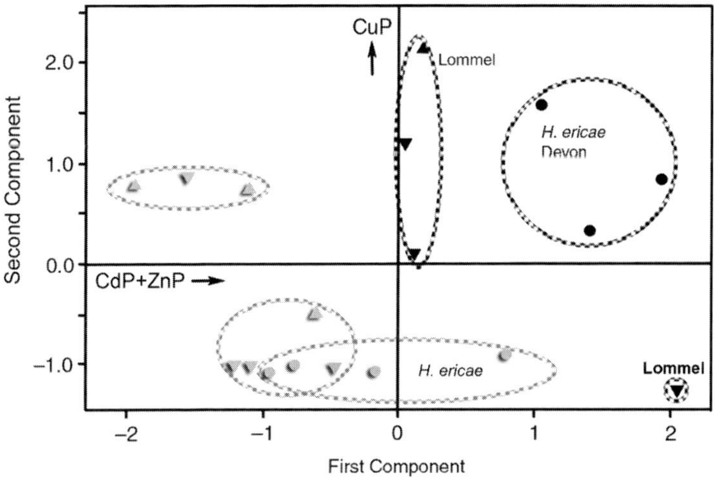

Figure 13.4 Principal component analysis of the solubilization of zinc, copper and cadmium phosphates by strains of the ericoid mycorrhizal fungus *Hymenoscyphus ericea* (circles) and ectomycorrhizal fungi of the genera *Paxillus* (triangles up) and *Suillus* (triangles down) isolated from metal-polluted (**black** symbols) and non-polluted habitats (**grey** symbols) (adapted from Fomina *et al.*, 2005c). Copper-tolerant isolates of the ericoid mycorrhizal fungus *Hymescyphus ericae* from the Devon Consol copper mine area (circled by **black dotted** line) demonstrated a much higher ability to solubilize cadmium, copper and zinc phosphates than isolates from non-polluted areas (circled by **grey dotted** line). Zinc-tolerant isolates of ectomycorrhizal fungi from the Lommel zinc smelter location (circled by **black dotted** lines) demonstrated a higher ability to dissolve zinc and cadmium phosphates than isolates from non-polluted soils (circled by **grey dotted** lines).

non-mycorrhizal roots. This indicates that these biogeochemical activities of ectomycorrhizas are conditional and could be altered by additives, e.g. phosphates, and the physico-chemical conditions. The nutritional status of the metal-polluted environment could therefore shift toxic metal transformation processes from metal mobilization to immobilization and vice versa.

Acidification or protonolysis was the main mechanism of toxic metal mineral solubilization by the majority of mycorrhizal fungi because most strains examined were not able to excrete strong chelating agents such as oxalate and citrate (Gadd, 1993; Fomina *et al.*, 2004, 2005c). Abiotic tests with agar medium amended with different toxic metal minerals clearly demonstrate the efficiency of oxalate-mediated dissolution (Fomina *et al.*, 2005b) (Fig. 13.5). It is noteworthy (see below) that oxalic acid was the only tested agent to give a clear solubilization zone for pyromorphite.

Pyromorphite or lead chlorophosphate ($Pb_5(PO_4)_3Cl$) is a geochemically-stable lead mineral that forms in urban and industrially-contaminated soils (Cotter-Howells & Caporn, 1996). Because of the very low solubility product of pyromorphite, the formation of such a form of lead phosphate

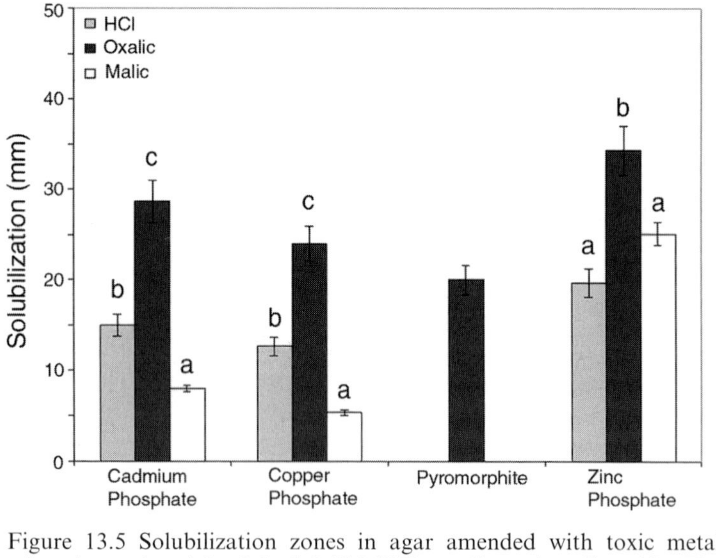

Figure 13.5 Solubilization zones in agar amended with toxic metal minerals produced by 0.1 ml 0.5 M HCl, oxalic and malic acids. The different letters denote significant differences at the 5% level using Fisher's LSD test (one-way ANOVA). Bars indicate ±SEM. (Adapted from Fomina *et al.*, 2005b; Fomina & Gadd, unpublished.)

in lead-contaminated soil (phosphate-induced metal stabilization (PIMS)) has been proposed as a potentially cost-effective *in situ* remediation technology (Cotter-Howells & Caporn, 1996; Conca, 1997; Zhang *et al.*, 1997; Brown *et al.*, 2004). However, pyromorphite, as the reaction product of PIMS-remediation, can be solubilized and transformed by several saprotrophic and mycorrhizal fungi (Sayer *et al.*, 1999; Fomina *et al.*, 2004, 2005b, c). A correlation between pH and lead mobilization from pyromorphite by mycorrhizal fungi in liquid medium confirmed that acidification was the main mechanism of pyromorphite dissolution for the majority of tested fungi (Fomina *et al.*, 2004). However, the ability of certain other fungi to over-excrete such strong chelators as oxalic acid shifted the mechanism of pyromorphite dissolution from proton-promoted to ligand-promoted and led to a dramatic increase in lead mobilization (Fomina *et al.*, 2004). Lead accumulation by fungal biomass grown on pyromorphite-containing agar also provided evidence that pyromorphite solubilization by mycorrhizal fungi occurred even in the absence of any clearing zone-visualization of this process (Fomina *et al.*, 2005c) (Fig. 13.6). The ability of free-living and mycorrhizal fungi to transform pyromorphite should clearly be taken into account in risk assessments of the long-term environmental consequences of *in situ* chemical remediation techniques.

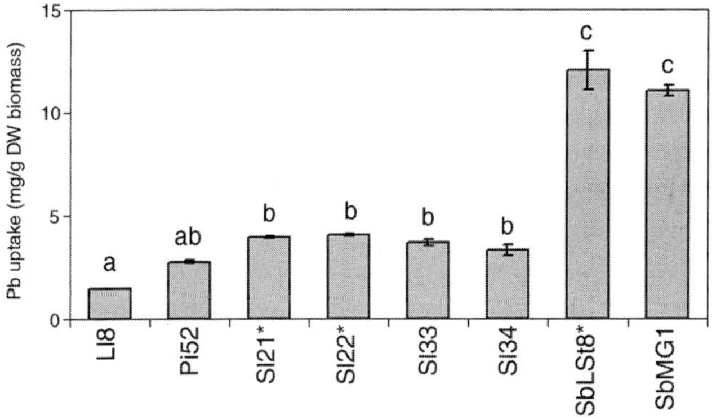

Figure 13.6 Lead bioaccumulation by biomass of ectomycorrhizal fungi grown on agar containing pyromorphite. Fungal strains tested were Ll8 – *Laccaria laccata* 8; Pi52 – *Paxillus involutus* 52; Sl21, Sl22, Sl33, Sl34 – *Suillus luteus* strains 21, 22, 33, 34; SbLSt8, SbMG1 – *Suillus bovinus* strains LSt8 and MG1. The different letters (a, b, c) denote significant differences at the 5% level using Fisher's LSD test (one-way ANOVA). Bars indicate ±SEM. The asterisk denotes strains isolated from zinc-polluted Lommel soil. (Adapted from Fomina *et al.*, 2005c.)

Immobilization

Metals can be immobilized by fungi via biosorption, which can be defined as metabolism-independent processes of metal uptake by fungal cell walls and other cell components (Gadd, 1990, 1999; Tobin, 2001). Biosorption occurs via a variety of mechanisms of a chemical and physical nature including ion exchange, complexation, hydrogen bonding, hydrophobic and van der Waals forces, and entrapment in fibrillar capillaries and spaces of the polysaccharide network (Gadd, 1990, 1999; Tobin, 2001). Certain pretreatments and immobilization of fungal biosorbents may make metal sorption more efficient. For example, growth of melanin-producing fungi in medium containing bentonite significantly enhanced the copper sorption ability of the resultant fungal biomineral sorbent (Fomina & Gadd, 2002). Fungal biomass has a high metal sorption affinity (Morley & Gadd, 1995; Fomina & Gadd, 2002). Potential binding sites on fungal biomass can include acetoamino groups from chitin, amino groups from proteins, sulfhydryl groups from proteins and peptides, hydroxyl groups, phosphate groups, and carboxyl groups from organic acids, polysaccharides, (poly)phenols/quinones, and melanin (Sarret *et al.*, 1998; Gadd, 1999; Tobin, 2001; Fomina & Gadd, 2002). It has been found that phosphate and carboxyl groups are of principal importance in metal biosorption to fungal cell walls. Phosphate groups were responsible for 95% of the lead bound to *Penicillium chrysogenum*, carboxyl groups were responsible for 55% of the zinc sorbed to *P. chrysogenum* and 70% of the zinc sorbed to *Trichoderma reesei* (Fourest *et al.*, 1996; Sarret *et al.*, 1999).

Bioaccumulation of mobile metal species occurs in living fungi. Identification of metal speciation within fungal biomass is a significant challenge because metals are likely to be fixed to biomolecules forming compounds of poor-crystallinity or of a non-crystalline nature. Synchrotron-based X-ray absorption spectroscopy (XAS) techniques provide a novel means to study the speciation of metals accumulated by fungi (Sarret *et al.*, 1998, 1999, 2002). XAS-studies of copper, lead and zinc speciation within the biomass of mycorrhizal and saprotrophic fungi grown in the presence of copper and zinc phosphates, and pyromorphite, confirmed the importance of phosphate and carboxyl groups in metal coordination (Fomina *et al.*, 2006b). For example, coordination of lead by oxygen-containing ligands fitting phosphate coordination was observed in the majority of ericoid- and ectomycorrhizal fungi grown on pyromorphite (Fig. 13.7).

Amorphous toxic metal precipitation can also occur in the microenvironment surrounding fungal hyphae. Non-crystalline precipitation of copper, cadmium and strontium was observed for saprotrophic (*Fennelia flavipes*

(=*Aspergillus flavipes*), *Trichoderma virens, Cladosporium cladosporioides*) and mycorrhizal (*Hymenoscyphus ericae, Rhizopogon rubescens*) fungi penetrating agar tiles amended with soluble toxic metal species (Fomina & Gadd, unpublished). Scanning electron microscopy coupled with energy dispersive X-ray microanalysis and elemental mapping demonstrated that toxic metals were concentrated and immobilized within precipitates: precipitated copper was associated with phosphorus (Fig. 13.8).

Some fungi are able to precipitate toxic metals within true mycogenic minerals with a well-crystallized structure, e.g. metal oxalates (Sayer *et al.*, 1997, 1999; Gadd, 2000; Fomina *et al.*, 2005b) (Fig. 13.9). When grown on media amended with cadmium-, copper-, zinc- and lead-containing minerals, oxalic acid over-excreting fungi can produce toxic metal oxalates (Fomina *et al.*, 2005b) (Fig. 13.9A–C). On nitrate-containing medium, ectomycorrhizal fungi (*R. rubescens* and *Suillus collimitus*) were able to transform zinc and copper phosphates into zinc and copper oxalate (Fig. 13.9D–F). Zinc-containing tetragonal crystals have also been observed within the extramatrical mycelium of the ectomycorrhizal association between *Pinus sylvestris* and *R. rubescens* grown with zinc phosphate (Fomina & Gadd, unpublished) (Fig. 13.9G–J).

Conclusions

To summarize the biogeochemical significance of mycorrhizal and free-living fungi, a simple model of metal mineral transformations in the mycorrhizosphere must include the following processes: (1) interaction between the mycorrhizal fungus and the toxic metal mineral and its dissolution via proton- and ligand-promoted mechanisms; (2) the release of anionic (e.g. phosphate) nutrients and metal cations as a result of mineral dissolution; (3) the uptake of anionic nutrients by the host plant, fungus, and other soil macro- and microbiota; (4) interaction of the fungus with

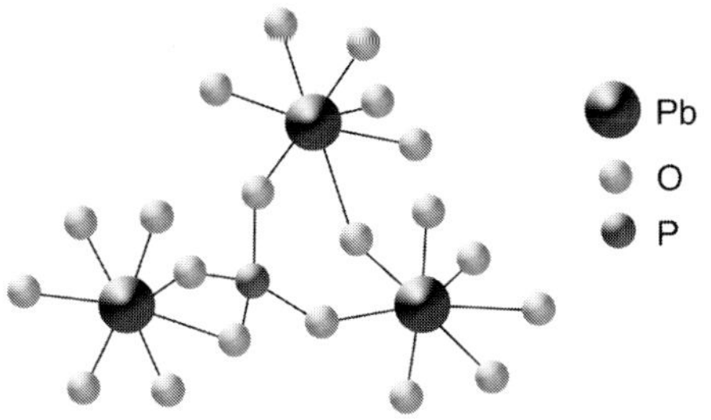

Figure 13.7 Model of phosphate coordination of lead accumulated within biomass of a variety of mycorrhizal and saprotrophic fungi grown in the presence of pyromorphite (adapted from Fomina *et al.*, 2006b).

mobilized metal species via intra- and extracellular processes; (5) the formation of mycogenic minerals, e.g. oxalates; (6) interactions of toxic metals mobilized by fungi with soil constituents, e.g. clay minerals, metal oxides and humic substances. Because fungi are perfectly suited for biogeochemical activity, usually comprise the largest bulk biomass in soil and play a significant role in the establishment and survival of host plants via mycorrhizal associations, fungal-based bioremediation (or mycoremediation)

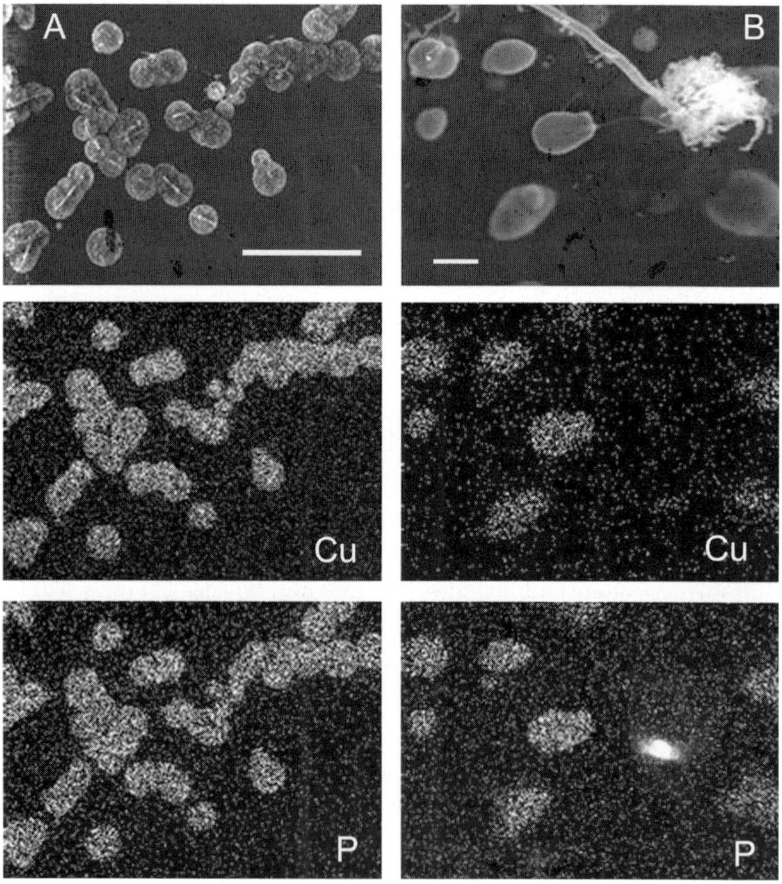

Figure 13.8 Visualization of extracellular amorphous copper precipitation by (A) the ectomycorrhizal fungus *Rhizopogon rubescens* and (B) *Fennelia flavipes* (=*Aspergillus flavipes*) penetrating agar tiles amended with soluble toxic metal species. Scale bars are (A) 100 μm and (B) 20 μm. The boxes below represent the results of elemental mapping of copper and phosphorus, which suggest association of copper with phosphate. (Fomina & Gadd, unpublished.)

approaches cannot be neglected. Any phytoremediation approach or *in situ* chemical remediation should be coupled with complementary mycoremediation techniques to assist either toxic metal extraction from polluted matrices or metal stabilization/immobilization. The mycobiota may compromise the efforts of some chemical remediation techniques as shown for pyromorphite (Sayer *et al.*, 1999). Phytoremediation

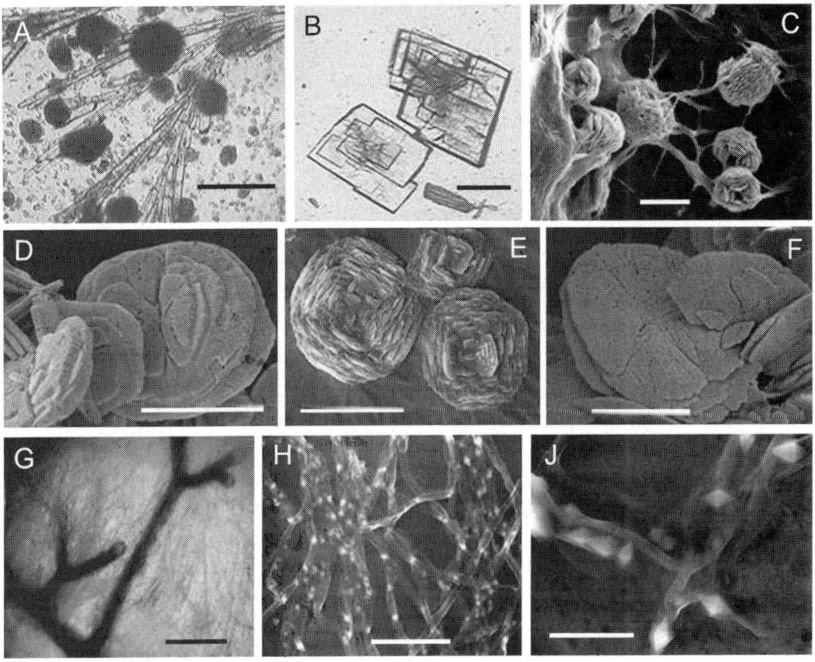

Figure 13.9 Precipitation of mycogenic toxic metal-containing minerals. (A–F) Formation of metal oxalate crystals by (A–C) *Beauveria caledonica* and ectomycorrhizal fungi (D, E) *Rhizopogon rubescens* and (F) *Suillus collimitus*. (A) Light microscopic images of the transformation of pyromorphite particles into lead oxalate; (B) cadmium oxalate crystals precipitated in agar amended with cadmium phosphate; (C) dry mode environmental scanning electron microscope (ESEM) image of air-dried Au/Pd-coated copper oxalate (moolooite) precipitated on fungal mycelium (adapted from Fomina *et al.*, 2005b) (D, F) SEM images of air-dried Au/Pd-coated crystals of (D, F) zinc oxalate; and (E) moolooite extracted from agar medium after growth of ectomycorrhizal fungi on corresponding toxic metal phosphates (Fomina & Gadd, unpublished). (G–J) Tetragonal zinc-containing crystals precipitated within extramatrical mycelium of *R. rubescens*/*Pinus sylvestris* ectomycorrhizal association grown on zinc phosphate-containing medium (Fomina & Gadd, unpublished). Scale bars are (A, B) 200 μm; (C–F, H) 20 μm; (G) 1 mm; (J) 5 μm.

techniques, especially revegetation assisted by mycorrhizal inoculation, still attracts attention (Chen *et al.*, 2004a, b; Christie *et al.*, 2004). There have been successful examples of revegetation using native trees, shrubs and grasses inoculated with specific ectomycorrhizal and arbuscular mycorrhizal species, e.g. the Ohio Abandoned Minelands Program, US, 1982–1999 (www.mcrcc.osmre.gov/PDF/Forums/Reforestation/Session%204/4-9.pdf). A complex approach combining all the relevant remediation techniques and which includes the study of fungal community structure and function in contaminated soil could provide significant improvements in the restoration of metal polluted environments.

Acknowledgements

The authors' research was funded by the BBSRC/BIRE programme (94/BRE13640), BNFL and CCLRC Daresbury SRS (SRS user grant 40107), which is gratefully acknowledged. We are also grateful to Drs John Charnock, Lorrie Murphy and Bob Bilsborrow (Stations 7.1 and 16.5, CCLRC Daresbury SRS, UK) for assistance with X-ray absorption spectroscopy scanning and analysis, and to Mr Martin Kierans (Centre for High Resolution Imaging and Processing (CHIPs), School of Life Sciences, University of Dundee) for assistance with scanning electron microscopy.

References

Andriaensen, K., Van der Lelie, D., Van Laere, A., Vangrosveld, J. & Colpaert, J. V. (2003). A zinc-adapted fungus protects pines from zinc stress. *New Phytologist*, **161**, 549–55.

Banfield, J. P., Barker, W. W., Welch, S. A. & Taunton, A. (1999). Biological impact on mineral dissolution: application of the lichen model to understanding mineral weathering in the rhizosphere. *Proceedings of the National Academy of Sciences USA*, **96**, 3404–11.

Blaylock, M. J. & Huang, J. W. (2000). Phytoextraction of metals. In *Phytoremediation of Toxic Metals: Using Plants to Clean-up the Environment*, eds. I. Raskin & B. D. Ensley. New York: John Wiley & Sons, Inc, pp. 53–70.

Brandl, H. (2001). Heterotrophic leaching. In *Fungi in Bioremediation*, ed. G. M. Gadd. Cambridge: Cambridge University Press, pp. 383–423.

Brown, M. T. & Wilkins, D. A. (1985). Zinc tolerance of mycorrhizal *Betula*. *New Phytologist*, **99**, 101–6.

Brown, S., Chaney, R., Hallfrisch, J., Ryan, J. A. & Berti, W. R. (2004). *In situ* soil treatments to reduce the phyto- and bioavailability of lead, zinc, and cadmium. *Journal of Environmental Quality*, **33**, 522–31.

Burford, E. P., Fomina, M., Gadd, G. M. (2003). Fungal involvement in bioweathering and biotransformation of rocks and minerals. *Mineralogical Magazine*, **67**, 1127–55.

Burgstaller, W. & Schinner, F. (1993). Leaching of metals with fungi. *Journal of Biotechnology*, **27**, 91–116.

Castaldi, P., Santona, L. & Melis, P. (2005). Heavy metal immobilization by chemical amendments in a polluted soil and influence on white lupin growth. *Chemosphere*, **60**, 365–71.

Chen, X.-B., Wright, J. V., Conca, J. L. & Peurrung, L. M. (1997). Evaluation of heavy metal remediation using mineral apatite. *Water, Air and Soil Pollution*, **98**, 57–78.

Chen, B. D., Shen, H., Li, X. L., Feng, G. & Christie, P. (2004a). Effects of EDTA application and arbuscular mycorrhizal colonization on growth and zinc uptake by maize (*Zea mays* L.) in soil experimentally contaminated with zinc. *Plant and Soil*, **261**, 219–29.

Chen, B. D., Liu, Y., Shen, H., Li, X. L. & Christie, P. (2004b). Uptake of cadmium from an experimentally contaminated calcareous soil by arbuscular mycorrhizal maize (*Zea mays* L.). *Mycorrhiza*, **14**, 347–54.

Christie, P., Li, X. L. & Chen, B. D. (2004). Arbuscular mycorrhiza can depress translocation of zinc to shoots of host plants in soils moderately polluted with zinc. *Plant and Soil*, **261**, 209–17.

Coleman, N. J., Lee, W. E. and Slipper, I. J. (2005). Interactions of aqueous Cu^{2+}, Zn^{2+} and Pb^{2+} ions with crushed concrete fines. *Journal of Hazardous Materials*, **121**, 203–13.

Colpaert, J. V. & Van Assche, J. A. (1992). Zinc toxicity in ectomycorrhizal *Pinus sylvestris*. *Plant and Soil*, **143**, 201–11.

Colpaert, J. V., Muller, L. A. H., Lambaerts, M., Andriaensen, K. & Vangronsveld, J. (2004). Evolutionary adaptation to Zn toxicity in populations of Suilloid fungi. *New Phytologist*, **162**, 549–59.

Conca, J. L. (1997). Phosphate-induced metal stabilization (PIMS). *Final Report to the US Environmental Protection Agency 68D60023*, Res. Triangle Park, NC.

Cotter-Howells, J. D. & Caporn, S. (1996). Remediation of contaminated land by formation of heavy metal phosphates. *Applied Geochemistry*, **11**, 335–42.

Fomina, M., and Gadd, G. M. (2002). Metal sorption by biomass of melanin-producing fungi grown in clay-containing medium. *Journal of Chemical Technology and Biotechnology*, **78**, 23–34.

Fomina, M. A., Alexander, I. J., Hillier, S. & Gadd, G. M. (2004). Zinc phosphate and pyromorphite solubilization by soil plant-symbiotic fungi. *Geomicrobiological Journal*, **21**, 351–66.

Fomina, M., Burford, E. P. & Gadd, G. M. (2005a). Toxic metals and fungal communities. In *The Fungal Community: Its Organization and Role in the Ecosystem*, eds. J. Dighton, J. White & P. Oudemans. Boca Raton, FLA: CRC Press, Taylor & Francis Group, pp. 733–58.

Fomina, M., Hillier, S., Charnock, J. M., Melville, K., Alexander, I. J. & Gadd, G. M. (2005b). Role of oxalic acid over-excretion in toxic metal mineral transformations by *Beauveria caledonica*. *Applied and Environmental Microbiology*, **71**, 371–81.

Fomina, M. A., Alexander, I. J., Colpaert, J. V. & Gadd, G. M. (2005c). Solubilization of toxic metal minerals and metal tolerance of mycorrhizal fungi. *Soil Biology and Biochemistry*, **37**, 851–66.

Fomina, M., Charnock, J. M., Hillier, S., Alexander, I. J. & Gadd, G. M. (2006a). Zinc phosphate transformations by the *Paxillus involutus*/pine ectomycorrhizal association. *Microbial Ecology*, **52**, 322–33.

Fomina, M., Charnock, J., Bowen, A. D. & Gadd, G. M. (2006b). X-ray absorption spectroscopy (XAS) of toxic metal mineral transformations by fungi. *Environmental Microbiology* (in press).

Fourest, E., Serre, A. & Roux, J. C. (1996). Contribution of carboxylic groups to heavy metal binding sites in fungal cell walls. *Toxicological and Environmental Chemistry*, **54**, 1–10.

Gadd, G. M. (1990). Fungi and yeasts for metal binding. In *Microbial Mineral Recovery*, eds. H. Ehrlich & C. L. Brierley. New York: McGraw-Hill, pp. 249–75.

Gadd, G. M. (1993). Interactions of fungi with toxic metals. *New Phytologist*, **124**, 25–60.

Gadd, G. M. (1999). Fungal production of citric and oxalic acid: importance in metal speciation, physiology and biogeochemical processes. *Advances in Microbial Physiology*, **41**, 47–92.

Gadd, G. M. (2000). Bioremedial potential of microbial mechanisms of metal mobilization and immobilization. *Current Opinion in Biotechnology*, **11**, 271–9.

Gadd, G. M. (2001). Metal transformations. In *Fungi in Bioremediation*, ed. G. M. Gadd. Cambridge: Cambridge University Press, pp. 250–382.

Gadd, G. M. (2004). Mycotransformation of organic and inorganic substrates. *Mycologist*, **18**, 60–70.

Gadd, G. M. (2005). Microorganisms in toxic metal polluted soils. In *Microorganisms in Soils: Roles in Genesis and Functions*, eds. F. Buscot & A. Varma. Berlin: Springer-Verlag, pp. 325–56.

Gadd, G. M. & Griffiths, A. J. (1978). Microorganisms and heavy metal toxicity. *Microbial Ecology*, **4**, 303–17.

Gadd, G. M., Fomina, M. & Burford, E. P. (2005). Fungal roles and function in rock, mineral and soil transformations. In *Microorganisms in Earth Systems*, eds. G. M. Gadd, K. T. Semple & H. M. Lappin-Scott, Cambridge: Cambridge University Press, pp. 201–31.

Garrido, F., Illera, V. & Garcia-Gonzalez, M. T. (2005). Effect of the addition of gypsum- and lime-rich industrial by-products on Cd, Cu and Pb availability and leachability in metal-spiked acid soils. *Applied Geochemistry*, **20**, 397–408.

Garcia-Sanchez, A., Alastuey, A. & Querol, X. (1999). Heavy metal adsorption by different minerals: application to the remediation of polluted soils. *Science of the Total Environment*, **242**, 179–88.

Ghosh, M. & Singh, S. P. (2005). A comparative study of cadmium phytoextraction by accumulator and weed species. *Environmental Pollution*, **133**, 365–71.

Gorbushina, A. A., Krumbein, W. E., Hamann, R., Panina, L., Soucharjevsky, S. & Wollenzien, U. (1993). On the role of black fungi in colour change and biodeterioration of antique marbles. *Geomicrobiology Journal*, **11**, 205–21.

Hartley-Whitaker, J., Cairney, J. W. G. & Meharg, A. A. (2000). Sensitivity to Cd and Zn of host and symbiont of ectomicorrhizal *Pinus sylvestris* L. (Scots pine) seedlings. *Plant and Soil*, **218**, 31–42.

Hoffland, E., Giesler, R., Jongmans, T. & van Breemen, N. (2002). Increasing feldspar tunneling by fungi across a north Sweden podzol chronosequence. *Ecosystems*, **5**, 11–22.

Hu, Z. Q., Yang, X. H., Zhu, Y. G. & Zhang, Y. C. (2004). Remediation of Cu^{2+} contaminated soil with Na-bentonite. *Rare Metal Materials and Engineering*, **33**, 92–5.

Jentschke, G. & Godbold, D. L. (2000). Metal toxicity and ectomycorrhizas. *Physiologia Plantarum*, **109**, 107–16.

Jones, M. D. & Hutchinson, T. C. (1988). Nickel toxicity in mycorrhizal birch seedlings infected with *Lactarius rufus* or *Scleroderma flavidum*. II. Uptake of nickel, calcium, magnesium, phosphorus and iron. *New Phytologist*, **108**, 461–70.

Khan, A. G. (2005). Role of soil microbes in the rhizospheres of plants growing on trace metal contaminated soils in phytoremediation. *Journal of Trace Elements in Medicine and Biology*, **18**, 355–64.

Knox, A. S., Seaman, J. C., Mench, M. J. & Vangronsveld, J. (2000). Remediation of metal- and radionuclides-contaminated soils by *in situ* stabilization techniques. In *Environmental Restoration of Metals-Contaminated Soil*, ed. I. K. Iskandar. Boca Raton, FLA: Lewis Publishers, pp. 21–61.

Krämer, U. (2005). Phytoremediation: novel approaches to cleaning up polluted soils. *Current Opinion in Biotechnology*, **16**, 133–41.

Krupa, P. & Kozdroj, J. (2004). Accumulation of heavy metals by ectomycorrhizal fungi colonizing birch trees growing in an industrial desert soil. *World Journal of Microbiology and Biotechnology*, **20**, 427–30.

Kumar, R. & Kumar, A. V. (1999). *Biodeterioration of Stone in Tropical Environments: An Overview*. USA: The J. Paul Getty Trust.

Lesage, E., Meers, E., Vervaeke, P., Lamsal, S., Hopgood, M., Tack, F. & Verloo, M. (2005). Enhanced phytoextraction: II. Effect of EDTA and citric acid on heavy metal uptake by *Helianthus annuus* from a calcareous soil. *International Journal of Phytoremediation*, **7**, 143–52.

Leyval, C. & Joner, E. J. (2001). Bioavailability of heavy metals in the mycorrhizosphere. In *Trace Elements in the Rhizosphere*, ed. G. R. Gobran, W. W. Wenzel & E. Lombi. Boca Raton, FLA: CRC Press, pp. 165–85.

Lima, I. M., Marshall, W. E. and Wartelle, I. Y. H. (2004). Hardware-based granular activated carbon for metals remediation. *Journal of the American Water Works Association*, **96**, 95–102.

Martino, E., Perotto, S., Parsons, R. & Gadd, G. M. (2003). Solubilization of insoluble inorganic zinc compounds by ericoid mycorrhizal fungi derived from heavy metal polluted sites. *Soil Biology and Biochemistry*, **35**, 133–41.

Meers, E., Lesage, E., Lamsal, S., Hopgood, M., Vervaeke, P., Tack, F. & Verloo, M. (2005). Enhanced phytoextraction: I. Effect of EDTA and citric acid on heavy metal mobility in a calcareous soil. *International Journal of Phytoremediation*, **7**, 129–42.

Meharg, A. A. (2003). The mechanistic basis of interactions between mycorrhizal associations and toxic metal cations. *Mycological Research*, **107**, 1253–65.

Meharg, A. A. & Cairney, J. W. G. (2000). Co-evolution of mycorrhizal symbionts and their hosts to metal-contaminated environments. *Advances in Ecological Research*, **30**, 69–112.

Money, N. P. (2004). The fungal dining habit – a biomechanical perspective. *Mycologist*, **18**, 71–6.

Morley, G. F. & Gadd, G. M. (1995). Sorption of toxic metals by fungi and clay minerals. *Mycological Research*, **99**, 1429–38.

Olsson, P. A., Wallander, H. (1998). Interactions between ectomycorrhizal fungi and the bacterial community in soils amended with various primary minerals. *FEMS Microbiology Ecology*, **27**, 195–205.

Panfili, F. R., Manceau, A., Sarret, G., Spadini, L., Kirpichtchikova, T., Bert, V., Laboudigue, A., Marcus, M. A., Ahamdach, N. & Libert, M.-F. (2005). The effect of phytostabilization on Zn speciation in a dredged contaminated sediment using scanning electron microscopy, X-ray fluorescence, EXAFS spectroscopy, and principal components analysis. *Geochimica et Cosmochimica Acta*, **69**, 2265–84.

Paris, F., Bonnaud, P., Ranger, J. & Lapeyrie, F. (1995). *In vitro* weathering of phlogopite by ectomycorrhizal fungi I. Effect of K^+ and Mg^{2+} deficiency on phyllosilicate evolution. *Plant and Soil*, **177**, 191–201.

Perotto, S. & Martino, E. (2001). Molecular and cellular mechanisms of heavy metal tolerance in mycorrhizal fungi: what perspectives for bioremediation? *Minerva Biotechnologica*, **13**, 55–63.

Prasad, M. N. V. & Freitas, H. M. (2003). Metal hyperaccumulation in plants – biodiversity prospecting for phytoremediation technology. *Electronic Journal of Biotechnology*, **6**, 285–321.

Ryan, J. A., Zhang, P., Hesterberg, D., Chou, L. & Sayers, D. E. (2001). Formation of chloropyromorphite in a lead-contaminated soil amended with hydroxyapatite. *Environmental Science and Technology*, **35**, 3798–803.

Raskin, I., Smith, R. D. & Salt, D. E. (1997). Phytoremediation of metals: using plants to remove pollutants from the environment. *Current Opinion in Biotechnology*, **8**, 221–6.

Sarret, G., Manceau, A., Cuny, D., Van Haluwyn, C., Deruelle, S., Hazemann, J.-L., Soldo, Y., Eybert-Berard, L. & Menthonnex, J.-J. (1998). Mechanisms of lichen resistance to metallic pollution. *Environmental Science and Technology*, **32**, 3325–30.

Sarret, G., Manceau, A., Spadini, L., Roux, J. R., Hazemann, J. L., Soldo, Y., Eybert-Berard, L., Monthennex, J. J. (1999). Structural determination of Pb binding sites in *Penicillium chrysogenum* cell walls by EXAFS spectroscopy and solution chemistry. *Journal of Synchrotron Radiation*, **6**, 414–16.

Sarret, G., Saumitou-Laprade, P., Bert, V., Proux, O., Hazemann, J.-L., Traverse, A., Marcus, M. A. & Manceau, A. (2002). Forms of zinc accumulated in the hyperaccumulator *Arabidopsis halleri*. *Plant Physiology*, **130**, 1815–26.

Sayer, J. A., Kierans, M. & Gadd, G. M. (1997). Solubilization of some naturally occurring metal-bearing minerals, limescale and lead phosphate by *Aspergillus niger*. *FEMS Microbiology Letters*, **154**, 29–35.

Sayer, J. A., Cotter-Howells, J. D., Watson, C., Hillier, S. & Gadd, G. M. (1999). Lead mineral transformation by fungi. *Current Biology*, **9**, 691–4.

Sersen, F., Cik, G., Havranek, E. & Sykorova M. (2005). Bio-remediation by natural zeolite on plants cultivated in a heavy metal-contaminated medium. *Fresenius Environmental Bulletin*, **14**, 13–17.

Smith, S. E. & Read, D. J. (1997). *Mycorrhizal Symbiosis*. London, UK: Academic Press.

Sterflinger, K. (2000). Fungi as geologic agents. *Geomicrobiology Journal*, **17**, 97–124.

Tandy, S., Bossart, K., Mueller, R., Ritschel, J., Hauser, L., Schulin, R. & Nowack, B. (2004). Extraction of heavy metals from soils using biodegradable chelating agents. *Environmental Science and Technology*, **38**, 937–44.

Terzano, R., Spagnuolo, M., Medici, L., Tateo, F. & Ruggiero, P. (2005). Copper stabilization by zeolite synthesis in polluted soils treated with coal fly ash. *Environmental Science and Technology*, **39**, 6280–7.

Van Tichelen, K. K., Colpaert, J. V. & Vangronsveld, J. (2001). Ectomycorrhizal protection of *Pinus sylvestris* against copper toxicity. *New Phytologist*, **150**, 203–13.

Tobin, J. M. (2001). Fungal metal biosorption. In *Fungi in Bioremediation*, ed. G. M. Gadd. Cambridge: Cambridge University Press, pp. 424–44.

Turnau, K., Kottke, I. & Dexheimer, J. (1996). Toxic element filtering in *Rhizopogon roseolus*/*Pinus sylvestris* mycorrhizas collected from calamine dumps. *Mycological Research*, **100**, 16–22.

Verrecchia, E. P. (2000). Fungi and sediments. In *Microbial Sediments*, eds. R. E. Riding & S. M. Awramik. Berlin: Springer Verlag, pp. 69–75.

Watanabe, M. E. (1997). Phytoremediation on the brink of commercialization. *Environmental Science and Technology*, **31**, 182–6.

Wenzel, W. W., Lombi, E. & Adriano, D. C. (2004). Biogeochemical processes in the rhizosphere: role in phytoremediation of metal-polluted soils. In *Heavy Metal Stress in Plants: From Biomolecules to Ecosystems*, eds. M. Prasad & J. Hagemeyer. Berlin, Heidelberg, New York: Springer Verlag, pp. 273–303.

Whitelaw, M. A., Harden, T. J. & Helyar, K. R. (1999). Phosphate solubilization in solution culture by the soil fungus *Penicillium radicum*. *Soil Biology and Biochemistry*, **31**, 655–65.

Wu, L. H., Luo, Y. M., Xing, X. R. & Christie, P. (2004). EDTA-enhanced phytoremediation of heavy metal contaminated soil with Indian mustard and associated potential leaching risk. *Agriculture, Ecosystems and Environment*, **102**, 307–18.

Zhang, P. C., Ryan, J. A. & Bryndzia, L. T. (1997). Pyromorphite formation from goethite adsorbed lead. *Environmental Science and Technology*, **31**, 2673–8.

V

Fungal biocontrol of pests

14
Fungal control of subterranean pests

H. STRASSER, S. H. M. HUTWIMMER AND
R. ZELGER

Traditional crops are under increasing threat from a number of subterranean pests that have been proven particularly difficult to control. In Europe, these include the larvae of Scarabaeidae (examples include *Melolontha melolontha*, *Amphimallon solstitialis*, *A. majale*, *Phyllopertha horticola*, *Hoplia philanthus*), a number of different larvae of Elateridae (for example, *Agriotes lineatus*, *A. obscurus*, *Limonius* spp.) and Curculionidae (*Bothynoderes punctiventris*, *Otiorhynchus sulcatus*, for example), *Daktulosphaira vitifoliae* (Phylloxeridae), as well as the new exotic pest *Diabrotica virgifera* (Chrysomelidae). Such pests are difficult to control not only because of their hidden niches, but also because of the inherent difficulties of penetrating their habitat with appropriate control agents. In most of the agricultural systems affected in Europe, the application of chemical insecticides is undesirable or impossible. Currently, the use of virulent and ecologically competent strains of insect-pathogenic fungi appears to be the best approach. Fungal pathogens are endemic in pest populations, and they also fulfil the key criteria for biological control agents: effectiveness, auto-dissemination and persistence. In this article, examples of successful use of entomopathogenic fungi *Beauveria* spp. and *Metarhizium anisopliae* as preventive control approaches in the subterranean pest control in European agriculture are discussed.

Introduction

Scarabaeidae, Elateridae, Curculionidae, Phylloxeridae, Chrysomelidae and other subterranean pests are of increasing importance in Europe. These pests – especially in their larval stages – cause damage of several billion euros annually to a wide range of economically important crops throughout Europe (Keller & Zimmermann, 2005). Major problems

Exploitation of Fungi, eds. G. D. Robson, P. van West and G. M. Gadd. Published by Cambridge University Press.

in the control of such pest organisms are their hidden niche in the soil and the inherent difficulties of penetrating their habitat with appropriate control agents. Early methods of subterranean pest control relied on non-chemical and cultural practices, depending on the pest species. Adults of large species were hand collected or caught using traps. In the absence of effective controls, management practices have been modified to reduce the risk of pest attack. By the 1950s, modern synthetic chemicals for pest control were introduced and the chemical age in agriculture was launched, whereby many of the old control methods were forgotten. Organochlorine insecticides were widely used until the problems with residues and persistence forced their prohibition in most countries. Organophosphates, which followed organochlorines, are still under discussion because most of their soil formulations have not been reported by the European Industry in accordance with Article 5, Commission Regulation (EC/No 1112/2002). The main argument not to notify the authorities is the high costs of notification, the reason being the toxicity and lack of persistence of the active ingredient.

Currently, a new emphasis has been placed on integrated control. The integrated pest management (IPM) strategy involves the use of a range of control measures in an appropriate way to reduce the level of pests to below an economic threshold. During the last two decades, large research teams developed biological control methods against subterranean pests. BIPESCO (FAIR6-CT98-4105), ERBIC (FAIR-CT97-3489) and RAFBCA (QLK1-2001-01391) are examples of several EU funded projects, which aimed to develop safe crop protection strategies using macrobial and/or microbial control agents and reduce pesticide inputs in food crops.

Considerable progress has been made in the development of fungal biocontrol agents (BCAs) to be used in IPM programmes including the use of the insect-pathogens *Beauveria* spp. and *Metarhizium anisopliae* (Pickett *et al.*, 1995; Butt & Copping, 2000; Inglis *et al.*, 2001; Butt, 2002). Both genera have a number of advantages as biocontrol agents over chemical products. They are mass produced relatively easily, strains can be selected with appropriate levels of virulence and specificity, and formulations for increased shelf-life and field efficacy are possible (Milner, 2000). Reviews dealing with the successful use of *Beauveria* spp. and *M. anisopliae* and their development as microbial BCAs include those of Ferron (1985); McCoy, Samson & Boucias (1988); Evans (1989); Ferron, Fargues & Riba (1991); Glare & Milner (1991); Glare (1992); Roberts & Hajek (1992); Tanada & Kaya (1993); Hajek & St. Leger (1994); Boucias & Pendland (1998); Wraight & Carruthers (1999); Inglis *et al.* (2001) and Butt (2002).

Although some products based on these anamorphic fungi are already commercially available (Copping, 2004), in 'Green' Europe their real potential is still not exploited in a sufficient way. Currently, fungal agents are only in use where (i) chemicals are banned because of environmental damage and/or toxicity (for example organochlorines, methyl bromide); (ii) pesticides are ineffective due to pesticide resistance or absorption into organic soils; (iii) humans and other non-target organisms are at risk from chemicals (botanic gardens, golf courses, other public areas); and (iv) chemical pesticides cannot be used in sustainable organic systems and in crop agricultures where bees are essential for pollination. However, this state of affairs will hopefully change when the RENDER4 evaluation (Sanco/10377/2003) is completed in 2006, and both *Beauveria* spp. and *M. anisopliae* are included in Annex I of the amended Commission Directive 2001/36/EC. This will open the application of these genera to a broad spectrum of pests, thereby pushing the market towards a preventive biological soil treatment.

This article shows examples of successful applications of *Beauveria* and *Metarhizium* strains in the control of selected subterranean pests, and recommends the use of such well characterized entomopathogens for the control of new major pest species (such as *Diabrotica virgifera*) in Europe.

Subterranean pests and their economic relevance

Scarabaeidae

The target species is the European cockchafer or May beetle, *Melolontha melolontha* L. (Coleoptera: Scarabaeidae). The cockchafer has a three-year development (a four-year cycle is found at high altitudes and in the northern parts of its natural distribution area). Adults swarm between the end of April and the beginning of June. During this period the females deposit their eggs in the soil at a depth of 10–25 cm (or deeper in sandy soils). Larvae hatch about six weeks later and pass through three larval instars. After a development of two years, larvae are fully grown and pupate after a prepupal period in July/August. Adults hatch in September, but remain at the spot where they hatched. The following April/May they leave the soil to feed on trees, mainly at the borders of forests. Although forests can become completely defoliated, the adults are not considered pests. Conversely, larvae that feed on roots of nearly all cultivated plants (including trees) can cause considerable damage in grassland, arable crops, arboriculture and viticulture (Strasser, 2004).

Phyllopertha horticola L. (garden chafer): in contrast to the May and June beetles (*Melolontha melolontha* L. and *Amphimallon solstitialis* L.) the developmental cycle of the garden chafer is a one-year cycle (Strasser, 2000). The female chafer lays on average 30 to 40 eggs, individually, in cavities in the soil at a depth of approximately five to sixteen centimetres (Milne, 1957). Larvae in the first stage (L1) hatch after four to six weeks and develop very quickly (within three weeks) by feeding on roots. Once the skin has been shed, the second larval stage (L2) begins, which lasts about four weeks. After another shedding, the third larval stage begins usually in the last week of August, which is characterized by an eight- to ten-week feeding period. In this late summer period the infested areas of meadows and lawns, sports facilities, recreation areas, golf courses and reforestation areas will be devastated not only by the larvae themselves, but also through the predators feeding on the larvae (for example fox, badger, crow, wild boar). The affected areas can not only be found in the valley, but they also reach up to regions in the highlands. The larvae's hibernation begins in the third week of October at the earliest, when they migrate deeper (approximately 20 centimetres) and food intake is discontinued until the swarming. By the end of March the larvae pupate. After the pupa skin breaks open, the chafer remains motionless in the soil for one to seven days. Because of a hardened cuticle, the chafer requires approximately two days to reach the soil surface where it remains in the turf to await favourable weather conditions for swarming (end of May to the beginning of July). Within 3 to 38 days (the lifespan of the adult chafer) the chafer causes further damage feeding heavily in orchards, on deciduous trees, roses and other flowering shrubs.

Curculionidae

The black vine weevil, *Otiorhynchus sulcatus* Fabricius (Coleoptera: Curculionidae), is a serious pest of many horticultural crops in most temperate regions, including Europe, North America, Japan and Australia (Smith, 1932; Masaki, Ohmura & Ichinoe, 1984; Moorhouse, Charnley & Gillespie, 1992). Adult weevils are all females and reproduce by thelytokous parthenogenesis (Smith, 1932), which means that unfertilized *O. sulcatus* eggs develop into females, and the daughters are genetically identical to their mothers. Due to this parthenogenetic reproduction and due to the long reproductive life span of the adults, a single *O. sulcatus* female can establish a new population, which can reach a very high density in just a few generations. Under outdoor conditions, the weevil is univoltine with considerable overlap of life stages (Schread, 1972). Data on the timing of appearance and the duration of different stages is only approximate (Moorhouse *et al.*, 1992). *O. sulcatus* causes damage both above and below ground to over 140 plant species (Warner & Negley, 1976). Whereas larvae cause damage by feeding on root systems, which reduces plant vigour, stunts growth and – in severe infestations – causes death of the plant, foliar feeding adults are particularly damaging to ornamental plants due to the low damage threshold required in the affected crops.

Elateridae

Wireworms, the larvae of click beetles (Coleoptera: Elateridae), are worldwide pests of a large number of crops. During the past few years, wireworm damage has become an increasing problem for European potato growers; damage has even occurred in fields in all-arable rotations (i.e. no history of long-term grass). Up to 39 species from 12 genera have been recorded as pest species attacking potatoes (Jansson & Seal, 1994). By far the most important pest species of this group are those belonging to the genus *Agriotes*. Three species, *Agriotes lineatus* L., *Agriotes obscurus* L. and *Agriotes sputator* L. are responsible for the vast majority of attacks in England and most likely in Central Europe (Parker & Howard, 2001). Apart from *Agriotes*, other wireworm genera can sometimes be found in both grassland and arable fields, often in mixed populations with *Agriotes* species. Wireworm species in genera such as *Athous* (particularly *Athous haemorrhoidalis* F.) and *Corymbites* have long been known to cause damage to potato and other root crops (Anonymous, 1948). Recent surveys in western France (Blot *et al.*, 1999) indicate that *Athous* spp. wireworms were present in 19% of insecticide-treated fields in maize–wheat rotations.

Click beetles live above ground and feed on pollen (Keller & Schweizer, 2001). After mating in May or June, female click beetles deposit 100 to 300 spherical (0.5 mm diameter) translucent eggs just below the soil surface, either singly or in small clusters, and usually within the protection of grassy or weedy ground, which minimizes the risks of desiccation (Parker & Howard, 2001). Depending on the temperature, *Agriotes* spp. larvae hatch in four to six weeks (Furlan, 1996) passing one to three instars each year (Roberts, 1919, 1921, 1922; Pill, Keaster & Chippendale, 1976). The duration of larval development varies strongly between genera and individual sites. Usually, larvae run through eight stages, which last four to five years. Keller & Schweizer (2001) revealed that wireworms start to damage crops at the beginning of their fourth larval stage, when their size is about 10 mm. Wireworms usually reach maturity between July and September, at which point they burrow deep into the soil and hollow out small pupation cells 5–30 cm below the soil surface. After three to four weeks they become adults, but usually remain in the cell for hibernation unless disturbed, in which case they move upwards and hibernate elsewhere (Gratwick, 1989). Mature click beetles hatch the following spring, starting in April (Hill, 1987).

Phylloxeridae

Grape phylloxera, *Daktulosphaira vitifoliae* Fitch (Hemiptera: Phylloxeridae), is a serious pest of commercial grapevines worldwide. On the fresh roots of grapevines, the aphid-like grape phylloxera causes so-called nodosities, beak-like swellings, as a result of feeding activity. High populations of this pest can result in premature defoliation, reduced shoot growth, reduced yield, reduced quality of the crop and even crop shortfall.

Phylloxera has been reported to cause serious problems to Californian and Australian vineyards (Grannet, Timper & Lider, 1991; Corrie, 1998) and more recently to European vines (Presser, Schmid & Rühl, 1993; Remund & Boller, 1994; Leclant & Coer D'Acier, 1995; Redl, 1999; Porten & Huber, 2003). In Rheingau, Germany, for example, the affected area has almost doubled to 361 ha between 1994 and 1997 (Jung, 1998).

Grape phylloxera has a complex life-cycle (McLeod & Williams, 2005). Overwintering takes place either as winter eggs under the bark of older canes or trunks, or as nymphs on grapevine roots. The winter egg develops into a fundatrix (or stem mother), which feeds on shoot tips, thereby eliciting gall formation and becoming enclosed into a gall on the underside of the grape leaf. This parthenogenetic female is capable of producing several hundred eggs. First instar nymphs leave the galls, begin feeding, and initiate gall formation themselves. In North America, there are three to five generations of foliar phylloxera per season. Throughout the summer, certain portions of the foliar nymphs move to the soil and eventually reach grapevine roots. Phylloxera may also overwinter on grapevine roots as first or second instars nymphs. Increasing temperature triggers feeding on the roots, which results in nodosities and tuberosities. The latter eventually result in dead vines. From July to October, some root-infesting phylloxera develop wing buds and eventually become fully winged adults. Alates deposit male and female eggs, whose developed mature forms mate. The resulting eggs are deposited individually under the bark of old canes or trunks, which completes the life-cycle.

Chrysomelidae

The western corn root worm, *Diabrotica virgifera virgifera* LeConte (Coleoptera: Chrysomelidae), is the most important pest of maize in the USA and Canada causing yield loss and chemical control costs of up to one billion US dollars annually (Krysan & Miller, 1986; Chandler *et al.*, 2000). Just before 1992, when it was accidentally introduced near Belgrade, Serbia, its expansion into Europe took place at an estimated mean rate of 33 km $year^{-1}$ (Hemerik, Busstra & Mols, 2004) and the western corn root worm became a major new threat to European corn agriculture.

Diabrotica virgifera females deposit their white, oval eggs (0.1 mm long) in the soil during the summer. After hibernation, these eggs hatch in late May or in early June. The first larval stages feed on fine root hairs and burrow themselves into the root tips of corn, whereas larval stages two and three (10–18 mm) feed on primary roots. The root injury caused by larval feeding reduces the uptake of water and nutrients, thus affecting plant growth (Chiang, 1973). Extensive feeding makes plants susceptible to lodging (Kahler *et al.*, 1985). The larvae of the third instar pupate in the soil as white pupae and emerge as adults in July and August. Soon after their emergence, adults of *D. virgifera* mate and females start ovipositing after two weeks. The yellowish coloured adults are approximately 5–6 mm long and have a black stripe on the elytrum. They feed primarily on corn silk and pollen, but also on leaf tissue. The duration of each stage (egg, larvae instars 1–3 and pupae) depends mainly on the temperature. Members of *Diabrotica virgifera* complete one generation per year (Hemerik *et al.*, 2004).

Fungal biological control agents

Relatively few species of entomopathogenic fungi are effective against subterranean pest species. The two most common and well-studied genera of all entomopathogens are *Beauveria* and *Metarhizium*.

Historically, the first field trials were carried out with the fungus *Beauveria brongniartii* between 1891 and 1894 (Keller, 1986) and with *M. anisopliae* between 1884 and 1888 (Zimmermann, 1993). Although encouraging results were achieved, research was not continued. About 80 years later, the soil application technique of *Beauveria* and *Metarhizium* was revived and tested for the control of white grubs, weevils, sugarcane spittlebug, cockroaches, grasshoppers and locust, respectively, in France, Russia, China and Brazil (Ferron, 1981). Biological control of subterranean pests became a matter of research worldwide because it was quickly noted that the applied fungi establish themselves well in the soil, which is a significant advantage over chemical control agents.

Taxonomy of the genera *Beauveria* and *Metarhizium*

Anamorphic fungi, which mainly belong to the Ascomycota, are asexually or mitotically sporulating fungi, which are traditionally classified as the 'Fungi Imperfecti' or the form taxon 'Deuteromycotina' (Seifert & Gams, 2001). The genera *Beauveria* and *Metarhizium* are considered to be anamorphic stages of the order Clavicipitales (Ascomycota) and are commonly characterized by morphological features of hyphae, conidiophores and spores (Seifert & Gams, 2001).

Beauveria Vuill. (1912)

There has been one report from Japan (Shimazu, Mitsuhashi & Hashimoto, 1988) where an isolate of *B. brongniartii* (Sacc.) Petch has been associated with a sexual stage or teleomorph (*Cordyceps brongniartii* Shimazu). According to classical taxonomic criteria based on morphology of conidiophores and spores, historically three species were recognized within the genus *Beauveria* by De Hoog (1972): *B. bassiana* (Bals.-Criv.) Vuill., *B. brongniartii* (Sacc.) Petch and *B. alba* (Limber) Saccas.

To this day, mycologists have problems with the nomenclature and the systematics of *Beauveria* species. For instance, there have been different opinions concerning the separation of *B. bassiana* and *B. brongniartii*. This is why so many synonyms exist in the textbooks. Synonyms for *B. brongniartii* are: *Botrytis brongniartii, Botrytis melolonthae, Botrytis tenella, Beauveria densa, Beauveria epigaea, Beauveria melolonthae, Beauveria tenella, Isaria densa* and *Sporotrichum epigaeum* (De Hoog, 1972; Domsch, Gams & Anderson, 1980; *DOC 7846/VI/97 EEC)*. Synonyms for *B. bassiana* are: *Botrytis bassiana, Sporotrichium densum, Beauveria densa, Sporotrichium globuliferum, Beauveria globulifera, Beauveria bassiana* subsp. *tenella* (Domsch, Gams & Anderson, 1993).

Besides morphological criteria, physiological patterns (BIPESCO FAIR6 CT-98-4105), or molecular characterization such as RAPDs and rDNA RFLP can be used to separate *B. bassiana* and *B. brongniartii* (Couteaudier, Viaud & Riba, 1996; Cravanzola *et al.*, 1997; Glare *et al.*, 1996). The only restriction occurs when the species are degenerated. In that situation molecular methods will also fail (Gams, personal communication).

In 1972, *B. alba* was reclassified as *Engyodontium album* (Limber) de Hoog. Since that time, the human pathogenic fungus no longer belongs to the genus *Beauveria* but to the new genus *Engyodontium* (Gams, DeHoog & Samson, 1984).

Currently, the genus *Beauveria* contains eight species (*B. amorpha, B. bassian, B. brongniartii, B. caledonica, B. felina, B. epigea, B. velata,* and *B. vermiconia*). Of this genus, *B. bassiana* and *B. brongniartii* are potentially important entomopathogenic fungi.

Metarhizium Sorokin (1879)

One (not necessarily valid) species, *Metarhizium taii* Z. Q. Liang & A.Y. Liu, has been associated with its sexual stage, or teleomorph, *Cordyceps taii* Z.Q. Liang & A.Y. Liu (Liang, Liu & Liu, 1991). According to classical taxonomic criteria based on morphology of phialides and spores, only two species were recognized by Tulloch (1976): *M. flavoviride* W. Gams & Rozsypae and *M. anisopliae* (Metchn.) Sorokin. He divided the latter species into the short-spored 'var. *anisopliae*' and the long-spored 'var. *majus*'. Tulloch also provided access to the original nineteenth century descriptions by Metchnikoff and Sorokin, including a history of the extra 'r' (on occasions, *Metarhizium* has been spelt *Metarrhizium*). After various minor contributions to the taxonomy of this genus, increasing interest in the application of strains as biocontrol agents and increasing utility of molecular methodology led to a major revision of the genus by Driver, Milner and Trueman (2000). Their work is based on analysis of rDNA sequence data, and is generally accepted as definitive. The authors recognize three species: *M. anisopliae, M. flavoviride* and *M. album* Petch. The species *M. anisopliae* is further divided into four varieties: *anisopliae, majus* (both corresponding to Tulloch's original classification), *lepidiotum*, and *acridum* (two new varieties, the first affecting scarabs in Australia, New Zealand, Papua New Guinea and Fiji, and the second found in grasshoppers and locusts worldwide).

Applications and current research approaches

In the last few years, considerable progress has been made in the development of environmentally benign, effective new formulations of fungal biological control agents (BCAs) based on *Beauveria* and *Metarhizium* strains. Eighteen products, based on *B. bassiana, B. brongniartii* and *M. anisopliae*, have been registered in Europe (Table 14.1). Currently, five *Beauveria* and three *Metarhizium* products will be reported to the European Commission in accordance with Article 5 of Commission Regulation (EC) No 1112/2002 (Table 14.2). Most of the products used for subterranean pest control consist of cereal kernels colonized with

Table 14.1 *Registered* Beauveria *spp. and* Metarhizium anisopliae *products (Butt, Jackson & Magan, 2001; Copping, 2004).*

Fungus	Country	Product name	Intended use	Manufacturer
B. bassiana	USA	Naturalis	Cotton pests incl. boll-worms	Troy Biosciences
B. bassiana	Czech Republic	Boverol	Colorado beetle	Czechoslovakia
B. bassiana	France	Ostrinil	European corn borer, Asiatic corn borer	Calliope SAS* (formerly NPP)
B. bassiana	USA	CornGuard	European corn borer	Mycotech*
B. brongniartii	France	Betel	White grubs	Calliope SAS* (formerly NPP)
B. brongniartii	Switzerland	Engerlingspilz	Cockchafers	Andermatt
B. brongniartii	Switzerland	Schweizer	Cockchafers	Erich Schweizer
B. brongniartii	Austria	Melocont	Cockchafers	Agrifutur, Kwizda AGRO
M. anisopliae	USA	BioPath	Cockroaches	Eco-Science
M. anisopliae	USA	BioBlast	Termites	Eco-Science
M. anisopliae	Brazil	Metaquino	Sugar-cane spittlebug	Several
M. anisopliae	Brazil	Metabiol	Pasture spittlebug	Planterra Productos Biotechologicus
M. anisopliae	South Africa	Green Muscle	Locusts	BCP
M. anisopliae	CILSS	Green Muscle	Locusts and grasshoppers	NPP
M. anisopliae	Germany	Bio1020	Vine weevil	Earth BioScience, Inc.* (formerly Taensa, Inc.)
M. anisopliae	Australia	Green Guard	Locusts	SGM
M. anisopliae	Australia	BioGreen	Sugar-cane black grub	BioCare Technology
M. anisopliae	Venezuela	Cobican	Sugar-cane spittlebug	Probioagro

Note:
* Notified by company according to Commission Regulation (EC) No 2229/2004 of 3 December 2004.

mycelium. Standard application systems, such as seed drills, can be used directly or with minor adaptations in different crop systems. The application of fungal kernels or granules into the soil leads to spots with very high concentrations of spores, resulting in a high probability of infection of

Table 14.2 *The following notifiers go for* Beauveria *spp. and* Metarhizium anisopliae *as biocontrol agents (Sanco/10377/2003, 24 June 2003).*

Fungus	Notifier
Beauveria bassiana	Mycotech Europe Ltd (UK)
Beauveria bassiana	Agrichem, SA (Spain)
Beauveria bassiana	Calliope SAS (France)
Beauveria bassiana	Agribiotec srl (Italy)
Beauveria brongniartii	Calliope SAS (France)
Metarhizium anisopliae	Intrachem Bio Italia Spa (Italy)
Metarhizium anisopliae	Earth BioScience, Inc. (formerly Taensa, Inc.) (Germany)
Metarhizium anisopliae	Agrifutur srl (Italy)

soil dwelling larvae. Spores are distributed throughout the soil either by water movement or by invertebrate activity. Ferron (1979) defined a threshold concentration of $>2 \times 10^4$ spores g^{-1} dry weight in soil to ensure epidemic levels in grassland and agricultural systems. This suggested concentration is easily achieved by applying 30 to 50 kg of fungal BCAs twice.

Nowadays, liquid applications of BCAs to protect aerial plant parts are widespread (for example *Bacillus thuringiensis*, Baculoviruses, several fungal formulations). However, very poor vertical movement of the spores in the soil prevented the application of BIO 1020 spore suspension and dry conidia products until 2000 (Zimmermann, 1992).

Since the EU project BIPESCO, *Beauveria* and *Metarhizium* can be sold in two forms – either products based on water-dispersible granules, or a wettable powder product (Seger *et al.*, 2005). Mounted sprayers with special equipment are used for the injection of the liquid formulations of the spore-suspension into the soil. They are easy to handle and inexpensive. These products are now available and give users not only a greater flexibility in choosing application methods, but also allow the implementation of alternative control strategies. This is important because the adapted drilling machines could not be used in certain situations such as steep areas, closely spaced row crops, or in forest systems.

Because of the development of these new formulations and because of the already high acceptance of fungal biocontrol agents by farmers, a new offensive has been initiated in Europe to verify the success of *Beauveria* and *Metarhizium* for the control of subterranean pests. High throughput screening of *Beauveria*, *Metarhizium* and other new fungal isolates and products have been performed, using standardized bioassay methods that are already available (Laengle *et al.*, 2005). In addition, standard protocols have been developed to organize field experiments more efficiently, i.e. sampling, sample preparation, to demonstrate improved efficacy of pathogens developed for subterranean pest control, and to evaluate the narrow host range of the specific isolate.

Beauveria brongniartii control of the European cock chafer

The application of *B. brongniartii* using barley kernels in Austria provided sufficient suppression of *M. melolontha* populations after only two years of application (>20% prevalence of mycosis). Also, populations were reduced from >70 larvae m^{-2} to less than 22 larvae m^{-2} after five years. Since the spring of 1998, no relevant damage by *M. melolontha* has been reported at treated sites (Strasser, 2001). Results of an efficacy study with *B. brongniartii*, commercialized under the name Melocont®-Pilzgerste, for the control of *M. melolontha* in potato fields revealed mortality rates of 38% to 63%. The dose of Melocont®-Pilzgerste calculated to kill 50% *Melolontha* larvae (LD50) was 56 kg ha^{-1} (Laengle *et al.*, 2005). Similar results can be expected with the new formulations, Melocont®-WP (water dispensable powder) and Melocont®-WG (water dispensable granules), which were applied in apple orchards (Trentino, Italy, in 2004). With a single dose application of 1×10^{13} conidia per hectare, sprayed with a mounted sprayer system, a threshold concentration of $>5 \times 10^{3}$ spores/g dry weight in soil at a depth of 10 cm could be achieved. Bioassays of third instar *M. melolontha* larvae showed a high infection by *B. brongniartii* treated with Melocont®-Pilzgerste, Melocont®-WG, and Melocont®-WP (Fig. 14.1; Traugott, Weissteiner & Strasser, 2005).

Metarhizium anisopliae control of the garden chafer

An EPPO-based pilot study at a golf course in Igls-Rinn (Austria) tested the efficacy of *M. anisopliae* products against *Phyllopertha horticola* (Pernfuss *et al.*, 2005; Strasser *et al.*, 2005). Eight weeks post treatment with GRANMET®, *M. anisopliae* BIPESCO 5, caused 17% mortality of larvae, other granular formulations of *M. anisopliae* caused mortality rates of 14% and 16%. Positive long-term effects of an extensive application were observed, as one year after the treatment no damage was found in the treated areas, whereas severe damage caused by *P. horticola* larvae were observed in the surrounding areas. GRANMET® was fully registered in Austria in 2006.

Metarhizium anisopliae control of grape phylloxera

M. anisopliae products have also been tested for their potential abilities to control *Dactulosphaira vitifoliae* (Kirchmair *et al.*, 2004). In pot experiments, soil surrounding phylloxera-infested grapes was inoculated with barley colonized with *M. anisopliae*. After 32 days, in eight of ten *Metarhizium*-applied pots no fresh phylloxera infections could be observed. In contrast, in all untreated pots fresh nodosities with single (two of six plants) or multiple (four of six plants) occupation with phylloxera could be found.

Metarhizium anisopliae control of vine weevil

To control *Otiorhynchus sulcatus*, the *Metarhizium*-based product 'BIO 1020' (Bayer) had been available in Austria up to 2003, but registration was not renewed because of economical considerations. Currently, a new study with BIPESCO 5 (*M. anisopliae* grown on barley kernels) is being conducted in South Tyrol (Italy), which aims to achieve an indication-extension of already existing *Metarhizium* products. In this context, field trials in strawberry and vine cultures are being performed.

Metarhizium anisopliae control of western corn root worm

An EU-funded project was initiated in 1999 (QLRT-1999-01110) to determine the occurrence of indigenous natural enemies of *D. v. virgifera*

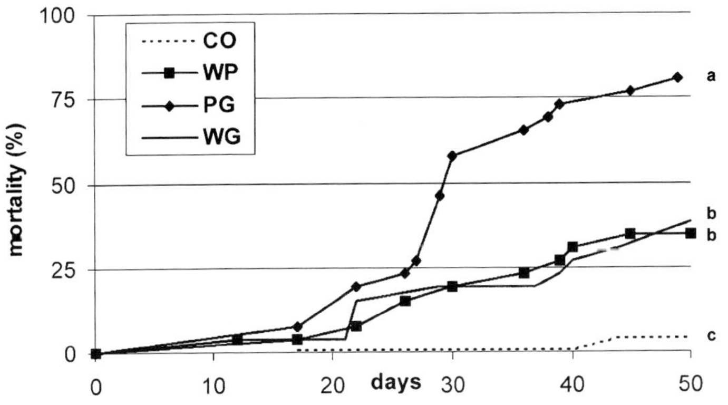

Figure 14.1 *Beauveria brongniartii* infection of third instar *Melolontha melolontha* larvae treated with Melocont®-Pilzgerste (PG), Melocont®-WP (WP), Melocont®-WG (WG) and within the control treatment (CO) during 50 d incubation time (n = 26 larvae per treatment). Letters denote significant differences in infection curves (Logrank test $P < 0.001$; Traugott *et al.*, 2005).

in Europe. Toepfer and Kuhlmann (2004) reported that the fungi *B. bassiana* and *M. anisopliae* attack adults and larvae, but at an extremely low level (<1%). However, no other entomopathogenic fungi, nematodes or parasitoids were found on eggs, larvae, pupae or adults. In spring 2005, research groups in Austria, Croatia and Switzerland launched projects to control *D. v. virgifera* with strains of *B. bassiana* and *M. anisopliae*. The research in Austria (Austrian Research Promotion Agency – FFG 809806) and Switzerland builds on earlier EU-funded projects (RAFBCA – QLK1-CT-2001-01391; BIPESCO – FAIR6 CT-98-4105), which aimed at assessing the risks and improving the efficacy of insect-pathogenic fungi for subterranean pest control.

Conclusion

The potential for fungal biocontrol agents against subterranean pests is enormous, but it is still not exploited to a sufficient extent. The technology of fungal biological control has reached the appropriate levels of maturity to allow demonstrational approaches. Therefore, such demonstrations will bridge the validation gap by transforming laboratory results into tangible technologies. The validation of technologies on a realistic scale should then facilitate the transfer of products to the market.

Acknowledgements

The authors express special thanks to the British Mycological Society for extending the invitation to present this article, and to Marietta Lou (Salem, USA) for kindly reviewing the manuscript. This work was supported by the Austrian Research Promotion Agency – FFG 809806, by Agrifutur srl (Italy) and by Kwizda Agro Gmbh (Austria).

References

Anonymous. (1948). *Wireworms and Food Production. A Wireworm Survey of England and Wales (1939–42)*. London: HMSO.

Blot, Y., Brunel, E. & Courbon, R. (1999). Enquete sue l'infestation de parcelles de blé at de mad's par les larves des taupins des genres Agriotes et Athous dans l'ouest de la France (Coleopteres: Elaterides). *Annales de la Société Entomologique de France*, **35**, 453–7.

Boucias, D. G. & Pendland, J. C. (1998). *Principles of Insect Pathology*. Boston, MA: Kluwer Academic Publishers.

Butt, T. M. (2002). Use of entomogenous fungi for the control of insect pests. In *The Mycota XI. Agricultural Applications*, ed. F. Kempken. Berlin: Springer Verlag, pp. 111–34.

Butt, T. M. & Copping, L. (2000). Fungal biological control agents. *Pesticide Outlook*, **11**, 186–91.

Butt, T. M., Jackson, C. & Magan, N. (2001). *Fungi as Biocontrol Agents: Progress, Problems and Potential.* Wallingford: CABI.

Chandler, L. D., Coppedge, J. R., Edwards, C. R., Tollefson, J. J. & Wilde, G. E. (2000). Corn rootworm area-wide management across the United States. In *Area-Wide Control of Fruit Flies and Other Insect Pests*, ed. K. H. Tan. Penang: Penerbit Universiti Sains Malaysia, pp. 159–67.

Chiang, H. C. (1973). Bionomics of the northern and western corn rootworms. *Annual Review of Entomology*, **18**, 47–72.

Corrie, A. (1998). Phylloxera problems in Australia and possible solutions. *19. Internationale Geisenheimer Rebveredlertagung*, Geisenheim.

Copping, L. G. (2004). *The Manual of Biocontrol Agents – New Edition.* Surrey, British Crop Protection Council.

Couteaudier, Y., Viaud, M. & Riba, G. (1996). Genetic nature, stability, and improved virulence of hybrids from protoplast fusion in *Beauveria. Microbial Ecology*, **32**, 1–10.

Cravanzola, F., Piatti, P., Bridge, P. D. & Ozino, O. I. (1997). Detection of genetic polymorphism by RAPD-PCR in strains of the entomopathogenic fungus *Beauveria brongniartii* isolated from the European cockchafer (Melolontha spp.). *Letters in Applied Microbiology*, **25**, 289–94.

De Hoog, G. S. (1972). The genera *Beauveria, Isaria, Tritirachium* and *Acrodontium* gen. nov. *Studies Mycol*, **1**, 1–41.

Domsch, K. H., Gams, W. & Anderson, T. H. (1980). *Beauveria* Vuill. 1912. In *Compendium of Soil Fungi*, eds. K. H. Domsch, W. Gams & T. H. Anderson. London: Academic Press, pp. 136–40.

Domsch, K. H., Gams, W. & Anderson, T. H. (1993). *Compendium of Soil Fungi.* IHW-Verlag Braunschweig, pp. 136–40.

Driver, F., Milner, R. J. & Trueman, J. H. W. (2000). Taxonomic revision of *Metarhizium* based on the phylogenetic analysis of ribosomal DNA sequence. *Mycological Research*, **104**, 134–50.

Evans, H. C. (1989). Mycopathogens of insects of epigeal and aerial habitats. In *Insect-Fungus Interactions*, eds. N. Wilding, N. M. Collins, P. M. Hammond & J. F. Webber. London, UK: Academic Press, pp. 205–38.

Ferron, P. (1979). Study of the virulence of some mutants of *Beauveria brongniartii* (*Beauveria tenella*) fungi infecting *Melolontha melolontha. Journal of Invertebrate Pathology*, **34**, 71–7.

Ferron, P. (1981). Pest control by the fungi *Beauveria* and *Metarhizium*. In *Microbial Control of Pests and Diseases 1970–1980*, ed. H. D. Burges. London: Academic Press, pp. 465–82.

Ferron, P. (1985). Fungal control. In *Comprehensive Insect Physiology, Biochemistry and Pharmacology*, eds. G. A. Kerkut & L. I. Gilbert. Oxford, UK: Pergamon Press, pp. 313–46.

Ferron, P., Fargues J. & Riba, G. (1991). Fungi as microbial insecticides against pests. In *Handbook of Applied Mycology*, eds. D. K. Arora, L. Ajello & K. G. Mukerji. NY: Marcel Dekker, pp. 665–706.

Furlan, L. (1996). The biology of *Agriotes ustulatus* Schäller (Col., Elateridae). I. Adults and oviposition. *Journal of Applied Entomology*, **120**, 269–74.

Gams, W., De Hoog, G. S. & Samson, R. A. (1984). The hyphomycete genus *Engyodontium* – a link between *Verticillium* and *Aphanocladium. Persoonia*, **12**, 135–47.

Glare, T. R. (1992). Fungal pathogens of Scarabs. In *Use of Pathogens in Scarab Pest Management*, eds. T. R Glare & T. A. Jackson. Andover: Intercept, pp. 63–77.

Glare, T. R. & Milner, R. J. (1991). Ecology of entomopathogenic fungi. In *Handbook of Applied Mycology, Vol. 2: Humans, Animals, and Insects*, eds. D. K. Arora, L. Ajello & L. G. Mukerji. NY: Marcel Dekker, pp. 547–612.

Glare, T., Inwood, A., van der Wielen, P. & Nelson, T. (1996). Identification of novel strains of *Beauveria* and *Metarhizium* in New Zealand. *Proc. 3rd International Workshop on Microbial Control Soil Dwelling Pests*. pp. 39–49.

Grannet, J., Timper, P. & Lider, L. (1991). Deadly insect pest poses increased risk to north coast vineyards. *California Agriculture*, **45**, 30–2.

Gratwick, M. (1989). *Potato Pests*. MAFF Reference Book 187. London: HMSO.

Hajek, A. E. & St. Leger, R. J. (1994). Interaction between fungal pathogens and insect hosts. *Annual Review of Entomology*, **39**, 293–332.

Hemerik, L., Busstra, C. & Mols, P. (2004). Predicting the temperature-dependent natural population expansion of the western corn rootworm, *Diabrotica virgifera*. *Entomologia Experimentalis et Applicata*, **111**, 59–69.

Hill, D. S. (1987). *Agricultural Insect Pests of Temperate Regions and their Control*. Cambridge, UK: Cambridge University Press.

Inglis, G. D., Goettel, M. S., Butt, T. M. & Strasser, H. (2001). Use of hyphomycetous fungi for managing insect pests. In *Fungal Biocontrol Agents: Progress, Problems and Potential*, eds. T. M. Butt, C. Jackson & N. Magan. Wallingford: CABI, pp. 1–69.

Jansson, R. K. & Seal, D. R. (1994). Biology and management of wireworm on potato. *Proceedings of the International Conference on 'Advances in Potato Pest Biology and Management'*. Jackson Hole, Wyoming.

Jung, C. (1998). Aktuelle Reblausprobleme im Rheingau. *Geisenheimer Berichte*, **40**, 12.

Kahler, A. L., Olness, A. E., Sutter, G. R., Dybing, C. D. & Devine, O. J. (1985). Root damage by western corn rootworm and nutrient content in maize. *Agronomy Journal*, **77**, 769–73.

Keller, S. (1986). Biologische Maikäferbekämpfung. In *Neuere Erkenntnisse über den Maikäfer*, eds. R. Büchi, E. Keller, S. Keller, W. Meier, A. Staub & Th. Wildbolz. Mitteilung der Thurgauischen Naturforschenden Gesellschaft. Frauenfeld, pp. 84–91.

Keller, S. & Schweizer, Ch. (2001). Ist das Drahtwurm-Problem ein Pilz-Problem? *Agrarforschung*, **8**, 248–51.

Keller S. & Zimmermann, G. (2005). Scarabs and other soil pests in Europe: Situation, perspectives and control strategies. *IOBC/wprs Bulletins*, **28**, (2): 9–12.

Kirchmair, M., Huber, L., Porten, M., Rainer, J. & Strasser, H. (2004). *Metarhizium anisopliae*, a potential agent for the control of grape phylloxera. *BioControl*, **49**, 295–303.

Krysan, J. L. & Miller, T. A. (1986). *Methods for the Study of Pest Diabrotica*. New York: Springer Verlag.

Laengle, T., Pernfuss, B., Seger, C. & Strasser, H. (2005). Field efficacy evaluation of *Beauveria brongniartii* (Sacc.) Petch against *Melolontha melolontha* L. in potato cultures. *Sydowia*, **57**, 54–93.

Leclant, F. & Coer D'Acier, A. (1995). Doit-on à nouveau craindre le phylloxéra? *Phytoma*, **477**, 20–5.

Liang, Z. Q., Liu, A. Y. & Liu, J. L. (1991). A new species of the genus *Cordyceps* and its *Metarhizium* anamorph. *Acta Mycologia Sin*, **10**, 257–62.

Masaki, M., Ohmura, K. & Ichinoe, F. (1984). Host range studies of the black vine weevil, *Otiorhynchus sulcatus* (Fabricius) (Coleoptera: Curculionidae). *Applied Entomology and Zoology*, **19**, 95–106.

McCoy, C. W, Samson, R. A. & Boucias, D. G. (1988). Entomogenous fungi. In *Handbook of Natural Pesticides, Vol. 5: Microbial Insecticides, Part A: Entomogenous Protozoa and Fungi*, eds. C. Ignoffo & N. B. Mandava. Boca Raton, FL: CRC Press, pp. 151–234.

McLeod, M. J. & Williams, R. N. Grape Phylloxera. HYG-2600-94. http://ohioline.osu.edu/hyg-fact/2000/2600.html 2005-10-01.

Milne, A. (1957). Biology and ecology of the garden chafer *Phyllopertha horticola* (L.). II – The cycle from egg to adult in the field. ed. Hall, W. J. *Bulletin of Entomological Research*, **56**, 23–42.

Milner, R. J. (2000). Current status of *Metarhizium* as a mycoinsecticide in Australia. *Biocont. News Information*, **21**: 47N–50N.

Moorhouse, E., Charnley, A. & Gillespie, A. (1992). A review of the biology and control of the black vine weevil, *Otiorhynchus sulcatus* (Coleoptera: Curculionidae). *Annals of Applied Biology*, **121**, 431–54.

Parker, W. E. & Howard, J. J. (2001). The biology and management of wireworms (*Agriotes* spp.) on potato with particular reference to the UK. *Agricultural and Forest Entomology*, **3**, 85–98.

Pernfuss, B., Zelger, R., Kron-Morelli, R. & Strasser, H. (2005). Control of the garden chafer *Phyllopertha horticola* with GranMet-P, a new product made of *Metarhizium anisopliae*. *IOBC/wprs Bulletin*, **28**, 47–50.

Pickett, J. A., Butt, T. M., Doughty, K. J., Wallsgrove, R. M. & Williams I. H. (1995). Minimising pesticide input in oilseed rape by exploiting natural regulatory processes. *Proceedings of the Ninth International Rapeseed Congress, Rapeseed Today and Tomorrow*. pp. 565–71.

Pill, B. A., Keaster, A. J. & Chippendale, G. M. (1976). Larval survival and pupation of the wireworm *Melanotus depressus* and *Limonius dubitans* in natural substrates. *Environmental Entomology*, **5**, 845–8.

Porten, M. & Huber, L. (2003). An assessment method for the quantification of *Daktulosphaira vitifoliae* (Fitch) (Hem., Phylloxeridae) populations in the field. *Journal of Applied Entomology*, **127**, 157–62.

Presser, C., Schmid, J. & Rühl, E. H. (1993). Die Reblaus – kein Problem mehr? *Das Deutsche Weinmagazin*, **23**, 22–5.

Redl, H. (1999). Die Reblaus auch in Österreich wieder (beängstigend?) im Kommen. *Der Pflanzenarzt*, **52**, 8–12.

Remund, U. & Boller, E. (1994). Die Reblaus – wieder aktuell? *Schweizerische Zeitschrift für Obst- und Weinbau*, **130**, 242–4.

Roberts, A. W. R. (1919). On the life history of 'wireworms' of the genus *Agriotes*, Esch., with some notes on that of *Athous haemorrhoidalis*, F. Part I. *Annals of Applied Biology*, **6**, 116–35.

Roberts, A. W. R. (1921). On the life history of 'wireworms' of the genus *Agriotes*, Esch., with some notes on that of *Athous haemorrhoidalis*, F. Part II. *Annals of Applied Biology*, **8**, 193–215.

Roberts, A. W. R. (1922). On the life history of 'wireworms' of the genus *Agriotes*, Esch., with some notes on that of *Athous haemorrhoidalis*, F. Part III. *Annals of Applied Biology*, **9**, 306–24.

Roberts, D. W. & Hajek, A. E. (1992). Entomopathogenic fungi as bioinsecticides. In *Frontiers in Industrial Mycology*, ed. G. F. Leatham. NY: Chapman and Hall, pp. 144–59.

Schread, J. C. (1972). The black vine weevil. New Haven, Connecticut, Connecticut Agricultural Experimental Station.

Seifert, K. A. & Gams, W. (2001). The taxonomy of anamorphic fungi. In *The Mycota: Systematics and Evolution VII Part A*, eds. D. J. McLaughlin, E. G. McLaughlin, & P. A. Lemke. Berlin, Heidelberg: Springer, pp. 307–47.

Seger, C., Sturm, S., Längle, T., Wimmer, W., Stuppner, H. & Strasser, H. (2005). Development of a sensitive HPLC-DAD assay for the detection and quantification of the fungal metabolite oosporein from fungal culture broth and biological pest control formulations. *Journal of Agricultural and Food Chemistry*, **53**, 1364–9.

Shimazu, M., Mitsuhashi, W. & Hashimoto, H. (1988). Cordyceps brongniartii sp. nov. the teleomorph of *Beauveria brongniartii*. *Transactions of the Mycological Society of Japan*, **29**, 323–30.

Smith, F. (1932). Biology and control of the black vine weevil. *USDA Technical Bulletin*, **325**.

Strasser, H. (2000). Searching for alternative biological pest control agents against *Phyllopertha horticola* (L.). Smith, P. H. Insect Pathogens and Insect Parasitic Nematodes. *IOBC/wprs Bulletin*, **23**, 23–7.

Strasser, H. (2001). Was tun gegen den Maikäfer? Maßnahmen zur Maikäferbekämpfung im Ackerbau und Grünland. *Der Pflanzenarzt*, **3**, 10–13.

Strasser, H. (2004). Assessment of scarab infestation in Austria. *Laimburg Journal*, **1**, 171–5.

Strasser, H., Zelger, R., Pernfuss, B., Laengle, T. & Seger, C. (2005). EPPO based efficacy study to control *Phyllopertha horticola* in golf courses. *IOBC/wprs Bulletin*, **28**, 189–92.

Tanada, Y. & Kaya, H. K. (1993). *Insect Pathology*. London, UK: Academic Press.

Toepfer, S. & Kuhlmann, U. (2004). Survey for natural enemies of the invasive alien chrysomelid, *Diabrotica virgifera virgifera*, in Central Europe. *Biocontrol*, **49**, 385–95.

Traugott, M., Weissteiner, S. & Strasser, H. (2005). Effects of the entomopathogenic fungus *Beauveria brongniartii* on the non-target predator *Poecilus versicolor* (Coleoptera: Carabidae). *Biological Contol*, **33**, 107–12.

Tulloch, M. (1976). The genus *Metarhizium*. *Transactions of the British Mycological Society*, **66**, 407–11.

Warner, R. E. & Negley, F. B. (1976). The genus *Otiorhynchus* in America north of Mexico (Coleoptera: Curculionidae). *Proceedings of the Entomological Society of Washington*, **78**, 240–62.

Wraight, S. P. & Carruthers, R. I. (1999). Production, delivery and use of mycoinsecticides for control of insect pests of field crops. In *Methods in Biotechnology, Vol. 5: Biopesticides: Use and Delivery*, eds. F. R. Hall & J. J. Menn. Totowa, NJ: Humana Press, pp. 233–69.

Zimmermann, G. (1992). Laborversuche zum Versickerungsverhalten des insektenpathogenen Pilzes *Metarhizium ansiopliae* (Metsch.) Sorok. in Standardböden und Sand. *Nachrichtenblatt Deutscher Pflanzenschutzdienst*, **44**, 19–23.

Zimmermann, G. (1993). The entomopathogenic fungus *Metarhizium anisopliae* and its potential as a biocontrol agent. *Pesticide Science*, **37**, 375–9.

15

Development of mycoherbicides and evaluation of potential risks (using *Stagonospora* as a model)

M. MAURHOFER, D. BOSS, M.O. AHONSI
AND G. DÉFAGO

Several weeds such as the bindweeds *Convolvulus arvensis* and *Calystegia sepium*, which belong to the economically most important weeds, are difficult to control with conventional physical and chemical methods. Biological control with fungal biocontrol agents (FBCA) offers a promising alternative. However, the employment of such FBCAs implies not only the identification of highly effective strains and the development of cheap formulation and production techniques, but also an extensive risk assessment including tracking of FBCAs and their toxic metabolites in the environment. *Stagonospora convolvuli* strain LA39 effectively controls field and hedge bindweed in the field. For this strain formulation and application techniques were developed and currently markers for environmental tracking of the fungus are being designed. A major metabolite produced by LA39 is elsinochrome A (ELA). To assess the risk of ELA production by LA39, different methods, such as toxicity tests, stability tests and methods for environmental monitoring, were developed. Our results revealed that ELA is toxic to a wide array of organisms. However, ELA was never detected in LA39 treated bindweed or crop plants, and the ELA content in the applied biocontrol product is far too low to result in a toxic concentration in the environment. Furthermore, ELA is quickly degraded by sunlight. In conclusion, LA39 is a promising candidate for the development of a successful biocontrol product against bindweed. Harmful side effects of ELA production can be excluded.

Introduction

Field bindweed (*Convolvulus arvensis* [L.] R. Br.) and hedge bindweed (*Calystegia sepium* L.) are noxious perennial weeds in Europe and many agricultural areas all over the world (Weaver & Riley, 1982; Maillet,

Exploitation of Fungi, eds. G. D. Robson, P. van West and G. M. Gadd. Published by Cambridge University Press.

1988). Bindweeds cause serious problems in grapes, raspberries, beans, maize, wheat and also in amenity areas such as gardens, parks and cemeteries. Field bindweed can reduce crop yield to 50–60% when infestations are dense (Callihan *et al.*, 1990) and was ranked the twelfth worst weed in the world (Holm *et al.*, 1977). Bindweeds reduce crop value through competition for nutrients and water, and by interference with harvest procedures. They can reduce the available soil moisture in the top 60 cm of soil to below the wilting point for many plant species (Weaver & Riley, 1982). Bindweeds are deep-rooted and display a vigorous vegetative reproduction capacity. These properties enable them to escape most mechanical and chemical control methods.

Although some herbicides like 2,4-D and glyphosate are effective against bindweed, they do not provide sufficient control, especially if a bindweed population is already well established. In addition, even repeated defoliation does not destroy bindweed population because of abundant reserves stored in the rhizomes and roots of bindweeds. Chemical control is also restricted by the fact that chemicals effective against bindweeds often affect many other plant species. Furthermore, there is a growing public concern regarding environmental problems associated with the use of chemical pesticides. For these reasons, biological control appears to be a promising alternative against bindweeds.

In recent years, efforts to find suitable biocontrol organisms have been undertaken by several research groups. Insect and gall mites have been considered (Boldt & Sobhian, 1993; Rosenthal & Carter, 1977; Rosenthal & Platts, 1990) as well as several fungal bindweed pathogens such as *Phoma proboscis*, *Phomopsis convolvulus* and *Stagonospora convolvuli* (Pfirter *et al.*, 1997). Among the fungi, especially isolates of *S. convolvuli* have evolved as highly promising candidates for the biocontrol of hedge and field bindweed. Here we use *S. convolvuli* strain LA39 as a model to evaluate not only the potential of such a mycoherbicide but also possible risks connected with its use. In the first part of this chapter, we describe the development, use and efficacy of an LA39-based mycoherbicide. In the second part, we address the role of toxin production in *Stagonospora* biocontrol agents, and in the last part we outline a strategy for the evaluation of potential risks of a fungal biocontrol agent (FBCA) to the environment and the consumer of treated crops.

Use and efficacy of *Stagonospora convolvuli* LA39 as a biocontrol agent against field and hedge bindweed

Stagonospora convolvuli LA39 was selected as a potential biocontrol agent among 600 fungal isolates. This strain was isolated from infected field bindweed leaves collected in Long Ashton (UK) in 1994. Its host range is restricted strictly to *Convolvulaceae* (Pfirter & Défago, 1998). Within the *Convolvulaceae* strain LA39 is highly aggressive to field and hedge bindweed, causes far less severe symptoms on other plant species belonging to this family and is not at all pathogenic to sweet potato (Pfirter & Défago, 1998). The possibility of LA39 application causing damage to non-target plants can therefore be excluded. Applied to field bindweed, strain LA39 causes severe leaf necrosis resulting in extensive defoliation (Pfirter & Défago, 1998). With LA39 infected hedge bindweeds show leaf necrosis and partial defoliation. Although the total biomass is not reduced, LA39 infection leads to lower carbohydrate reserves in the rhizomes and thereby influences winter survival and subsequent vegetative propagation (Défago *et al.*, 2001). Thus, strain LA39 might cause depletion in starch reserves in the long term leading to an efficient control of bindweeds.

Mass production, formulation and storage

A prerequisite for the development of a successful mycoherbicide is to optimize mass production and to develop a formulation that facilitates application, infection process and storage of the biocontrol agent. The mass production of LA39 spores is complicated by the fact that sporulation is induced by light and does not occur in liquid medium. Testing many different solid substrates revealed that optimal conidia production ($5 \times 10^8 – 1 \times 10^9$ conidia/g) occurs on couscous (cracked hard wheat, *Triticum durum* Desf; Pfirter *et al.*, 1999).

LA39 is applied as a conidia suspension using standard spraying techniques (for example flat fan nozzles operating at 2 bars) at 300–600 l/ha, which corresponds to $1.5–3 \times 10^{12}$ conidia/ha (Guntli *et al.*, 1998; Guntli *et al.*, 1999; Pfirter *et al.*, 1997). Pfirter *et al.* (1998) reported that the use of a 10% vegetable oil emulsion significantly increased the biocontrol capacity of LA39. The addition of oil to the spore suspension reduces the water/humidity requirement during the infection process and improves the distribution of the spores on the leaves. Further formulation experiments revealed, however, that the addition of 10% rape seed oil can be replaced by the surfactants Greemax (0.01%, Stallen Bio Agro AG, Basel, Switzerland) or Genol (0.3%, Syngenta, Switzerland) without changing the effectiveness of the biocontrol agent. For all the mycoherbicide experiments presented here, conidia were applied at a concentration of 5×10^6/ml in a 0.01% Greemax/water emulsion. For field application, 2×10^{12} conidia/ha were applied.

In addition to optimized mass production and application techniques, an acceptable shelf life is another requirement for a good mycoherbicide.

S. convolvuli LA39 spores keep their viability and pathogenicity for over six months when stored on kaolin (1 g/10^9 conidia) at 3°C (Pfirter *et al.*, 1999).

Efficacy in the field

Between 1995 and 2004, *S. convolvuli* strain LA39 has been tested in several field experiments and its capacity for effective bindweed control has been validated. In 1995, application of LA39 to an experimental field plot planted with maize stopped the increase of ground coverage by bindweeds (Pfirter *et al.*, 1997). In 1996 and 1997, LA39 was applied to maize fields either as sole treatment or in combination with underseed of *Trifolium pratense* as a living green cover. LA39 application reduced the ground coverage with *C. arvensis* by 70–83% (Guntli *et al.*, 1999). The underseed of *T. pratense* did not affect bindweed control by strain LA39. LA39 not only successfully controls bindweeds in cereal production systems. In 1997, LA39 was tested in a cemetery where a ground cover of cotoneaster was heavily infested with field bindweed. Treatment of the infested areas with the fungus reduced the bindweed density to half of that of untreated control areas (Guntli *et al.*, 1998). In 2002, LA39 was applied to a meadow-like area in a cemetery with natural *C. arvensis* infection. In untreated control plots, the ground was weed covered by 17.4% by mid July, whereas bindweed coverage in LA39 treated plots was only 4.5% (Table 15.1). In 2004, the biocontrol fungus was used on amenity areas at two different sites. Applied to a meadow-like area heavily infested with *C. arvensis*, LA39 reduced the bindweed covered area by approximately 50% compared to the untreated control plot (Table 15.1). On a flower border with a severe *Calystegia* problem, the ground coverage by bindweed increased to more than 70% by mid July, whereas treatment with LA39 reduced weed coverage to 9% (Table 15.1).

In summary, *S. convolvuli* LA39 has the capacity, even at heavily bindweed infected sites, not to completely eradicate, but to reduce the noxious weed to an acceptable level. However, at sites or under conditions where bindweed occurs at an extremely high density as in 2004, between four and six mycoherbicide applications may be necessary, whereas under normal conditions two applications would be sufficient.

Production of toxins: a prerequisite for efficient biocontrol?

Biologically active substances, which are involved in the expression of disease symptoms and tissue colonization, have been isolated from most phytopathogenic fungi (Daub & Ehrenshaft, 1993, 2000). Like many

Table 15.1 *Reduction of bindweed coverage by application of* S. convolvuli *LA39 at different experimental sites.*

			Bindweed ground coverage (%) evaluation date		
Experimental site/ year	Natural infection with	LA39 treatment[d]	Mid June	Mid July	Mid August
Witikon cemetery, 2002[a]	Field bindweed	–	9.1	17.4	6.4
		+	16.5[e]	4.5[e]	2.0[e]
Eschikon meadow, 2004[b]	Field bindweed	–	47.1	31.6	37.7
		+	23.8[e]	12.7[e]	18.8
Witikon flower border, 2004[c]	Hedge bindweed	–	86.6	74.1	57.5
		+	41.7[e]	9.1[e]	15.7[e]

Notes:
[a] Meadow-like amenity area, mycoherbicide treatments: two between 10 June and 15 July.
[b] Meadow-like amenity area, mycoherbicide treatments: five between 10 May and 10 July.
[c] Amenity area covered with ornamental plants, mycoherbicide treatments: five between 10 May and 10 July.
[d] LA39 conidia were suspended in water containing 0.01% Greemax and were applied at a concentration of 5×10^9/l, 0.04 l/m^2.
[e] = significantly different from the untreated control at $P \leq 0.05$ (Fisher's protected LSD test).

necrotrophic fungi, *S. convolvuli* LA39 produces phytotoxic compounds in vitro, notably leptosphaerodione and elsinochrome A (ELA), as has been demonstrated by Nicolet & Tabacchi (1999). These authors also identified the toxin cercosporin in another isolate of *S. convolvuli*. Cercosporin and elsinochrome A are closely related photosensitizing perylenequinone toxins produced typically by species of *Cercospora* and *Elsinöe*, respectively (Chen *et al.*, 1966; Daub, 1982; Weiss *et al.*, 1987; Daub & Ehrenshaft, 2000). These compounds are chemical substances, which become energetically activated upon light absorption. They react with molecular oxygen to form both radical and non-radical species of activated oxygen, which have near-universal toxicity (Daub & Ehrenshaft, 1993, 2000). Cercosporin is the best described photosensitizing perylenequinone, and its toxicity to a wide array of organisms such as plants, fungi, bacteria, mice and human cells has been documented (Batchavora *et al.*, 1992; Daub, 1982, 1987; Daub & Ehrenshaft, 2000; Yamazaki *et al.*, 1975). Little is known about the effects of ELA on different organisms.

Ma *et al.* (2003) found that light-activated ELA caused apoptosis in human colorectal carcinoma cells.

About the toxicity and mode of action of leptosphaerodione, practically nothing is known. It was first isolated from the marine ascomycete, *Leptosphaeria oraemaris*, and described by Guerriero *et al.* (1991). Later, it was identified in culture filtrates of *Stagonospora convolvuli* LA39 and was found to be phytotoxic to field bindweed (*Convolvulus arvensis*) and tomato plants (*Lycopersicon esculentum*) by Nicolet and Tabacchi (1999).

Before a *Stagonospora* based mycoherbicide can be produced commercially, detailed information on the production of such substances by the biocontrol fungus and the potential risk its usage poses to the environment and to the consumer of treated crops are necessary. Hence, it is important to know if the production of these metabolites is widespread among *Stagonospora* pathogens of field and hedge bindweeds, and if there is a correlation between toxin production and aggressiveness. In order to minimize the risks connected with the application of fungal biocontrol agents it would of course be most convenient to select highly aggressive strains, which do not produce compounds with universal toxicity. Ahonsi *et al.* (2005) characterized 28 isolates of *Stagonospora* isolated from hedge or field bindweed for the quantitative production of ELA, cercosporin and leptosphaerodione as well as for their potential to cause disease on both bindweed species. In Table 15.2, a summary of these results is presented. Among the 28 tested isolates, 19 produce ELA, 12 leptosphaerodione and only 4 cercosporin. All the 13 isolates, which were classified as highly aggressive on both bindweed species, thus potential mycoherbicides, produce ELA, most of them but not all also leptosphaerodione and none of them cercosporin. These results suggest that the production of ELA is correlated with efficacy in biocontrol.

In order to evaluate the effect of ELA on bindweed leaves, ELA solutions were rubbed onto bindweed leaves dusted with carborundum. A concentration of 30 μM ELA was necessary to cause first visible necrosis on both bindweed species, and at a concentration of 300 μM about 60% of the leaf area was necrotic (Table 15.3). Interestingly, grapevine leaves were much less susceptible to ELA; leaves treated with 300 μM expressed a necrotic area of only 19%.

These results suggest that ELA might thus play a role in the infection process and disease development by generating active oxygen species, which destroy the membranes of host plants, as suggested for cercosporin by Daub & Ehrenshaft (2000). Yet, spraying a combination of purified ELA and LA39 conidia did not result in increased disease severity (data

Table 15.2 *Characterization of* Stagonospora *spp. isolates for the production of metabolites and for aggressiveness to* Calystegia sepium *and* Convolvulus arvensis.

	Metabolite production			
Isolate	Elsinochrome A	Leptosphaerodione	Cercosporin	Aggressiveness[a]
182 Caa	−	−	+	−
LA37	−	−	−	−
172 Caa	−	−	−	+/−
178 Caa	−	−	+	+/−
214 Caa	−	−	+	+/−
LA36	−	−	−	+/−
A7	+	+	−	+
174Coa	−	−	−	+
175 Caa	−	−	+	+
213 Coa	+	−	−	+
92 Coa	+	−	−	+
D6	−	−	−	+
LA10B	+	−	−	+
LA30B	+	−	−	+
B7	+	+	−	+
LA10A	[illegible]	−	−	++
C5	+	−	−	++
LA311	+	+	−	++
LA39	+	+	−	++
177 Cab	+	−	−	++
B4	+	+	−	++
LA24	+	+	−	++
B8	+	+	−	++
LA35	+	+	−	++
LA31	+	+	−	++
LA51	+	+	−	++
LA53	+	+	−	++
LA57	+	+	−	++

Notes:
[a] Aggressiveness to field and hedge bindweed: − = not infectious, +/− = weakly aggressive (maximal 30% leaf necrosis), + = moderately aggressive (= maximal 60% leaf necrosis); ++ = highly aggressive (more than 75% necrosis on one bindweed species, or more than 60% necrosis on both bindweed species).

not shown). In summary, the role of elsinochrome A in the pathogenesis of *S. convolvuli* still remains unclear. However, ELA production seems to be connected with high aggressiveness and thus it seems unlikely that a screening for efficient biocontrol strains, which do not produce ELA would be successful.

Table 15.3 *Toxicity of elsinochrome A to bindweeds and vines.*

	Necrotic or chlorotic leaf area (%)			
Plant species	0 μM	3 μM	30 μM	300 μM
Field bindweed	0.9 ± 0.3	1.3 ± 0.6	19.1 ± 39.9	62.1 ± 8.9
Hedge bindweed	0.9 ± 0.2	0.5 ± 0.2	35.7 ± 10.1	57.8 ± 13.5
Grapevine	0.2 ± 0.1	1.2 ± 0.7	4.6 ± 4.8	18.8 ± 9.1

Notes:
40 μl of differently concentrated elsinochrome A solutions were rubbed onto carborundum-dusted leaves. After seven days, leaves were assessed for necrosis and chlorosis.
The means and standard deviations of three individual experiments with six plants per concentration are presented.

Risk assessment of fungal biocontrol agents using *Stagonospora convolvuli* LA39 as a model

A prerequisite for the commercialization of a mycoherbicide is the assessment of risks connected with the application of a certain biocontrol organism in the field. When doing such a risk assessment the following major questions have to be addressed.

(1) The toxicity of the biocontrol product itself. Does the FBCA product contain any known toxic fungal metabolites, which could harm people handling the product or the environment?
(2) The production of harmful metabolites in the field. Is it possible that known fungal metabolites, which are produced by the BCA while infecting its target, enter the food chain or pose a risk to the environment? How stable are such metabolites in the environment?
(3) Dissemination and persistence. How long can the biocontrol agent survive in the environment and how fast or how far can it spread?

Here we present the results we have obtained having addressed the first and the second question. In order to answer the third question, about the survival and spread of LA39 after field application, we are currently developing molecular markers, which will enable us to monitor strain LA39 in the field.

The production of ELA seems to be a prerequisite for *Stagonospora* isolates for the successful control of bindweeds (see above). Therefore, ELA and strain LA39 were used as a model to develop a strategy for the assessment of potential risks associated with the use of a mycoherbicide producing a certain metabolite. In a first step, the toxicity of ELA to a broad spectrum of organisms was investigated. After that, ELA was quantified in the applied mycoherbicide product and the maximal possible environmental contamination was estimated. Furthermore, methods to detect ELA on the target of the mycoherbicide, that is the bindweeds, as well as on crop plants were developed, and it was investigated whether ELA can enter the food chain and pose a risk to the consumer. In the last step, the stability and degradation of ELA have been studied in order to estimate its persistence in the environment.

The data resulting from these experiments allow the estimation of the risks of ELA production by biocontrol strain LA39 for the environment and the consumer.

Toxicity of elsinochrome A

In order to gain knowledge about the toxicity of ELA, the effect of this compound on the viability of protozoa, on the growth of different species of fungi, oomycetes and bacteria, as well as on the germination of seeds of different plants, was investigated. Cells of the protozoan *Tetrahymena pyriformis* were suspended in water containing different amounts of ELA. After four hours of exposure the mortality of the protozoa was evaluated by estimating the percentage of dead cells. The fungi *Botrytis cinerea* and *Trichoderma atroviride*, as well as the oomycete *Pythium ultimum*, were grown on malt agar plates amended with different amounts of ELA. Fungal growth was evaluated after 1–4 days by measuring the radial mycelial growth. The bacteria *Pseudomonas syringae*, *Escherichia coli* and *Pseudomonas aeruginosa* were grown in Luria Bertani broth containing different ELA concentrations. After 24 h bacterial growth was evaluated by measuring the optical density at 600 nm. In order to investigate the effect of ELA on the germination of plant seeds, seeds of *C. arvensis*, *Lycopersicum esculentum* (tomato), *Lepidium sativum* (cress) and *Triticum aestivum* (wheat) were germinated on water agar with different concentrations of ELA. After 2–3 days the length of the roots was measured.

All the toxicity tests were performed under white light in order to activate ELA. For all the tested organisms the minimal effective ELA concentration (MEC) was determined, and the EC_{50} (concentration which causes 50% of growth inhibition or mortality) was calculated. The results are presented in Table 15.4.

ELA seems to be toxic to a broad spectrum of organisms, since it inhibited the growth of all tested species and caused cell death in the case of *T. pyriformis*. As most sensitive species emerged the fungi *B. cinerea* and

Table 15.4 *Toxicity of elsinochrome A to different bacterial, fungal, plant and protozoan species.*

Organism	MEC (μM)[a]	EC_{50} (μM)[b]
Tetrahymena pyriformis[c]	0.30[g]	0.65[g]
Pseudomonas syringae[d]	0.07[g]	0.14[g]
Escherichia coli[d]	25.0	45.40
Pseudomonas aeruginosa[d]	25.0	25.80
Trichoderma atroviride[e]	0.01	0.10
Botrytis cinerea[e]	0.03	2.90
Pythium ultimum[e]	0.17	11.40
Convolvulus arvensis[f]	3.0	3.90
Lycopersicum esculentum[f]	30.0	>30.0
Lepidium sativum[f]	0.03	2.78
Triticum aestivum[f]	3.0	>30.0

Notes:
[a] MEC = minimal effective concentration = lowest concentration at which toxic effects were observed.
[b] EC_{50} = concentration which caused 50% cell death (*T. pyriformis*), 50% growth inhibition (bacteria and fungi) or 50% inhibition of seedling root elongation (plants). EC_{50} values were calculated with PriProbit 1.63.
[c] *T. pyriformis* cells were suspended in water and exposed to different ELA concentration. After 4 h incubation with light the percentage of dead cells was counted.
[d] Bacteria were grown in Luria Bertani broth containing different amounts of ELA at 27°C with light. After 24 h bacterial growth was evaluated by measuring the optical density at 600 nm.
[e] Fungi were grown on malt agar amended with different amounts of ELA at 20°C with light. After 1–4 days the radial mycelial growth was measured.
[f] Plant seeds were germinated on water agar amended with different amounts of ELA at 20°C and light. After 2–3 days the length of the roots was measured.
[g] Skrobek *et al.*, 2006.

T. atroviride, the bacterium *P. syringae* and cress seedlings (Table 15.4). They were already affected by ELA at concentrations as low as 0.01–0.07 μM. The protozoan *T. pyriformis*, the oomycete *P. ultimum* and the bindweed *C. arvensis* were less sensitive to ELA, with EC_{50} values ranging from 0.7–11.4 μM. The least sensitive organisms were the bacteria *E. coli* and *P. aeruginosa* as well as tomato and wheat plants, which were only severely affected at ELA concentrations higher than 25 μM (Table 15.4). These results show that ELA displays a general toxicity and is not specifically toxic to a certain group of organisms, because within all tested organism groups (bacteria, fungi and plants) there were highly and less sensitive species.

Since ELA is known to form radical and non-radical species of activated oxygen upon reaction with light, it is not surprising that this compound seems to be of near universal toxicity. The assessment of the risks of using an ELA producing biocontrol agent thus becomes the more essential.

Evaluation of the risk of a known fungal metabolite to enter the food chain or contaminate the environment

In order to calculate the risks of ELA entering the food chain or contaminating the environment at first it was tested, whether ELA was present in the mycoherbicide product. The LA39 product consists of conidia produced on couscous, which are suspended in water amended with a surfactant (see above). Eleven samples of conidia, harvested from 300 g of couscous each, were extracted with ethyl acetate and the extracts were analyzed by HPLC. The detection limit of this method was 0.57 μmol ELA per kg couscous culture. The mean ELA concentration detected in the samples was 1.19 μmol per kg couscous taking an extraction efficiency of 65% into account. Out of 1 g couscous culture, conidia for 200 ml product can be harvested. In the final applied product, the average ELA concentration would thus be 5.95 nM. In the environment, however, the ELA concentration would be much lower because of immediate dilution. Therefore, we conclude that ELA concentrations in the environment would be far too low to pose any risk to non-target organisms, since the EC_{50} value for the most sensitive organism (*T. atroviride*) was 0.1 μM.

For the consumer the risks would also be negligible. Even if fruit cultures, for example strawberries, were treated with LA39 product containing 5.95 nmol ELA/l immediately before harvest the estimated maximal content in the fruits would be 0.01 nmol per kg fruit. Taking into account that normally the last mycoherbicide treatment is applied several weeks before harvest and that ELA is rapidly degraded by sunlight (see below), any risk for the consumer of treated strawberries can be excluded.

The presence of ELA in the mycoherbicide product is one question to be addressed. The second is the production of ELA by the BCA after its application. In order to find out whether ELA is produced in the environment and to estimate the probability of ELA entering the food chain, methods were developed to detect ELA in the target bindweed plants as well as in strawberry fruits. Bindweed leaves and strawberry fruits were ground with liquid nitrogen, were extracted with ethyl acetate and the extracts were analyzed using HPLC. The ELA detection limit was 1.7 μmol per kg (fresh weight) for field bindweed, 3.3 μmol per kg (fresh weight) for hedge bindweed and 1.3 μmol per kg (fresh weight) for strawberry fruits.

In order to test whether ELA is produced on the target plants, field and hedge bindweed plants were treated with normally concentrated or tenfold concentrated LA39 product. The experiments were performed in the greenhouse as well as in the field. At the time when the leaf infection just became visible as well as later when the disease was fully developed, bindweed leaves were collected. Samples of 150 leaves were extracted and analyzed by HPLC. In none of the treatments was any ELA detected. Therefore, if the compound was produced in planta at all, the concentrations would be below the detection limits (1.7–3.3 μmol/kg). Concentrations close to the detection limits would thus be the worst-case scenario. In the environment, actual concentrations, for example in the soil or in ground water, would be much smaller due to high dilution factors and thus be far too low to pose any risk to non-target organisms.

For the consumer the question is, could ELA-contaminated bindweed get into the crop. In the case of berries, this is hardly possible regarding the harvesting techniques; in the case of cereals, this could be possible. However, the bindweed content in harvested cereals can be estimated to be below 100 ppm. Thus, the maximal ELA content in the crop, if indeed there was ELA production in bindweed close to the detection limit, would be below 0.33 nmol per kg and, consequently, far too low to pose any risk to the consumer.

The last question, which had to be answered, was the possibility of ELA being produced in the crop directly. Strain LA39 is strongly specific and only infects bindweeds (Pfirter & Défago, 1998). Therefore, the probability of ELA production on a crop plant is minimal. Nevertheless, strawberry plants were treated with LA39 at different growth stages from flowering up to the time of harvest. Plants and fruits were assessed for LA39 infection and fruits were extracted and checked for the presence of ELA. LA39 was not able to infect either the green parts of the plants or the fruits. The fungus was not even able to grow on fully ripe berries. As expected, ELA was never detected in extracts of strawberries.

We therefore conclude that the risk of ELA being present in animal or human food is negligible.

Stability of elsinochrome A in the environment

With the aim of studying the environmental stability of ELA, in a first set of experiments the compound was exposed to different temperatures, ranging from 3°C to 27°C in the dark, either as pure substance or dissolved in water. In both cases ELA was stable at any temperature tested and degradation could not be observed after even three months (our

Table 15.5 *Degradation of elsinochrome A dissolved in water under natural light conditions.*

	Degradation (%)			
	Experiment 1[a]		Experiment 2[a]	
Time of exposure	with daylight	without light	with daylight	without light
1 day	88.6 b[b]	0 a	3.4 a	0 a
2 days	100 b	0 a	nd	nd
3 days	100 b	0 a	35.7 b	0 a
4 days	100 b	7.9 a	nd	nd
5 days	100 b	9.9 a	61.2 c	0 a
10 days	nd	nd	82.0 d	0 a
Calculated half-life	14.1 h		99.2 h	

Notes:
[a] Experiment 1 was performed from 30 July to 5 August; experiment 2 was performed from 25 August to 1 September.
[b] Means of three replicates are presented. Values for the same experiment followed by the same letter are not significantly different at $P < 0.05$ (Fisher's protected LSD test). nd = not determined.

unpublished results). Furthermore, ELA seems to be heat stable. Even when ELA was exposed to 100°C for 2 h, 100% of the initially used ELA amount could still be recovered (unpublished data).

In a second set of experiments, ELA was exposed to daylight for five days. During experiment one (30 July–5 August 2004) the weather was sunny, whereas during experiment two (25 August–1 September 2004) the weather was unstable and partly cloudy. In experiment one, ELA exposed to daylight was rapidly degraded with a half-life of only 14 hours (Table 15.5). Contrarily, control samples sheathed from daylight by being wrapped in aluminium foil remained unchanged throughout the five-day experimental period. The results obtained in experiment two were similar. However, ELA was degraded more slowly with a half-life of 99 hours (Table 15.5). The reason for a slower degradation compared to experiment one was probably that the sunlight reaching the samples during this experiment was far less intensive.

Taken together, ELA does not seem to be affected by temperatures up to 100°C, but is degraded by sunlight. Rapid degradation by sunlight, furthermore, diminishes the risks of ELA contamination of the environment and the crop.

Conclusions

During the last decade, *S. convolvuli* LA39 has confirmed its effectiveness against field and hedge bindweed in many field experiments at different sites and under different conditions. Strain LA39 has thus the potential for a successful mycoherbicide against field and hedge bindweeds. However, the application method still has to be optimized. So far, in some years with weather conditions favouring bindweed development, up to six mycoherbicide applications have had to be made. Optimizing the timing of the sprayings by better adapting it to the weather conditions of the respective year will certainly reduce the number of applications.

The formulation of the FBCA in water amended with a surfactant, the mass-production of conidia on couscous, as well as the storage of the conidia on kaolin, seem to be appropriate methods for the development of a commercial product since they are efficient, easy and inexpensive.

A big question to be solved before thinking of commercializing an FBCA is the assessment of potential risks such a fungus could pose to the environment or the consumer. In the case of strain LA39 there is the problem of toxin production. The results presented here have shown that ELA is a compound with a broad-spectrum toxicity. However, it also seems to be related to high aggressiveness towards bindweeds. Thus, ELA production probably is essential for effective bindweed control and the risks connected to it have to be studied. With newly developed methods for ELA detection in the mycoherbicide product as well as on the target bindweeds, it could be estimated if ELA produced by LA39 poses a risk for the environment or enters the food chain. In the applied product the ELA concentration is far too low to affect any organisms which could be in contact with it. Furthermore, ELA has never been detected in bindweed leaves, not to mention in the crop. Consequently, the risk of ELA entering the food chain is negligible. Last but not least, stability tests have shown that ELA is rapidly degraded by sunlight and, therefore, is most probably not stable in the environment. All these studies reveal that it can be ruled out that ELA produced by the fungal biocontrol strain LA39 poses a risk to the environment or the consumer.

The bindweed problem is a serious one, especially because commercial herbicides have clear limits in controlling *C. sepium* and *C. arvensis*, and mechanical control is too labour-intensive. The need for alternatives such as FBCAs is growing and, additionally, the public demand for environmentally friendly methods further supports the market for a mycoherbicide. A *Stagonospora*-based product would certainly fit that niche and its integration into integrated pest management systems would be easy to accomplish. The mycoherbicide approach thus offers a promising solution to the bindweed problem.

References

Ahonsi, M. O., Maurhofer, M., Rüegg, D. & Défago, G. (2005). Relationship between aggressiveness of *Stagonospora* sp. isolates on field and hedge bindweeds, and *in vitro* production of fungal metabolites cercosporin, elsinochrome A and leptosphaerodione. *European Journal of Plant Pathology*, **111**, 203–15.

Batchvarova, R. B., Reddy, V. S. & Benett, J. (1992). Cellular resistance in rice to cercosporin, a toxin of *Cercospora*. *Phytopathology*, **82(6)**, 642–6.

Boldt, B. E. & Sobhian, R. (1993). Release and establishment of *Aceria malherbae* (Acari: Eriophydiae) for control of field bindweed in Texas. *Environmental Entomology*, **22**, 234–7.

Callihan, R. H., Eberlein, C. V., McCaffrey, J. P. & Thill, D. C. (1990). Field bindweed: Biology and management. University of Idaho, Cooperative Extension System, College of Agriculture, Bulletin, # 719.

Chen, C-T., Nakanishi, K. & Natori, S. (1966). Biosynthesis of elsinochrome A, the perylenequinone from *Elsinöe* spp. 1. *Chemical and Pharmaceutical Bulletin*, **14**, 1434–9.

Daub, M. E. (1982). Cercosporin, a photosensitizing toxin from *Cercospora* species. *Phytopathology*, **72**, 370–4.

Daub, M. (1987). Resistance of fungi to the photosensitizing toxin, cercosporin. *Phytopathology*, **77** (11), 1515–20.

Daub, M. E. & Ehrenshaft, M. (1993). The photoactivated toxin cercosporin as a tool in fungal photobiology. *Physiologia Plantarum*, **89**, 227–36.

Daub, M. E. & Ehrenshaft, M. (2000). The photoactive *Cercospora* toxin cercosporin: contributions to plant disease and fundamental biology. *Annual Review of Phytopathology*, **38**, 461–90.

Défago, G., Ammon, H. U., Cagán, L., Draeger, B., Greaves, M. P., Guntli, D., Hoeke, D., Klimes, L., Lawrie, J., Moënne-Loccoz, Y., Nicolet, B., Pfirter, H. A., Tabacchi, R. & Tóth, P. (2001). Towards the biocontrol of bindweeds with a mycoherbicide. *BioControl*, **46**, 157–73.

Guerriero, A., Dambrosio, M., Cuomo, V. & Pietra, F. (1991). A novel, degraded polyketidic lactone, leptosphaerolide, and its likely diketone precursor, leptosphaerodione: Isolation from cultures of the marine ascomycete *Leptosphaeria oraemaris* (Linder). *Helvetica Chimica Acta*, **74**, 1445–50.

Guntli, D., Burgos, S., Heeb, M., Pfirter, H. A. & Défago, G. (1999). Biological control of hedge bindweed (*Calystegia sepium*) with mycoherbicide *Stagonospora convolvuli* strain LA39 in combination with competition from red clover (*Trifolium pratense*). *Biological Control*, **15**, 252–8.

Guntli, D., Pfirter, H. A., Moënne-Loccoz, Y. & Défago, G. (1998). *Stagonospora convolvuli* LA39 for biocontrol of field bindweed infesting cotoneaster in a cemetery. *HortScience*, **33**, 860–1.

Holm, L. G., Plucknett, D. L., Pancho, J. V. & Herberger, J. P. (1977). In *The World's Worst Weeds: Distribution and Biology*. Honolulu: University Press of Hawaii.

Ma, L., Tai, H., Li, C., Zhang, Y., Wang, Z. H. & Ji, W. Z. (2003). Photodynamic inhibitory effects of three perylenequinones on human colorectal carcinoma cell line and primate embryonic stem cell line. *World Journal of Gastroenterology*, **9(3)**, 485–90.

Maillet, J. (1988). Les liserons. Phytoma. *Défense de Culture*, **399**, 11–15.

Nicolet, B. & Tabacchi, R. (1999). Secondary metabolites produced by *Stagonospora* sp., a potential biocontrol agent against bindweeds. In *Modern Fungicides and Antifungal Compounds II*, eds. H. Lyr, P. E. Russell & H. D. Sisler. Andover, United Kingdom: Intercept Limited, pp. 469–76.

Pfirter, H. A., Ammon, H. U., Guntli, D., Greaves, M. P. & Défago, G. (1997). Towards the management of field bindweed (*Convolvulus arvensis*) and hedge bindweed (*Calystegia sepium*) with fungal pathogens and cover crops. *Integrated Pest Management Reviews*, **2**, 61–9.

Pfirter, H. A. & Défago, G. (1998). The potential of *Stagonospora* sp. as a mycoherbicide for field bindweed. *Biocontrol Science and Technology*, **8**, 93–101.

Pfirter, H. A., Guntli, D., Ruess, M. & Défago, G. (1999). Preservation, mass production and storage of *Stagnospora convolvuli* a bioherbicide candidate for field bindweed (*Convolvulus arvensis*). *BioControl*, **44**, 437–47.

Rosenthal, S. S. & Carter, J. (1977). Host specificity and biology of Galeruca rufa, a potential biocontrol agent for field bindweed. *Environmental Entomology*, **6**, 155–8.

Rosenthal, S. S. & Platts, B. E. (1990). Host specificity of *Aceria (Eriophyes) malherbe*, (Acari: Eriophyidae), a biological control agent for the weed *Convolvulus arvensis* (Convolvulaceae). *Entomophaga*, **35**, 459–63.

Skrobek, A., Boss, D., Butt, T. M., Défago, G. & Maurhofer, M. (2006). Evaluation of different biological test systems to assess the toxicity of metabolites from fungal biocontrol agents. *Toxicology Letters*, **161(1)**: 43–52.

Weaver, S. E. & Riley, W. R. (1982). The biology of Canadian weeds. 53. *Convolvulus arvensis* L. *Canadian Journal of Plant Science*, **62**, 461–72.

Weiss, U., Merlini, L. & Nasini, G. (1987). Naturally occurring perylenequinones. In *Progress in the Chemistry of Organic Natural Products* 52, eds. W. Herz, H. Grisebach, G. W. Kirby & C. H. Tamm. Vienna: Springer Verlag, pp. 1–71.

Yamazaki, S., Okubo, A., Akiyama, Y. & Fuwa, K. (1975). Cercosporin, a novel photodynamic pigment isolated from *Cercospora kikuchii*. *Agricultural and Biological Chemistry*, **39(1)**, 287–8.

16

A novel understanding of the three-way interaction between *Trichoderma* spp., the colonized plant and fungal pathogens

S. L. WOO, M. RUOCCO, R. CILIENTO, P. AMBROSINO, R. MARRA, F. VINALE, S. LANZUISE, D. TURRÀ, S. FERRAIOLI, I. SORIENTE, S. GIGANTE, F. SCALA AND M. LORITO

Abstract

Trichoderma-based biofungicides are a reality in agriculture, with more than 50 formulations available today as registered products worldwide. Several strategies have been applied to identify the main genes and compounds involved in this complex cross-talk between the fungal antagonist and the microbial pathogen, as mediated by the plant. Proteome and genome analysis have greatly enhanced our ability to conduct holistic and genome-based functional studies. We have identified and determined the role of a variety of novel genes and gene-products, including ABC transporters, enzymes and other proteins that produce or act as novel elicitors of induced systemic resistance, proteins recognized by the plant as avirulence factors, as well as molecules that generally activate the antagonistic activity in *Trichoderma* spp. We have cloned mycoparasitism-related promoters and used them in combination with GFP and other markers to study the interaction in vivo and *in situ* between *Trichoderma* and the fungal pathogen or the plant. Finally, we have transgenically improved the ability of the antagonist to kill other microbes and to activate plant defence mechanisms.

Introduction

Plant diseases caused by pathogenic fungi infections represent a major limiting factor for the cultivation and the conservation of agricultural plants of interest. The consequences of parasite attack result in both quantitative and qualitative reduction of crop production, large

Exploitation of Fungi, eds. G. D. Robson, P. van West and G. M. Gadd. Published by Cambridge University Press economic losses and represent a risk for human and animal health due to the accumulation of residues in the environment and mycotoxin contaminants in food products. Chemical control of phytopathogenic fungi is limited not only by economical cost, but also by concerns about the impact of continued applications on consumer health and the environment. In addition, applications of methyl bromide, the most frequently used soil fumigant (also causing a reduction of the ozone level in the atmosphere), will be completely banned in Europe in the next few years, which provides a strong impulse for the development of 'biological' methods based on, for instance, antagonistic microbes and soil solarization for disease management. Alternatively, new technologies for plant defence can be proposed that are based on the utilization of transgenic plants and microorganisms, the application of antagonist microbes (fungi or bacteria) capable of controlling important diseases or of useful compounds originating from these microorganisms.

In particular, fungi of the genus *Trichoderma* are among the most effective biopesticides applied for plant protection against fungal diseases, both in traditional and organic farming. The fungi are commercially produced, and several patents protect their use (Harman & Kubicek, 1998; Kubicek & Harman, 1998). Biopesticides based on antagonistic *Trichoderma* strains are used for biocontrol of phytopathogenic fungi that cause diseases including root and crown rot of vegetables, seedling damping-off, vascular wilt, take-all of cereals, root infection of fruit trees, mildew of grapes and other crops, monilia disease, and soft rot of pome and stone fruit etc. The availability and diffusion of these products is much more widespread than commonly known, and *Trichoderma* is also listed both in Europe and USA as a pesticide permitted for use in organic farming. The size of the biopesticide market based on antagonistic fungi is modest compared to chemicals, but it is steadily growing and is expected to increase significantly with the introduction of these products in India, China, northern Africa and central Asia. In addition to the registered products (a few dozen formulations), a large variety of non-registered bioprotectants, biostimulators and biofertilizers based on *Trichoderma* or other antagonistic fungi are being sold everywhere (Lorito *et al.*, 2004). For instance, commercial products based on the fungus *Trichoderma* spp. include Binab T™ (*T. harzianum* and *T. polysporum* registered in Sweden, UK, USA), Bio Fungus™ (*Trichoderma* spp. in Belgium), EcoSOM™ (*T. viride* in India), RootShield™ T-22G, T-22 Planter Box, Bio-Trek (*T. harzianum* in USA and north Europe), Root Pro™ (*T. harzianum* in Israel), Supresivit™ (*T. harzianum* in the

Czech Republic), Trichodex™ (*T. harzianum* in Italy, Israel and USA), Trieco™ (*T. viride* in India), Trichoderma 2000™ (*Trichoderma* spp. in Israel), Trichopel™ Trichojet, Trichodowels, Trichoseal (*T. harzianum* and *T. viride* in New Zealand) etc.

***Trichoderma* as a versatile biocontrol agent**

Trichoderma spp. have the remarkable ability to interact both as a parasite and symbiont with different substrates and living organisms, including plants and other microbes. These fungi can utilize a variety of nutrient sources. They are not only some of the most resistant microbes to natural and man-made chemicals and toxins, but they are also able effectively to degrade some compounds, including hydrocarbons, chlorophenolic compounds, polysaccharides and xenobiotic pesticides (Harman *et al.*, 2004). Many strains of *Trichoderma* are strong opportunistic invaders, fast growing, producers of enormous quantities of spores and powerful antibiotics. These properties make these fungi ecologically very successful, which is reflected in the fact that they are widely distributed in diverse ecosystems of agriculture, native prairie, forest, salt marsh and desert soils of all climatic zones, as well as in lake water, dead plant material, living roots of virtually any plant species, seeds and air.

These saprophytic fungi are commonly isolated from the soil, which is their natural habitat, but they are also present in other surroundings. The genus *Trichoderma*, according to recent systematics, is comprised of more than 30 species, whereby the majority exhibit only the asexual life-cycle, although in rare cases the sexual form *Hypocrea* is also found. *Trichoderma* fungi have a tremendous amount of genetic diversity and are biologically versatile with the ability to perform numerous different biological functions. For example, there are strains that are strong antagonistics towards phytopathogenic fungi, some that are effective soil colonizers and biodegraders, and others that are producers of important metabolites. Therefore, *Trichoderma* spp. are of substantial interest for studies on the biology of fungi and good candidates for applications in agriculture and industry.

Many *Trichoderma* strains are 'rhizosphere competent', having the ability to colonize the plant root, and they may also be capable of promoting plant growth and root development. Recent studies on the interaction of *Trichoderma* with plants and other microbes indicate that these fungi may induce systemic resistance in plants, thus increasing the plant defence response to attack from a variety of pathogens (Harman, 2000; Harman *et al.*, 2004).

Molecules produced by *Trichoderma* also have potential for promoting plant growth (Inbar *et al.*, 1994). Plants treated with *T. harzianum* showed improved seed germination, increased overall plant size and augmented leaf area and weight (Altomare *et al.*, 1999). The possibility of combining systemic biofungicides and plant growth promoters has a great market potential if the molecular basis of the mechanisms involved can be understood.

Natural and genetically modified strains of *T. reesei* are widely used in industry, typically for protein production, whereas isolates of *T. harzianum*, *T. atroviride* or *T. virens* are used in agriculture as biopesticides, bioprotectants, biostimulants and biofertilizers on a large variety of crops in the field and greenhouse. It is possible to use *Trichoderma* over a wide range of applications because its propagules can be produced economically, in high concentrations, both in liquid and dry formulations, and can be stored for many months. Today, more than 50 different *Trichoderma*-based agricultural products can be found registered in different countries across five continents, where they are sold as biocontrol products for protection and yield improvement of numerous crops.

The laboratory techniques that can be applied at present provide the possibility to genetically modify these fungi in a very specific way, permitting improvement of the ability to secrete given enzymes, inhibit or kill plant pathogens, or stimulate plant growth and resistance to disease attack (Brunner *et al.*, 2005). The research conducted on *Trichoderma* spp. in the past 20 years has revealed numerous molecular and genetic mechanisms involved in the many beneficial or useful biological processes involving these fungi. However, regardless of the obvious potential there are some difficulties that limit the development and application of these biopesticides. In addition to the lack of effective strains and the availability of correct formulations, accessibility is often limited in: (1) basic information required for product registration, such as a sufficient knowledge of the mechanisms of action; (2) efficacy testing in the geographic regions of the country where the product is to be registered; (3) methods for monitoring the production of eventual toxic metabolites produced by these fungi that permit an evaluation of possible risks derived from a large scale application.

Trichoderma a beneficial fungal antagonist

The mechanisms utilized by *Trichoderma* in the interaction with the plant and other microbes are very complex and include mycoparasitism (necrotrophic direct fungus–fungus interaction), competition for nutrients, antibiosis, inhibition of pathogen enzymes, antagonism against nematodes, colonization of the rhizosphere, phyllosphere and spermosphere, promotion of plant growth and root development, and induction of systemic resistance against different pathogens (Harman, 2000). Several cell wall degrading enzymes (chitinases and glucanases) (Limon *et al.*, 1996; Lorito and Woo, 1998; Lorito *et al.*, 2001; Woo *et al.*, 1999) and antibiotics have been included among the factors that play a main role in this complex biological process (Lorito *et al.*, 1996; Schirmböck *et al.*, 1994). Different types of molecules functioning as 'biocontrol/antagonism inducers' or elicitors for both *Trichoderma* and the plant have been found to be released by the antagonist and by the activity of its lytic enzymes on the cell wall of the fungal host (see below). In addition, *Trichoderma* sends a variety of biochemical signals to the plants including avirulence factors, as discussed below.

Several cell wall degrading enzymes are being used today, both directly as proteins or through gene expression, for the development of new defence strategies against phytopathogenic fungi (Lorito *et al.*, 2001; Lorito *et al.*, 1998). Especially promising is the use of enzyme mixtures capable of degrading the fungal cell wall that have been found to be active against a wide range of fungi. To date, no resistant phytopathogenic fungi have been discovered, and these fungal enzymes are considered to be amongst the most potent antifungal compounds known (in comparison to similar plant and bacterial products). *Trichoderma* spp. produce large amounts of these lytic enzymes, which are stable at room temperature, have efficacy levels similar to those of commercial fungicides and maintain biological activity even if applied ectopically to plants in greenhouse or produced in post-harvest (see Lorito & Scala, 1999 for a review). Moreover, these fungal enzymes are strongly synergistic with synthetic fungicides of different classes, in particular azole and other cell membrane affecting compounds (Lorito *et al.*, 1996). Moreover, *Trichoderma* strains produce a large number of antibiotics, including gliotoxin and viridin, alpha-pyrones, terpenes, polyketides, isocyanide derivatives, piperacines and complex families of peptaibols (Sivasithamparam & Ghisalberti, 1998). Many of these antibiotics are synergistic when combined with cell wall degrading enzymes (those originating from *Trichoderma* and other sources) and produce a strong inhibitory effect on many fungal plant pathogens (Fogliano *et al.*, 2002; Lorito *et al.*, 1996; Schirmböck *et al.*, 1994).

Competition for space or specific infection sites, nutrients and other factors necessary for growth in the rhizosphere may also be another mechanism through which antagonistic microbes control plant pathogens.

Although competition for space has not been clearly demonstrated in the case of *Trichoderma*, there is evidence showing that it plays a role in biocontrol when these antagonists establish their dominance in a specific environmental niche. *T. harzianum* is able to control the attack of *Botrytis cinerea* on grapes by colonizing blossom tissue and excluding the pathogen from its infection site (Gullino, 1992). The mechanisms of the competition for nutrients such as carbon, nitrogen and iron, root and seed exudates have been better defined. Competition for carbon and nitrogen between *T. harzianum* and *F. oxysporum* f. sp. *melonis* on rhizosphere colonization has been identified as the most important mechanism involved in the biocontrol of this vascular disease. It was demonstrated that all the other possible modes of action played little or no role in this case (Sivan & Chet, 1989). Mechanisms by which antagonists parasitize other fungi have been thoroughly studied by using *Trichoderma* spp. Many studies, recently performed by using different gene expression reporter systems, have visualized the presence and the activity of propagules of the antagonist in different fungal structures of the pathogen, including sclerotia (Lu *et al.*, 2004; Mach *et al.*, 1999).

Many studies indicate that *Trichoderma* spp. act not only as antagonists against plant pathogens, but also as agents to alter plant disease susceptibility. The ability of *Trichoderma* species to stimulate defence responses in the host plant has been clearly observed: in cotton with strains of *T. virens* (Howell *et al.*, 2000), in cucumber with some strains of *T. harzianum*, in pepper (Sid Ahmed *et al.*, 2000) and in tobacco (Chang *et al.*, 1997) against various phytopathogens including *Rhizoctonia solani* and some *Phytophthora* species. The *Trichoderma* treatments to different plant hosts caused the increase of ex-novo synthesis of compounds involved in active defence mechanisms, such as the production of terpenoids (Howell *et al.*, 2000), capsidiol (Sid Ahmed *et al.*, 2000), as well as peroxidases, beta-glucosidase, cellulase and chitinase activities (Yedidia *et al.*, 2000). For example, in grape cell suspension cultures, an elicitor produced by *T. viride* was able to activate defence responses resulting in cellular necrosis and an increase in resveratrol concentration, an important phytoalexin produced following *B. cinerea* and *Plasmopara viticola* infections (Calderon *et al.*, 1993; Dai *et al.*, 1995).

There are many aspects of the molecular mechanisms that *Trichoderma* utilizes in its relationship with plants and other microbes that are not understood due to the complexity of the processes involved, but also because many of the recent and potent tools available for investigating molecular-genetic and biochemistry techniques (such as EST sequencing, DNA array and miniarray or proteome analysis) have yet to be fully implemented to study the mechanisms of biocontrol of fungal antagonists.

Interactions among plants, pathogens and antagonists

In comparison to the relationship between plants and antagonists, the mechanisms involved in the interaction between the plant and fungal pathogens have been investigated extensively and they are much better understood. The involvement of different genes and other factors have been identified as having a role in various aspects of avirulence and pathogenicity on the part of the microorganisms or in the process of resistance on the side of the plant. The molecular cross-talk that involves pathogen elicitors and plant receptors has been very well studied (Dangl & Jones, 2001). Clearly, in order to identify genetic and molecular factors that regulate the three-way plant–pathogen–antagonist interaction requires an holistic approach aimed at investigating the changes in gene expression that occur in each of the three partners involved in this interaction, singly and then in all combinations. This approach parallels the natural crop conditions where plants are exposed contemporaneously to a multitude of microorganisms, pathogens and antagonists (in addition to symbionts), and disease development is a consequence of this complex, multivariate interaction.

The means for studying this three-way relationship can be determined by examining the information available from work involving the mechanisms of action in both the plant–pathogen and the plant–antagonist interactions, and using this information for developing new genetic and molecular tactics to analyze the multicomponent interaction. Present techniques used in post-genomic analysis with the development of EST sequences, gene library, arrays, proteomic maps and transcriptomic analysis are opportune for identifying and characterizing new genes, proteins and signal molecules involved in the process, which have potential utility in developing novel strategies for plant protection (Lorito *et al.*, 2001; Rey *et al.*, 2004).

Pathogen attack of the plant instigates various mechanisms of resistance that generally involves the activation of several localized responses at the infection site, such as the 'oxidative' and 'nitrosative' burst, characterized by a rapid accumulation of reactive oxygen species (ROS) and nitric oxide (NO) (Sandermann, 2000), the induction of genes involved in phytoalexin biosynthesis or the encoding of several pathogenesis-related proteins (PR proteins), the production of mechanical structures to reinforce plant tissues or block pathogen movement, and the activation of the hypersensitive response (HR) (Keen, 2000).

Plant perception of pathogen presence is mediated by signals produced by the pathogen (elicitors) that are perceived by specific plant receptors. It is likely that similar molecular mechanisms to activate specific genes also occur during the interactions between antagonist–pathogen and antagonist–plant. Recently, it has been demonstrated that there is a production of mycoparasitism inducers that are released from the pathogen or the plant and are 'detected' by the antagonist, although the specific molecules and the genes involved in the process have yet to be identified. Furthermore, several preliminary observations suggest that specific oligosaccharides derived from the pathogen or plant cell wall (or even from *Trichoderma* itself) are able to activate the process of biocontrol in *Trichoderma* spp., and induce at the same time systemic resistance in the plant. Potentially, these molecules (proteins and low molecular weight compounds) may be applied as 'elicitors' of biocontrol to stimulate resident fungal antagonist populations and defence mechanisms in the plant, thus improving the efficacy of biopesticides in the field as well as the production of enzymes or biomass in commercial fungal strains at the industrial level.

Interaction between *Trichoderma* and the plant

Fungal antagonists such as *Trichoderma* do not perform only, as previously described, by physically and chemically attacking phytopathogens, but they can also control diseases by interacting directly with the plant. Recently, numerous studies have demonstrated that there is an induction of localized and systemic resistance in plants treated with fungal antagonists, including *Trichoderma* spp. These plant defence responses to pathogen attack or other injuries caused by physical or insect damage, application of chemicals, etc., are very complex and involve signal molecules such as salicylic acid, jasmonic acid, ethylene, nitrous oxide, etc. These signals activate a cascade of reactions involving the production of metabolites (phytoalexins etc.) and proteins (PR proteins) that are components of plant disease resistance.

Recently, the role of rhizobacteria in plant resistance has been partially clarified (Pieterse & van Loon, 1999). These microbes activate a specific type of induced resistance, particular to the rhizobacteria, known as induced systemic resistance (ISR). In ISR, PR proteins are not expressed by the plant, and plants react more strongly to pathogen attack because of an increased sensitivity to jasmonic acid and ethylene (Kuc, 2001). Similarly, some fungi, such as *Fusarium*, *Rhizoctonia*, symbiontic mycorrhizal fungi and *Trichoderma*, have this capacity to promote plant growth and to induce resistance to pathogens. The first report of ISR caused by *Trichoderma* was observed on tomato, lettuce, pepper, bean and tobacco plants treated with *T. harzianum* strain T39 in the biocontrol of *B. cinerea* (De Meyer *et al.*, 1998). Since then, numerous studies on the induction of systemic resistance in various plants to a variety of pathogens by the application of different *Trichoderma* species has been reported, including *R. solani* and some *Phytophthora* spp. (see Harman *et al.*, 2004 for a review). The treatments with the biocontrol agent were found to alter the production of substances implicated in the active defence mechanisms; an increase in the production of substances including peroxidases, beta-glucosidases, cellulases and chitinases activities (Yedidia *et al.*, 2000), or the ex-novo synthesis of compounds such as terpenoids (Howell *et al.*, 2000) and capsidiol (Sid Ahmed *et al.*, 2000). In grape cell cultures, an elicitor produced by *T. viride* activated defence responses that resulted in cellular necrosis and accumulation of resveratrol (Calderon *et al.*, 1993; Dai *et al.*, 1995).

In order to discern if the biocontrol effect of *T. virens* on soilborne diseases was due to localized or systemic acquired resistance, as an alternative to direct mycoparasitism by the antagonist, it was necessary to use mutants of *Trichoderma* with different antibiotic or mycoparasitic capabilities. The biocontrol of *R. solani* on cotton seedlings was found to depend upon the ability of the antagonist strain to induce the production of phytoalexins in the plant such as hemoglossypol, deoxyhemoglossypol and glossypol (Howell *et al.*, 2000). These results provide a new insight into the mechanisms involved in biocontrol and suggest that induced resistance may, in some cases, play a more important role than direct action by an antagonist on pathogen control (Harman *et al.*, 2004). The activation of induced resistance may confer protection for a longer period of time and against a wider range of pathogens, including viruses (Harman *et al.*, 2004). For example, an application of *T. harzianum* to tomato roots was found to reduce significantly leaf infection caused by *Alternaria solani* almost three months later.

Induced resistance in plants caused by the interaction with *Trichoderma* may involve different molecules (Fig. 16.1). It was noted that these fungi produced a xylanase of 22 kDa that induced a localized resistance reaction (Lotan & Fluhr, 1990). Several peptides and proteins produced by *T. virens* were found to have elicitor activity and to induce the biosynthesis of terpenoid phytoalexin and peroxidase activity in cotton (Howell *et al.*, 2000). Our preliminary results indicate that *T. harzianum* produces proteins that may be homologous to the products of the avirulence genes *Avr4* and *Avr9* found in the tomato pathogen *Cladosporium fulvum*. These proteins, acting as race or pathovar specific elicitors, induce resistance responses in plants having the matching resistance genes (Dangl & Jones, 2001).

An additional genre of molecules that are capable of inducing resistance may originate from the interaction *Trichoderma* has with pathogens and plants. Cell wall degrading enzymes secreted by the antagonist have activity both on the cell walls of the fungal pathogens as well as the plants,

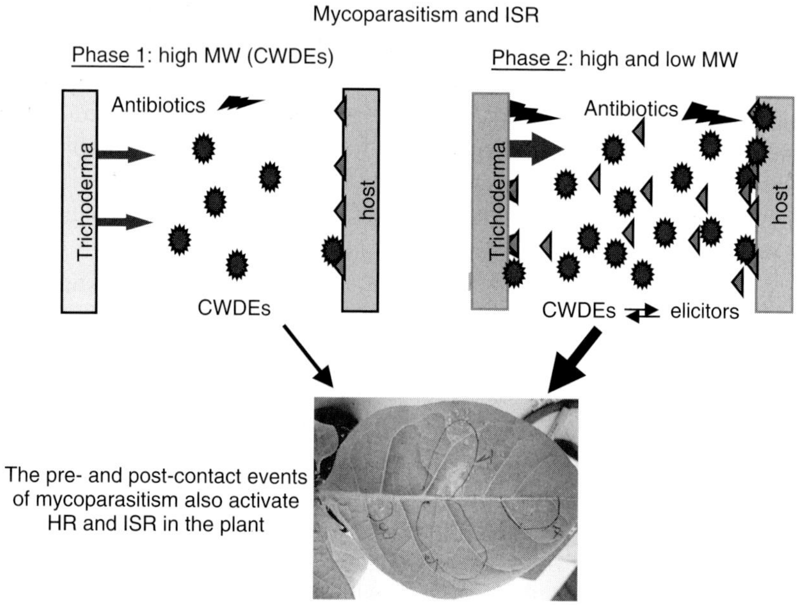

Figure 16.1 Model for a correlation between mycoparasitism and induced systemic resistance (ISR) during multiple interactions between *Trichoderma*, plants and pathogenic fungi. In the fungus–fungus interaction there may be first a pre-contact phase 1 in which the mycoparasite secretes antibiotics and cell wall degrading enzymes (ovals) to search the environment for a living or a dead host/substrate. This phase can be inhibited by blocking the movement of high molecular weight compounds, found to be chitinases, glucanases and other lytic enzymes. When a host is found (phase 2), the degradation products from the host cell walls (triangles) reach the antagonist and mediate the activation of the mycoparasitism-related genes, which induces stronger secretion of enzymes and antibiotics, chemiotactic growth toward the host, coiling and host penetration, etc. This phase can be inhibited by blocking the movement of low molecular weight compounds, found to be oligosaccharides with an amino acid moiety (Lorito *et al.*, unpublished). Arrows indicate that the plant has developed the ability to 'sense' these molecular interactions (especially during phase 2) occurring in the rhizosphere and use these signals to activate defence responses involving hypersensitive reaction (shown infiltration of low MW compounds in tobacco leaf) and ISR.

which consequently causes the release of oligosaccharides that may activate both the mechanisms of antagonism in *Trichoderma* and defence responses in plants (Fig. 16.1). Recent results indicate that *Trichoderma* culture filtrates contain macromolecules and low molecular weight compounds that induce strong changes in cytosolic Ca^{2+} levels in plant cell cultures and may cause programmed cell death (Mariani & Lorito, unpublished data). Rhizosphere competent strains of *Trichoderma* are able to colonize plant roots and penetrate the root epidermis, thus causing a change in plant metabolism, such as an increase in the levels of plant enzymes involved in the defence mechanisms (Harman *et al.*, 2004). The means by which *Trichoderma* limits penetration only to the outer cortex of the plant tissue are not yet known, although there may be a correspondence to its ability to activate the plant defence system by producing avirulence (avr) like proteins and other types of elicitors. Furthermore, it has also been reported that there is production and accumulation of phytoalexins; in the case of cucumber, root colonization increased the accumulation of phenolic compounds in the leaves that are inhibitory towards *Pseudomonas syringae* pv. *lachrymans*, *Agrobacterium tumefaciens*, *F. oxysporum* f. sp. *melonis*, *B. cinerea* and other fungi (Harman *et al.*, 2004). It appears that many *Trichoderma* strains, which maintain close relationships with the plant, may have the intrinsic ability to act as phytopathogens, and probably have all the required molecular tools to behave as such, but instead establish beneficial interactions with the plant by stimulating root development and increasing plant resistance to pathogenic microbes. In exchange, the plant also provides some mutual positive effects for *Trichoderma* by increasing the available nutrient base through plant root exudates and limiting competition with other microbes found in the rhizosphere.

Accumulating evidence indicates that molecular cross-talk occurs between *Trichoderma* and the plant that it colonizes. When the fungus transfers molecules to receptive plants, these compounds are recognized and able to activate various processes such as defence to biotic attack by microorganisms and insects, and accelerate development or resistance to abiotic stresses caused by drought, nutrient deficiency, etc.

Today, the mechanism by which *Trichoderma* is able to protect the plant is considered more important, at least in some applications, than 'classical' direct mycoparasitism and competition that occurs with the phytopathogenic host. Numerous researchers in the field support the hypothesis that ISR and not antagonistic ability may be the mechanism that produces the better wide-spectrum performances of some commercial biocontrol strains of *Trichoderma*. However, much is yet to be discovered about the molecules and genes involved in the process of ISR, and similar mechanisms involved in plant interactions with *Trichoderma*, mycorrhizal fungi, rhizobacteria or microbial pathogens. Different molecular genetic techniques that have been utilized to study *Trichoderma* spp., including homologous recombination, use of reporter genes, use of constitutive and inducible promoters, and secretion of heterologous proteins, will be very useful in examining its dual interaction with the plant and pathogens (Mach *et al.*, 1999; Lu *et al.*, 2004; Woo *et al.*, 1999).

Molecular cross-talk occurring between *Trichoderma* and the plant

Only recently, with the advent of new technologies and a concerted effort of several research groups, have some of the molecules that control the *Trichoderma*–plant interaction, and in particular the induced resistance effect, been identified. *Trichoderma* is able to produce at least three different types of elicitors, those comprising of (1) enzymes or peptides; (2) avr-like proteins; and (3) oligosaccharides or low-molecular-weight compounds released by the action of specific lytic enzymes on fungal and plant cell walls.

The elicitors made up of different peptides and proteins vary in size from 6.2 to 42 kDa (possibly including a serine protease and a xylanase) and induce the biosynthesis of terpenoid phytoalexins and peroxidase in the plant (Harman *et al.*, 2004). Data from our laboratory suggest that among these elicitors there are proteins that *Trichoderma* uses for its mycoparasitic and antagonistic activities, such as chitinases and glucanases. These are secreted initially at low levels and then profusely when the presence of a host is detected (Fig. 16.1). The plant recognizes these compounds as foreign fungal molecules (regardless of their enzymatic mode of action) probably by using specific receptors and resistance genes, similar to those utilized in pathogen recognition. Perhaps the detection of the *Trichoderma* metabolites serves as an indicator for the plant of the ongoing antagonist–pathogen interactions occurring in close proximity and the necessity to activate the defence system because of the risk of pathogen attack. This hypothesis is supported by the observation that the increased resistance mediated by *Trichoderma* in the plant is based on a pre-activation of the biochemical response rather than on the over-accumulation of defence proteins.

Our recent results indicate one possibility of how some *Trichoderma* strains elicit plant defences by the use of avirulence proteins. The biocontrol strain *T. atroviride* P1 was transformed with the gene encoding for the Avr4 protein of *Cladosporium fulvum*, a leaf pathogen of tomato, under a strong constitutive or inducible promoter. The Avr4-secreting *Trichoderma* mutants over-expressing the transgene were applied to the seeds of tomato cultivars containing and not containing the corresponding *Cf4* resistance gene. Fewer plants emerged from the *Cf4* containing seeds, and these seedlings were stunted and less healthy than those emerging from the non-*Cf4*-containing seeds, which did not differ from the untreated controls. When *Trichoderma* avr mutants were applied to the roots of mature *Cf4* plants, necrotic zones rapidly appeared on the root and leaf surface that resembled HR (Ruocco & Lorito, unpublished data). From a biotechnological viewpoint, *avr* genes from biocontrol fungi are attractive because they can potentially provide protection against a large spectrum of pathogens, by transforming plants or beneficial microbes with the *Trichoderma avr* genes, or ectopically applying the avr products to stimulate plant resistance.

The third type of elicitors are molecular compounds released from the pathogen and the plant cell wall by the action of *Trichoderma* cell wall degrading enzymes. These molecules not only stimulate antagonistic activity in the fungus by activating the mycoparasitic gene expression cascade, but they also act as elicitors of plant defence when plant cells are exposed to them or when injected in the root and leaf surface (Woo & Lorito, unpublished data, Figure 16.1). Some of these compounds were found to be comprised of short oligosaccharides made of two types of monomers either with or without an amino acid residue. This provides a novel possibility of using the enzymes secreted by *Trichoderma*, rather than the fungus itself, to induce ISR in the plant. This enzyme treatment may stimulate the activity of resident or introduced *Trichoderma* populations by the release of fungal inducers and activate the ISR system in the plant by the release of plant elicitors, as well as having an antimicrobial effect in the application zone.

Finally, the understanding of the mechanism of interaction between *Trichoderma* and the plant has provided for the first time the opportunity to genetically increase the ability of a *T. atroviride* strain to induce ISR. We were able to over-express a glucose oxidase gene from *Aspergillus niger* in *Trichoderma* strain P1, and obtained mutants that, in vitro and in vivo experiments, performed much better than the wild type both as a mycoparasite and as an ISR inducing agent (Brunner *et al.*, 2005). In this case, bean seeds treated with the spores from the mutant produced plants that were more resistant to leaf infection by *B. cinerea* when compared to those treated with the wild type *Trichoderma*. Probably, the high glucose oxidase production by the genetically-improved *Trichoderma* catalyzed the production of hydrogen peroxide and reactive oxygen species that systemically induced the plant defence mechanisms.

Effect of *Trichoderma* on plant growth improvement

Many microorganisms, such as rhizobacteria, are capable of producing beneficial effects to plants when they colonize the roots resulting in increased growth and yields (Figs. 16.2, 16.3 and 16.4). The promotion of both root development and growth in a wide range of plants has been noted with rhizosphere competent strains of *Trichoderma* (Harman *et al.*, 2004). In particular, the effects of some commercial strains, such as *T. harzianum* T22, have been well studied on important crops like maize, whereby seeds treated with the antagonist produced plants that had much

Figure 16.2 Plant growth promotion effect of *Trichoderma* spp. strains on greenhouse cultivated lettuce. T (and boxed zone) = *Trichoderma* seed-treated plants, c = untreated controls.

Figure 16.3 Plant growth promotion effects of *Trichoderma* spp. strains on greenhouse cultivation of tomato (A) and zucchini (B). In each case, left-hand rows contain untreated controls and right-hand rows contain plants treated with *Trichoderma*.

deeper rooting compared to untreated controls and were also more tolerant to drought (Harman & Björkman, 1998). In addition, it was shown that this effect is accentuated when other beneficial root colonizing microorganisms, such as mycorrhyzal fungi, were also applied. Increased root development caused by the antagonists may also produce a significant increase in product yield, as observed in maize (Harman & Björkman, 1998). *Trichoderma* may provide an advantage to crops that are grown in less than optimum field conditions with various biotic stresses and thus substantially improve production. However, it should be noted that the results differed with the lines of maize tested, and in some cases *Trichoderma* even caused a reduction in plant growth and yield (Harman & Björkman, 1998). This was also found to be the case with tomato in which the different genetic background of the eight tested plant varieties reacted differently to a strain of *T. atroviride* and one of *T. harzianum* (Fig. 16.4 C and D) (Ruocco & Lorito, unpublished). This suggests that more studies are required to clarify the mechanisms involved in the

Figure 16.4 Plant growth promotion effects of different *Trichoderma* spp. strains on pot-grown tomato (T = *Trichoderma*-treated seeds, c = untreated controls). (A) and (B), effect of *T. harzianum* strain T22 on stem and root development. (C) and (D), effect of *T. atroviride* strain P1 and *T. harzianum* strain T22 on two different tomato cultivars.

Trichoderma–plant interaction to identify the most useful strains and crop combinations that produce desirable results.

The mechanisms that *Trichoderma* uses to increase root and shoot growth and yields may depend on several physiological processes, such as its ability to make nutrients more available for plant uptake, or degrade toxic compounds. *Trichoderma* are able to degrade hydrogen cyanide (HCN) either present in soil or produced by some deleterious rhizobacteria (Harman *et al.*, 2004). They are also able to produce a diffusible metabolite that has plant hormone characteristics (Windham *et al.*, 1986). Further, *Trichoderma* is able to solubilize minor nutrients like Mn^{4+}, Cu^{2+} and Fe^{3+} by reducing them to the soluble forms Mn^{2+}, Cu^{1+} and Fe^{2+}, making these minerals available for the plant (Altomare *et al.*, 1999). Similar results were observed with nitrogen, whereby *Trichoderma* treated plants produced maximum yields even when 40% less nitrogen-containing fertilizer was supplied. Therefore, *Trichoderma* may be able to improve nitrogen-uptake efficiency of plants and, consequently, reduce nitrogen pollution in soil and water (Harman *et al.*, 2004).

Conclusions

The mechanisms by which *Trichoderma* spp. are able to control plant diseases or stimulate resistance in the plant are just starting to be elucidated. In the complex association between this fungal antagonist, the plant and pathogens there is continuous molecular cross-talk maintained among the participants. The presence or absence of a member will alter the secretion of different compounds, biochemical and biological activities, perception or receptivity to the released by-products and consequent responses of antagonism by *Trichoderma* and/or defence response by the plant. The beneficial effects to the plant are not limited to disease control, but also include increased growth and development. The potential of applications utilizing *Trichoderma* as a bioagent (fungicide, stimulant, protectant, etc.), and its by-products, alone or in combination with existing technology and commercial products, is expanding as our knowledge increases about the mechanisms involved in this three-way relationship.

References

Altomare, C., Norvell, W. A., Björkman, T. & Harman, G. E. (1999). Solubilization of phosphates and micronutrients by the plant-growth promoting and biocontrol fungus *Trichoderma harzianum* Rifai strain 1295–22. *Applied and Environmental Microbiology*, **65**, 2926–33.

Brunner K., Zeilinger, S., Ciliento, R., Woo, S. L., Lorito, M., Kubicek, C. P. & Mach, R. L. (2005). Genetic improvement of a fungal biocontrol agent to enhance both antagonism and induction of plant systemic disease resistance. *Applied and Environmental Microbiology*, **71**, 3959–65.

Calderon, A. A., Zapata, J. M., Munoz, R., Pedreno, M. A. & Barcelo, A. R. (1993). Resveratrol production as a part of the hypersensitive-like response of grapevine cells to an elicitor from *Trichoderma viride*. *New Phytologist*, **124**, 455–63.

Chang, P.-F. L., Xu, Y., Narasimhan, M. L., Cheah, K. T., D'Urzo, M. P., Damsz, B., Kononowicz, A. K., Abad, L., Hasegawa, P. M. & Bressan, R. A. (1997). Induction of pathogen resistance and pathogenesis-related genes in tobacco by a heat-stable *Trichoderma* mycelial extract and plant signal messengers. *Physiologia Plantarum*, **100**, 341–52.

Dai, G. H., Andary, C., Mondolot-Cosson, L. & Boubals, D. (1995). Involvement of phenolic compounds in the resistance of grapevine callus to downy mildew (*Plasmopara viticola*). *European Journal of Plant Pathology*, **101**, 541–7.

Dangl, J. L. & Jones, J. D. G. (2001). Plant pathogens and integrated defence responses to infection. *Nature*, **411**, 826–33.

De Meyer, G., Bigirimana, J., Elad, Y. & Hofte, M. (1998). Induced systemic resistance in *Trichoderma harzianum* T 39 biocontrol of *Botrytis cinerea*. *European Journal of Plant Pathology*, **104**, 279–86.

Fogliano, V., Ballio, A., Gallo, M., Woo, S. L., Scala, F. & Lorito, M. (2002). *Pseudomonas* lipodepsipeptides and fungal cell wall degrading enzymes act synergistically in biocontrol. *Molecular Plant Microbe Interactions*, **15**, 323–33.

Gullino, M. L. (1992). Control of *Botrytis* rot of grapes and vegetables with *Trichoderma* spp. In *Biological Control of Plant Diseases, Progress and Challenges for the Future*, eds. E. C. Tjamos, G. C. Papavizas & R. J. Cook. New York: Plenum Press, pp. 125–32.

Harman, G. E, Lorito, M. & Lynch, J. M. (2004). Uses of *Trichoderma* spp. to alleviate or remediate soil and water pollution. *Advances in Applied Microbiology*, **56**, 313–31.

Harman, G. E. (2000). The dogmas and myths of biocontrol. Changes in perceptions based on research with *Trichoderma harzianum* T-22. *Plant Disease*, **84**, 377–93.

Harman, G. E. & Kubicek, C. P. (1998). *Trichoderma and Gliocladium, Vol. 2, Enzymes, Biological Control and Commercial Applications*. London: Taylor and Francis.

Harman, G. E. & Björkman, T. (1998). Potential and existing uses of *Trichoderma* and *Gliocladium* for plant disease control and plant growth enhancement. In *Trichoderma and Gliocladium, Vol. 2*, eds. G. E. Harman & C. P. Kubicek. London: Taylor and Francis, pp. 229–65.

Harman, G. E., Howell, C. R., Viterbo, A., Chet, I. & Lorito, M. (2004). *Trichoderma* species – Opportunistic, avirulent plant symbionts. *Nature Reviews Microbiology*, **2**, 43–56.

Howell, C. R., Hanson, L. E., Stipanovic, R. D. & Puckhaber, L. S. (2000). Induction of terpenoid synthesis in cotton roots and control of *Rhizoctonia solani* by seed treatment with *Trichoderma virens*. *Phytopathology*, **90**, 248–52.

Inbar, J., Abramsky, M., Coen, D. & Chet, I. (1994). Plant growth enhancement and disease control by *Trichoderma harzianum* in vegetable seedlings grown under commercial conditions. *European Journal of Plant Pathology*, 337–46.

Keen, N. T. (2000). A century of plant pathology, a retrospective view on understanding host-parasite interactions. *Annual Review Phytopathology*, **38**, 31–48.

Kubicek, C. P. & Harman, G. E. (1998). *Trichoderma and Gliocladium Vol. 1, Basic Biology, Taxonomy and Genetics*. London: Taylor and Francis.

Kuc, J. (2001). Concepts and direction of induced systemic resistance in plants and its application. *European Journal of Plant Pathology*, **107**, 7–12.

Limón, M. C., Llobel, A., Pintor Toro, J. A. & Benítez, T. (1996). Overexpression of chitinase by *Trichoderma harzianum* strains used as biocontrol fungi. In *Chitin Enzymology, Vol. 2*, ed. R. A. A. Muzzarelli. Ancona, Italy: AtecEdizioni.

Lorito, M. & Scala, F. (1999). Microbial genes expressed in transgenic plants to improve disease resistance. *Journal of Plant Pathology*, **81**, 73–88.

Lorito, M. & Woo, S. L. (1998). Advances in understanding the antifungal mechanism(s) of *Trichoderma* and new applications for biological control. In *Molecular Approaches in Biological Control*, eds. B. Duffy, U. Rosenberger and G. Défago. Dijon, France: IOBC wprs Bulletin/Bulletin OILB srop, Vol. 21 (9), pp. 73–80.

Lorito, M., Scala., F., Zoina, A. & Woo, S. L. (2001). Enhancing biocontrol of fungal pests by exploiting the *Trichoderma* genome. In *Enhancing Biocontrol Agents and Handling Risks*, eds. M. Vurro & J. Gressel. Amsterdam: IOS Press, pp. 248–59.

Lorito, M., Woo, S. L., Garcia Fernandez, I., Colucci, G., Harman, G. E., Pintor-Toro, J. A., Filippone, E., Muccifora, S., Lawrence, C. B., Zoina, A., Tuzun, S. & Scala, F. (1998). Genes from mycoparasitic fungi as a source for improving plant resistance to fungal pathogens. *Proceedings of the National Academy of Sciences of USA*, **95**, 7860–5.

Lorito, M., Woo, S. L. & Scala, F. (2004). Le biotecnologie utili alla difesa sostenibile delle piante, i funghi. *Agroindustria*, **3**, 181–95.

Lorito, M., Woo, S. L., D'Ambrosio, M., Harman, G. E., Hayes, C. K., Kubicek, C. P. & Scala, F. (1996). Synergistic interaction between cell wall degrading enzymes and membrane affecting compounds. *Molecular Plant-Microbe Interactions*, **9**, 206–13.

Lotan, T. & Fluhr, R. (1990). Xylanase, a novel elicitor of pathogenesis-related proteins in tobacco, uses a non-ethylene pathway for induction. *Plant Physiology*, **93**, 811–17.

Lu, Z., Tombolini, R., Woo, S. L., Zeilinger, S., Lorito, M. & Jansson, J. (2004). *In vivo* study of *Trichoderma*-pathogen interactions using constitutive and inducible green fluorescent protein reporter systems. *Applied and Environmental Microbiology*, **70**, 3073–81.

Mach, R. L., Peterbauer, C. K., Payer, K., Jaksits, S., Woo, S. L., Zeilinger, S., Kullnig, C. M., Lorito, M. & Kubicek, C. P. (1999). Expression of two major chitinase genes of *Trichoderma atroviride* (*T. harzianum* P1) is triggered by different regulatory signals. *Applied and Environmental Microbiology*, **65**, 1858–63.

Pieterse, C. M. J. & van Loon, L. C. (1999). Salicylic acid-independent plant defense pathways. *Trends in Plant Science*, **4**, 52–8.

Rey, M., Llobell, A., Monte, E., Scala, F. & Lorito, M. (2004). Genomics of *Trichoderma*. In *Applied Mycology and Biotechnology, Vol. 4, Fungal Genomics*, eds. D. K. Arora & G. G. Khachatourians. Elsevier Science, pp. 225–48.

Sandermann, H. (2000). Active oxygen species as mediators of plant immunity: three case studies. *Biological Chemistry*, **381**, 649–53.

Schirmböck, M., Lorito, M., Wang, Y.-L., Hayes. C. K., Arisan-Atac, I., Scala, F., Harman, G. E. & Kubicek, C. P. (1994). Parallel formation and synergism of hydrolytic enzymes and peptaibol antibiotics, molecular mechanisms involved in the antagonistic action of *Trichoderma harzianum* against phytopathogenic fungi. *Applied and Environmental Microbiology*, **60**, 4364–70.

Sid Ahmed, A., Sánchez, C. P. & Candela, M. E. (2000). Evaluation of induction of systemic resistance in pepper plants (*Capsicum annuum*) to *Phytophthora capsici* using *Trichoderma harzianum* and its relation with capsidiol accumulation. *European Journal of Plant Pathology*, **106**, 817–24.

Sivan, A. & Chet, I. (1989). The possible role of competition between *Trichoderma harzianum* and *Fusarium oxysporum* on rhizosphere colonization. *Phytopathology*, **79**, 198–203.

Sivasithamparam, K. & Ghisalberti, E. L. (1998). Secondary metabolism in *Trichoderma* and *Gliocladium*. In *Trichoderma and Gliocladium, Vol. 2*, eds G. E. Harman & C. P. Kubicek. London, UK: Taylor and Francis Ltd., pp. 139–91.

Windham, M. T., Elad, Y. & Baker, R. (1986). A mechanism for increased plant growth induced by *Trichoderma* spp. *Phytopathology*, **76**, 518–21.

Woo, S. L., Donzelli, B., Scala, F., Mach, R., Harman, G. E., Kubicek, C. P., Del Sorbo, G. & Lorito, M. (1999). Disruption of the *ech42* (endochitinase-encoding) gene affects biocontrol activity in *Trichoderma harzianum* P1. *Molecular Plant-Microbe Interactions*, **12**, 419–29.

Yedidia, I., Behamou, N., Kapulnik, Y. & Chet, I. (2000). Induction and accumulation of PR proteins activity during early stages of root colonization by the mycoparasite *Trichoderma harzianum* strain T-203. *Plant Physiology and Biochemistry*, **38**, 863–73.

17

Fungal parasites of invertebrates: multimodal biocontrol agents?

L. V. LOPEZ-LLORCA AND H.-B. JANSSON

Introduction

Nematophagous and entomopathogenic fungi (NEF) comprise an important group of fungal parasites of invertebrates (FPI). NEF belong to a wide range of fungal taxa, but most of them are anamorphic fungi and facultative parasites. These fungi can infect, kill and digest nematodes and insects, respectively, which we will call their canonical, or normal, hosts. These hosts have barriers to the environment (egg-shells and cuticles) that have common structural features. Therefore, the infection cycles share common strategies (e.g. adhesion to the host) or metabolites (e.g. proteases and chitinases for host penetration). Some species (e.g. *Lecanicillium lecanii*) can even be isolated from both infected nematodes and insects. The NEF may also infect other organisms (other fungi and plants) apart from their canonical hosts in a similar or different mode. We will use the term multimodal to describe the mode of action of these biological activities (Fig. 17.1). However, to date, the main emphasis in research has covered their mode of action on their canonical hosts (e.g. nematodes for nematophagous fungi). Many of these fungi are used for biological control of plant-parasitic organisms.

In this review we will describe the NEF and their hosts in general terms (both canonical and non-canonical) at biological, ecological and physiological-molecular levels. We will also analyze the reasons for this multitrophic behaviour, trying to use a comparative approach of both types of hosts (canonical and non-canonical) and pathogens (nematophagous and entomopathogenic fungi) under an evolutionary perspective. For recent reviews of the NEF see Charnley (2003); Morton, Hirsh & Kerry (2004); Jansson & Lopez-Llorca (2004).

Exploitation of Fungi, eds. G. D. Robson, P. van West and G. M. Gadd. Published by Cambridge University Press.

Fungal parasites of invertebrates

Most entomopathogenic and nematophagous fungi are facultative parasites. This way of life usually implies low host specificity and, consequently, a wide host range. They can also colonize a wide array of habitats, and their main species can be found worldwide. This versatile behaviour allows FPI to colonize (and eventually infect) hosts other than their canonical.

Entomopathogenic and nematophagous fungi bear multiple similarities. The most important species of both fungal groups have been described as soil inhabitants (Domsch, Gams & Anderson, 1993) where they spend most of the saprophytic phase of their life-cycles. Soil is – anyway – a reservoir of both organic matter and dead canonical hosts. It is also the environment of nearly all plant-parasitic nematodes and of soil-dwelling insects, such as pests of roots or other underground plant organs. The main parasitic phase of NEF on their canonical hosts will be briefly described, as an introduction to the multimodal behaviour.

Nematophagous fungi

Living stages of nematodes can be infected by several types of nematophagous fungi (Jansson & Lopez-Llorca, 2001). Dead nematodes may also be invaded saprophytically by nematophagous as well as other (saprophytic) fungi, but these latter are not regarded as proper nematode-destroying fungi. For instance, dead vermiform nematodes may be

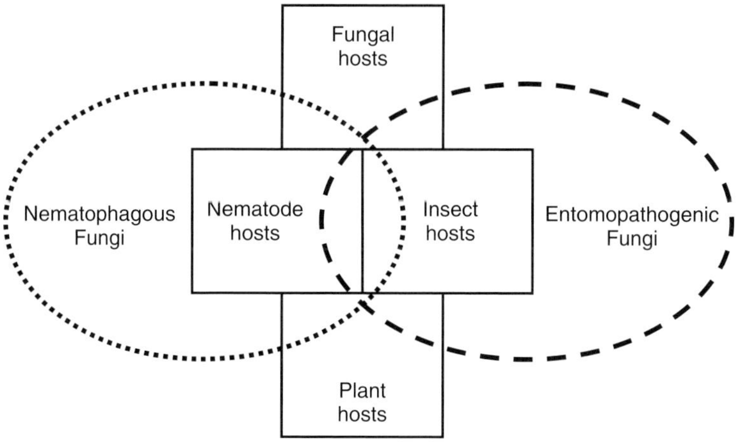

Figure 17.1 Multimodal behaviour of nematophagous and entomopathogenic fungi (NEF) on different hosts. Dashed ovals indicate host range of nematophagous and entomopathogenic fungi, respectively.

invaded by nematode-trapping fungi, which enter the natural openings (mouth, anus, etc.) of the nematodes but never penetrate their cuticles (Nordbring-Hertz & Stålhammar-Carlemalm, 1978). Most nematophagous fungi are facultative parasites and exist in both saprophytic and parasitic stages induced by external and/or internal signals.

The nematophagous fungi are divided into groups depending on their mode of infecting nematodes: nematode-trapping, endoparasitic, egg- and female-parasitic, and toxin-producing fungi (Jansson, Tunlid & Nordbring-Hertz, 1997). Nematode-trapping fungi capture vermiform nematodes in special trapping organs formed on the hyphae (Barron, 1977). The traps can have either an adhesive function, where the nematodes stick to the trap, or a mechanical function. These fungi, e.g. *Arthrobotrys oligospora*, are characterized by low host specificity and lower parasitic ability. The endoparasites, e.g. *Drechmeria coniospora*, use their spores to infect the nematode hosts. These fungi have high host specificity and are obligate parasites. The toxin-producing fungi, e.g. *Pleurotus ostreatus*, immobilize their victims using toxins prior to penetration. These three groups of fungi all attack vermiform nematode stages. Females of sedentary endoparasitic nematodes, as well as their eggs, can be infected by facultative egg-parasites, e.g. *Pochonia chlamydosporia* (syn. *Verticillium chlamydosporium*). As has been shown, the nematophagous habit has examples in the main taxa of the Fungal Kingdom. This has been explained as a multiple origin of the nematophagous habit in fungal evolutionary terms (Barron, 1992).

In practical terms, although some species of nematophagous fungi (mainly egg-parasites and trapping fungi) have experimentally been used for biocontrol of plant parasitic nematodes, no widely available biocontrol products based on these organisms are yet available.

Entomopathogenic fungi

Nearly 1000 fungal species are pathogenic on insects. Entomopathogenic fungi, like nematophagous fungi, belong to most taxonomic groups within the Fungal Kingdom (Charnley, 1997). This fact, shared to some extent with nematophagous fungi, gives entomopathogenic fungi a big biological potential due to the multiplicity of modes of reproduction/multiplication, characteristic of the taxa they belong to. This biological plasticity allows them to infect insects under different ecological conditions (Charnley, 1997).

Entomopathogenic fungi include specialist and generalist groups. The former includes mostly Entomophthorales, which are not commercialized for insect control, although they may cause epizootics in insects under natural conditions. The generalists are mostly anamorphic facultative parasites and they have been (and are being) developed as biological control agents of insects (Butt, 2002).

Chytridiomycetes include genera such as *Coelomomyces*, with species parasitic on mosquitoes. Within Ascomycetes, *Cordyceps* includes more than 300 species parasitizing most insect orders. Most entomopathogenic fungi belong to the anamorphic fungi. Amongst the most common we have the following genera: *Aschersonia, Beauveria, Culicinomyces, Hirsutella, Metarhizium, Nomuraea, Paecilomyces, Tolypocladium* y *Lecanicillium* (= *Verticillium*) (Charnley, 1997; Gams & Zare, 2001).

Some of these species are currently being exploited or developed for crop or forest protection. Failures or inconsistencies in biocontrol suggest that increased knowledge on the biology of the fungal species used is of paramount importance for future improvement (Bidochka, Kamp & de Croos, 2000). Commercialization of entomopathogenic and nematophagous fungi for pest and pathogen control requires better understanding of physiological aspects of growth, metabolism or genetic basis of virulence as well as their ecological performance. Some tools derived from biotechnology, and recently from genomics and proteomics approaches, may facilitate such studies (Burges, 1998; Tunlid & Talbot, 2002).

Physical barriers for FPI to overcome

Fungal pathogens obtain nutritional or other resources from damaging another organism (its host). Potential hosts have evolved barriers to shield themselves from the harsh environment and from pathogens. Host barriers are to be breached by attacking pathogens. As will be described in this chapter, the multimodal action of FPI implies that these fungi may infect (naturally or artificially) very diverse hosts: nematodes, insects, other fungi or plants (Fig. 17.2). Although diverse, the external barriers (egg shells, cuticles, cell walls, etc.) share structural features. They have a matrix, mostly formed by proteins, embedding structural microfibrils consisting of common or different materials: collagen in nematode cuticles; chitin in nematode eggshells, insect cuticles and fungal cell walls; cellulose in plant cell walls. To further strengthen the structure, the matrix proteins are often cross-linked and contain phenolic and other compounds, such as melanin and DOPA.

Canonical hosts

In spite of their differences, both insects and nematodes present a similar organization regarding composition and structure of their main barriers: cuticle and eggshell. The insect cuticle is formed by several layers (Andersen, 1979). The epicuticle is mainly made of lipoproteins and waxes, whereas the procuticle is mainly made of proteins and chitin. Other compounds (lipids, pigments and other small organic molecules and inorganic materials) are minoritary, athough they may affect the performance of entomopathogenic fungi. Chitin microfibrils (20–50% of procuticle composition) are embedded in a protein matrix (resilin) in different amounts, depending on the insect groups.

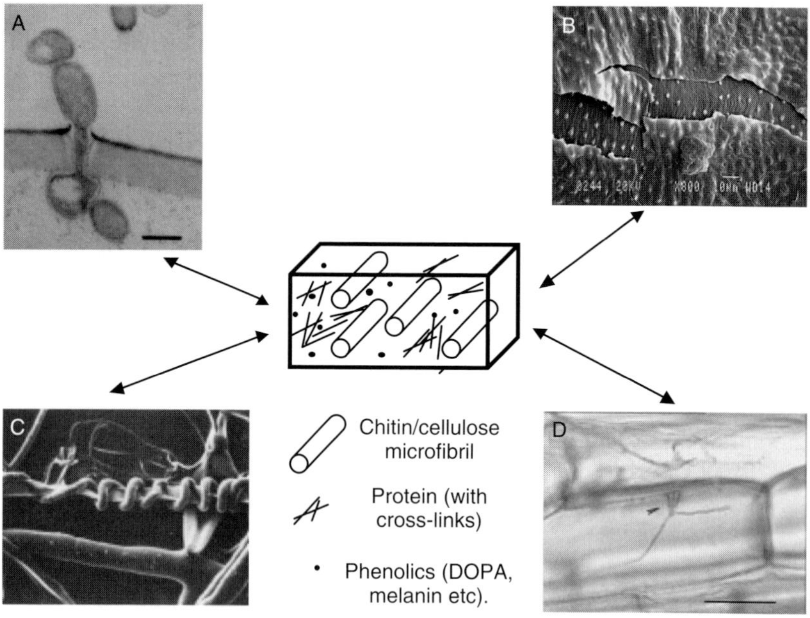

Figure 17.2 Similarities of external barriers of canonical and non-canonical hosts of NEF. (A) *Pochonia rubescens* hyphae penetrating eggshell of the cyst nematode *Globodera rostochiensis* (Lopez-Llorca & Robertson, 1992a, courtesy of *Nematologica*). (B) Degradation of insect (*Galleria mellonella*) cuticle by culture filtrate with proteolytic activity of the entomopathogen *Paecilomyces farinosus* (Lopez-Llorca, Carbonell & Gomez-Vidal, 2002, courtesy of *Mycological Progress*). (C) The nematode-trapping fungus *Arthrobotrys oligospora* coiling around its fungal host *Rhizoctonia solani* (courtesy of *FEMS, Yvonne Persson*). (D) *Pochonia chlamydospora* colonizing root cells of barley (Bordallo *et al.*, 2002, courtesy of *New Phytologist*).

The nematode surface, the cuticle, consists of several layers containing proteins (mainly collagen), lipids and carbohydrates (Bird & Bird, 1991). Externally to the cuticle a surface coat (or glycocalyx) consisting of glycoproteins is found (Bird & Bird, 1991). The surface coat is probably the part of the nematode surface most relevant to recognition and adhesion of nematophagous fungi, since proteolytic removal of this structure results in reduced adhesion of microorganisms (Bird, 1985; Jansson, 1993). The eggshell of plant parasitic nematodes is also a multilayered structure (Wharton, 1980; Bird & Bird, 1991). The thickest and most important is the chitinous layer, which like the insect cuticle is also made of chitin and proteins. Just as for insects, structure and composition varies with species. Regarding nematodes, chitin is, for instance, more abundant in *Meloidogyne* spp. eggshells, whereas protein is present in a higher degree in *Globodera* spp. Structurally, *Globodera* spp., have thicker egg-shells than those of *Heterodera* spp. (Lopez-Llorca & Robertson, 1992a). Molecular components of insect and nematode barriers suffer several types of modifications, e.g. protein cross-linking and melanization (Andersen, 1979; Lopez-Llorca & Fry, 1989). These facts can be highly relevant in the susceptibility of both nematodes and insects to FPI.

Non-canonical hosts

From the information given above, it becomes clear that regarding composition and structure the barriers of insects and nematodes bear striking similarities. In fact, the main barrier (cell wall) of plants (the most important hosts in nature of fungal pathogens) and also that of fungi shows some similarities too (Fig. 17.2). The plant cell wall contains cellulose microfibrils as the structural element. These are embedded in a matrix consisting of approximately 90% (dry weight) polysaccharides (mainly pectins, hemicelluloses and cellulose) and 10% (glyco)proteins (Fry, 2001). The cell wall components are strengthened by covalent and non-covalent cross-links, involvement of phenolics and, with age, an increasing lignification of the tissues. The cell wall of true fungi contains chitin microfibrils embedded in a matrix of proteins, also with cross-links, and carbohydrates, mainly various glucans (De Nobel *et al.*, 2001). Some authors (Bidochka *et al.*, 1999; Barron, 1992) suggest that FPI may have arisen from plant pathogens to escape competition via further specialization in different hosts. The capabilities of changing hosts between invertebrates (insects or nematodes), endophytism or mycoparasitism could then be the consequences of this evolutionary process favoured by some trends that hosts have developed in their barriers to defend themselves against environmental threats, such as the presence of parasites and pathogens.

Mode vs. multimodes of action: theoretical and practical implications

The diversity of modes of action and the subsequent extended host range of FPI merits some consideration before getting into the topic in more detail. Putting aside the fact of how frequent or relevant the non-canonical hosts are in nature, the fact that they exist has some theoretical and practical implications (Fig. 17.3). From the basic point of view, varying the hosts (even swapping them) in two groups of FPI and using biological tools at different scales will help in answering key questions: (i) how can this organism infect these hosts? or (ii) where did this capacity came from? It is clear that we are talking about the origin and evolution of pathogenicity, a question that even has implications in the medical field. The multimodal behaviour of FPI also has practical implications, both positive and negative (Fig. 17.3, (+) or (–) signs) regarding the use of FPI as biocontrol agents (BCA) of pests and diseases. Excluding plants as hosts, a potential expansion of the host range of FPI will increase the market of BCAs (a positive economic point). Conversely, it may affect their safety if non-targets (e.g. beneficial insects) are affected. This synergistic behaviour has, however, been used in chemical control where some insecticidal compounds are sold with claimed nematicidal action. The colonization of living plants (the hosts of pests and diseases) by FPI (and other BCA) has a wide array of benefits. FPI may modulate host defences and therefore add a new component in their canonical mode of action (infection of an insect/nematode). This may be positive if the modulation is to the right extent (i.e. not so high that it compromises the survival of the FPI, or so low that the FPI/BCA takes over the plant and becomes a pathogen).

How nematophagous fungi infect nematodes

The interactions between nematophagous fungi and their hosts involve several steps from recognition (attraction phenomena, contact), production of adhesives and lytic enzymes, differentiation of infection structures (appressoria and trapping organs) to host penetration and digestion (Tunlid, Jansson & Nordbring-Hertz, 1992). The nematode-trapping fungus *Arthrobotrys oligospora* forms so-called adhesive network traps on which vermiform nematodes are captured. Traps are essential for infection of living nematodes, and increase fungal attraction of nematodes (Jansson & Nordbring-Hertz, 1979).

After contact between the fungal trap and the nematode cuticle a possible contact recognition step occurs involving a fungal lectin binding to N-acetylgalactosamine (Gal-NAc) on the nematode surface (Nordbring-Hertz & Mattiasson, 1979). The trapping organ of *A. oligospora* contains

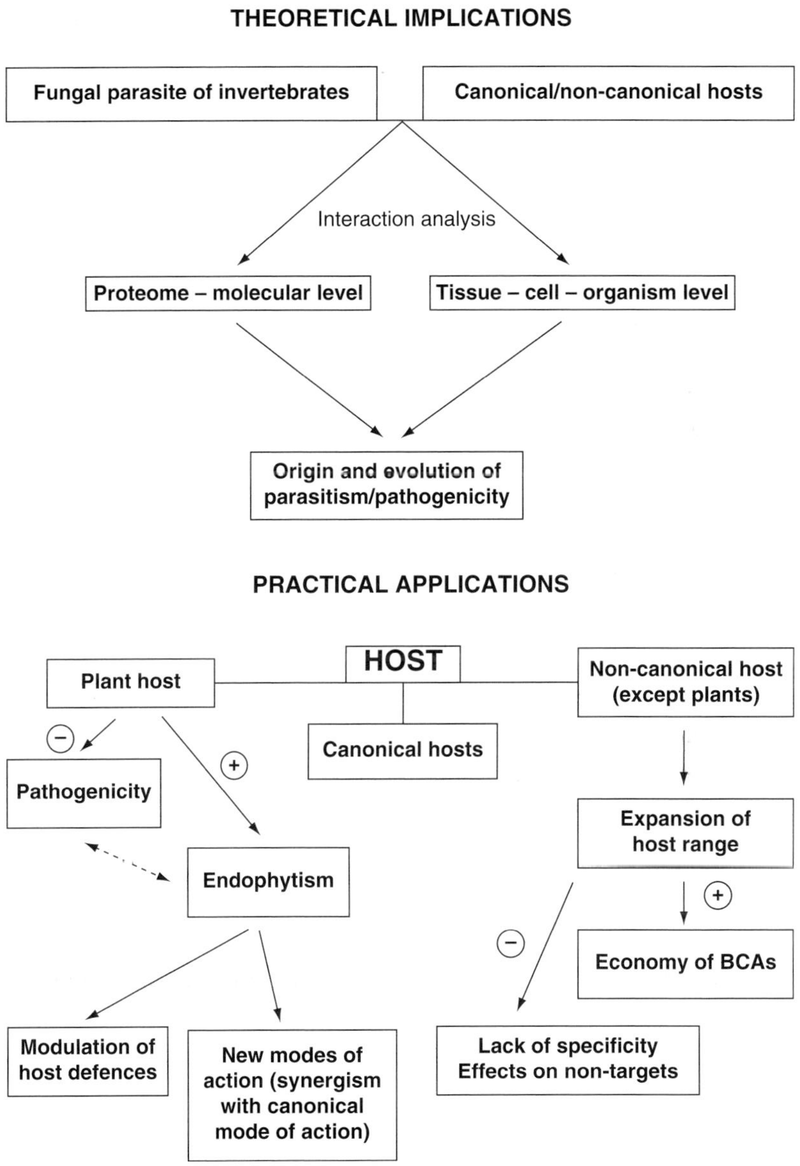

Figure 17.3 Mode vs. multimodes of action of FPI.

an adhesive material. Upon contact with the nematode surface a recognition step, possibly mediated, at least partly, by lectin binding, induces changes in the structure of the adhesive leading to capture of the nematode (Veenhuis, Nordbring-Hertz & Harder, 1985). The adhesive changes from amorphous material to fibrillar structure perpendicular to the nematode surface. This may anchor the nematode to the trap, thus facilitating infection. In contrast, the adhesive of the endoparasitic fungus *D. coniospora* always has a fibrillar structure even in the absence of nematodes (Jansson & Nordbring-Hertz, 1988). After attachment, *A. oligospora* penetrates the nematode cuticle and forms an infection bulb, from which trophic hyphae grow out to digest the nematode contents (Veenhuis *et al.*, 1985).

As in many other instances of fungal penetration of their hosts' surfaces, nematophagous fungi use both enzymatic and physical means. The nematode cuticle mainly contains proteins (Bird & Bird, 1991) and therefore the action of proteolytic enzymes may be important for penetration. A serine protease, PII, from *A. oligospora*, has been characterized, cloned and sequenced (Åhman *et al.*, 1996). The expression of PII is increased by the presence of proteins, including nematode cuticles (Åhman *et al.*, 1996). PII belongs to the subtilisin family and has a molecular mass of 32 kDa (for review see Jansson *et al.*, 1997).

Egg-destroying fungi, e.g. *Pochonia rubescens* and *P. chlamydosporia*, act on nematode eggs at two levels: directly as true parasites by penetrating and infecting eggs, and indirectly by causing distortions in the larvae or embryos they contain (Morgan-Jones & Rodriguez-Kábana, 1988). The former mode of action is well documented and is largely responsible for cases of soil supressiveness to nematodes.

Upon growth of germ tubes the hyphal tips swell and differentiate into appressoria on contact with nematode eggs (Lopez-Llorca & Claugher, 1990), as well as on artificial, especially hydrophobic, surfaces (Lopez-Llorca *et al.*, 2002b). An extracellular material (ECM) probably functions as adhesive, but possibly also seals the hole caused by the penetration hypha. This extracellular material can be labelled with the lectin Concanavalin A, indicating that the ECM contains mannose/glucose moieties probably on side chains of glycoproteins (Lopez-Llorca *et al.*, 2002b). Most ECMs of fungal hyphae consist of proteins and carbohydrates (Nicholson, 1996).

Nematode eggshells mostly contain protein and chitin (Clarke, Cox & Shepherd, 1967) organized in a microfibrillar and amorphous structure (Wharton, 1980). Therefore, a search for extracellular enzymes degrading those polymers was carried out. A 32 kDa serine protease (P32) was first purified and characterized from the egg parasite *P. rubescens* (Lopez-Llorca, 1990a). Involvement of the enzyme in pathogenesis was suggested by quick in vitro degradation of *Globodera pallida* eggshell proteins (Lopez-Llorca, 1990a), but most of all by its immunolocalization in appressoria of the fungus infecting *Heterodera schachtii* eggs (Lopez-Llorca & Robertson, 1992b). Although pathogenesis is a complex process involving many factors, inhibition of P32 with chemicals and polyclonal antibodies reduced egg infection and penetration (Lopez-Llorca *et al.*, 2002b). The similar species *P. chlamydosporia* also produces extracellular proteases (VcP1) (Segers *et al.*, 1994) which are immunologically related to P32 and similar enzymes from entomopathogenic fungi (Segers *et al.*, 1995). VcP1-treated eggs were more infected than untreated eggs, suggesting a role of the enzyme in eggshell penetration by egg-parasitic fungi. Recently, several chitinolytic enzymes of *Pochonia* spp. were detected. One of those accounting for most of the activity was a 43 kDa endochitiase (CHI43) (Tikhonov *et al.*, 2002). When *G. pallida* eggs were treated with both P32 and CHI43, damage to eggshell was more extensive than with each enzyme alone, suggesting a cooperative effect of both enzymes to degrade eggshells (Tikhonov *et al.*, 2002).

How entomopathogenic fungi infect insects

Infection of insects starts with the adhesion of spores to the surface of the cuticle. In this process extracellular polysaccharides, lectins and some extracellular enzymes play an important role (Clarkson *et al.*, 1998). For some fungal isolates, adhesion of spores to insect tegument is host specific. Some enzymes, including esterases and proteases, have been found in ungerminated conidia. Their role could be the modification of the surface of the insect cuticle (the infection court) before germination, to supply nutrients and help with adhesion. Therefore adhesion is not a mere union, but a complex process that modifies the surface of the host to allow further fungal development (St. Leger, 1993).

The germination process requires adequate moisture and nutrient availability for the production of the germ tube. Apart from these requirements, cuticle topography plays an important role in adhesion, germination and differentiation of appressoria (Butt, 2002). Failures in germination of spores of entomopathogens can be attributed to the presence of inhibitory compounds, such as free short-chain fatty acids, quinones and phenolics in the insect cuticle.

Penetration of the insect cuticle may take place in different ways. Germ tubes of *Verticillium lecanii* (= *Lecanicillium lecanii*, Zare & Gams, 2001) can directly penetrate the insect cuticle. On the contrary, *Metarhizium anisopliae* and *Beauveria bassiana* usually form special penetration structures (appressoria) (Bidochka *et al.*, 2000). The appressorium firmly links the fungus to the epicuticle of the insect and allows cuticle penetration through mechanical and enzymatic processes (Leucona, 1996; Clarkson *et al.*, 1998). In many fungi the appressorium secretes mucilage, thus helping the anchorage of the fungus (Butt, 2002). Penetration of the host cuticle is favoured mainly by extracellular proteases and other enzymes (e.g. lipases and chitinases) secreted by entomopathogenic fungi (Charnley, 1997; Butt, 2002).

Once the fungus has succesfully penetrated the insect cuticle and reached the subcuticular epidermis, it then invades the insect's haemocoel (Clarkson *et al.*, 1998). The fungus then produces spherical-ovoid bodies named blastoconidia or blastospores. These propagules have a cell wall different from that of mycelium, to avoid the immune system of the insect. Furthermore, some secondary metabolites produced by entomopathogenic fungi interfere with cell and humoral insect defences (Bidochka *et al.*, 2000; Butt, 2002).

Insect death occurs as a result of the invasion of vital organs, massive tissue destruction, the shortage of nutrients, or by the production of toxins by the fungus (Bidochka *et al.*, 2000; Butt, 2002). The insect death marks the end of the parasitic phase. The fungus then continues to grow saprotrophically in all insect tissues (Lecuona, 1996).

The life-cycle is completed when the fungus sporulates over the insect cadaver. Under adequate conditions, high relative moisture in particular, the fungus breaks the insect cuticle producing aerial spores. These are dispersed by the wind, rain or by other arthropods. The fungus may produce resistant spores within the dead insect that allow long survival periods under adverse conditions (Charnley, 1997; Butt, 2002).

Antibiosis phenomena of the pathogenic fungus in the dying insect, limits the growth of saprotrophic microorganisms of the insect. These are, in general, potential competitors of the entomopathogenic fungus. After total colonization, the insect cadaver is transformed into a mummy, which serves as a fungus reservoir to withstand adverse climatic conditions (Lecuona, 1996).

Finally, many traits of the host plant may influence, directly or indirectly, the survival of entomopathogenic fungi and its efficacy in biological control programmes. Plant growth, morphology and especially chemical composition are highly relevant. Plant chemical compounds, depending on their nature and concentration, may influence conidia viability and/or insect susceptibility to biocontrol fungi (Butt, 2002). Crucifers influence germination of conidia of *M. anisopliae* through the presence of stimulating or inhibitory compounds in epicuticular leaf waxes (Inyang *et al.*, 1999).

Mycoparasitic NEF

Several species of nematode-trapping fungi, including *A. oligospora*, can attack other (mostly plant-pathogenic) fungi, e.g. *Rhizoctonia solani* (Persson, Veenhuis & Nordbring-Hertz, 1985). This mycoparasitic behaviour takes place by coiling of the hyphae of the nematode-trapping fungi around the host hyphae (Fig. 17.2C), which results in disintegration of the host cell cytoplasm without penetration of the host. Although this phenomenon has never been observed in soil, it may increase the fitness of the nematode-trapping fungi in soil by reducing competition and providing nutrients. Moreover, it may extend the biocontrol capability of nematophagous fungi as biological control agents to fungal parasites as well as nematodes. Furthermore, *P. chlamydosporia* has been described infecting propagules of important plant pathogens, such as uredospores of rust fungi (Leinhos & Buchenauer, 1992), and oospores of *Phytophthora* and other oomycetes (Sneh, Humble & Lockwood, 1977).

We have found that *P. chlamydosporia* growing endophytically in roots (see NEF as endophytes) could reduce growth of the plant-pathogenic fungus *Gaeumannomyces graminis* var. *tritici* (take-all fungus, Ggt) in dual culture Petri dishes and in growth tube experiments. In pot experiments, *P. chlamydosporia* increased plant growth whether Ggt was present in the roots or not, suggesting a growth promoting effect by *P. chlamydosporia* (Monfort *et al.*, 2005).

The entomopathogen *Paecilomyces fumosoroseus* has been reported to infect the powdery mildew fungus *Sphaerotheca fuliginea* by colonizing conidia and hyphae (Kavkova & Curn, 2005). Furthermore, *Verticillium lecanii* has been reported as mycopathogen of *Penicillium digitatum* and *S. fuliginea* (Benhamou, 2004; Verhaar, Hijwegen & Zadoks, 1998). *Beauveria bassiana* has recently been shown to produce a fungitoxic compound active against the palm pathogen *Penicillium vermoesenii* (Asensio, 2004).

NEF as endophytes

Nematophagous fungi as root endophytes

Endophytic microorganisms colonize the interior of plants, especially leaves, branches, stems and roots, showing no apparent harm to the host, and the interaction involves metabolic exchange between endophyte and host. They may play an important role in host protection against pests and pathogens (Lodge, Fisher & Sutton, 1996; Azevedo *et al.*, 2000; Schulz *et al.*, 2002).

All groups of nematophagous fungi, apart from endoparasitic fungi (*Hirsutella rhossiliensis* and *Nematoctonus pachysporus*), have the ability to grow endophytically in roots (Lopez-Llorca *et al.*, 2006). *P. chlamydosporia* or *A. oligospora* colonized epidermis and root cortex of barley and the epidermis of tomato (Lopez-Llorca *et al.*, 2002a; Bordallo *et al.*, 2002). These fungi grew inter- and intracellularly and formed appressoria when penetrating plant cell walls of epidermis and cortex root cells, but never entered vascular tissues. In contrast to *Pochonia* spp., appressoria had previously never been observed in *A. oligospora*. Plant defence reactions (e.g. cell wall appositions) induced by nematophagous fungi could be detected. *P. chlamydosporia* grew extensively, especially in monocotyledonous plants, producing abundant mycelia and chlamydospores. Necrotic areas of the roots were observed at initial stages of colonization by the nematode-trapping *A. oligospora* and the toxin-producing *P. ostreatus*, indicating a different behaviour of the fungi. Roots colonized by *P. chlamydosporia* displayed higher proteolytic activity than non-inoculated control roots using immunochemical techniques (Monfort, 2004). The significance of this fact for biological control of root pathogens is under investigation in our laboratory.

Arthrobotrys dactyloides and *P. chlamydosporia* formed coiling structures in barley root cells and extensively colonized the roots. Such structures are also formed by other root endophytes, e.g. *Piriformospora indica* (Varma *et al.*, 1999), and presumably improve the exchange of metabolites.

Whether the root endophytic behaviour of nematophagous fungi is functional in their nematophagous habit is not known, although structures resembling trapping organs were observed in epidermal cells colonized by *A. oligospora* (Bordallo *et al.*, 2002). These traps may serve to capture nematodes in the roots. The internal root colonization by egg-parasitic fungi, e.g. *Pochonia* spp., may help egg infection of economically important endoparasitic nematodes, e.g. *Meloidogyne* spp., thus reducing their spread and root damage. The ability to colonize plant roots may also be a survival strategy of these fungi and could explain soil suppressiveness to plant-parasitic nematodes in nature. Root inoculation of nematophagous fungi may help us to circumvent the lack of receptivity of soil (even sand) to inoculum of nematophagous fungi (Monfort, 2004), which prevents *P. chlamydosporia* from controlling *Meloidogyne* spp. (Verdejo-Lucas *et al.*, 2003). This, in spite of the fact that the nematode is naturally found to be infected by *P. chlamydosporia* and other egg-parasitic fungi in similar Mediterranean agroecosystems (Verdejo-Lucas *et al.*, 2002; Olivares-Bernabeu & Lopez-Llorca, 2002).

Entomopathogenic fungi as leaf and stem endophytes

Insect pests are more abundant in aerial plant organs than in roots. Therefore endophytic behaviour of entomopathogenic fungi in these tissues seems a desirable trait for these FPI. To this respect, *Beauveria bassiana* inoculated in corn leaves and stems (Bing & Lewis, 1991) grew endophytically and controlled the European corn borer, *Ostrinia nubilalis* (Wagner & Lewis, 2000). The control of *O. nubilalis* achieved was due to the fungus' capacity to penetrate corn leafs, follow the apoplast pathway in all directions, and even reach the xylem (Wagner & Lewis, 2000). Perhaps, fungal movement within corn could be attributed to passive movement of *B. bassiana* within the xylem or to active development within the plant through mycelial growth (Bing & Lewis, 1991; Wagner & Lewis, 2000). Although *B. bassiana* is a ubiquitous entomopathogenic fungus in soil, certain isolates have been found colonizing plant tissue (Bing & Lewis, 1993; Boucias & Pendland, 1998). The capability of EF to colonize plant tissues is also documented by the fact that inoculum of entomopathogenic fungi can be efficiently produced on plant wastes. We have used almond mesocarp (Lopez-Llorca & Carbonell, 1998) or palm seed and leaves (Lopez-Lorca, Carbonell & Salinas, 1999) for production of entomopathogenic fungi, e.g. *B. bassiana* and *L. lecanii*. This endophytic behaviour of entomopathogenic fungi could be a choice for targeting armoured sap-sucking insects such as scale insects. *Phoenicoccus marlattii*, the red scale-insect of palms, could be a good example (Asensio *et al.*, 2005). We are currently investigating the endophytic behaviour of entomopathogenic fungi in date palm (*Phoenix dactylifera*) leaf tissue.

Beauveria bassiana, *Lecanicillium dimorphum* and *L.* c.f. *psalliotae* were inoculated into *P. dactylifera* leaf petioles. The entomopathogenic fungi colonized leaf petioles endophytically and were recovered up to 2 cm from the inoculation site. Fungi were detected inside the parenchymatic tissue and rarely within vascular tissue. The fungi survived for at least 30 days and colonized date palm tissues both in bioassays in laboratory and field experimental conditions, with no evidence of significant damage to date palm tissues (Gómez-Vidal *et al.*, 2006). *B. bassiana* was located in intercellular spaces and inside parenchymatic cells and was also occasionally found inside vascular bundles, mostly close to the inoculation site. In rare occasions, hyphal tip swellings resembling appressoria were found. We have found that the structures of fungal antagonists colonizing plant waste substrates (including leaves) by entomopathogenic and mycoparasitic fungi (Lopez-Llorca *et al.*, 1999) are very similar to those produced for canonical host penetration.

Hu and St. Leger (2002), using a gfp-transformed *Metarhizium anisopliae* strain, showed that the fungus can survive for over a year in soil, especially in the inner rhizosphere of cabbage plants. The authors concluded that rhizosphere soils are potential reservoirs for the fungus.

Similarities and differences between nematophagous and entomopathogenic fungi

Nematophagous and entomopathogenic fungi must breach the barriers of their hosts to get inside them and resume their infection process. These structures have been briefly described and compared earlier in this chapter (see Physical barriers for FPI to overcome). Some striking similarities have been stressed. The first step for a successful infection by FPI is the adhesion of the infective structures (conidia, hyphae) to the host barrier (cuticle, eggshell). In general, both groups of fungi produce adhesive substances for that purpose. For entomopathogenic fungi, with the best known adhesion process, the adhesion of conidia to the insect cuticle comprises three major steps (St. Leger, 1993; Hajek & St. Leger, 1994): adsorption (passive stage), consolidation of adhesion/germination, and fungus growth on the cuticle surface until penetration (active stage). For *Metarhizium anisopliae*, these stages are favoured by hydrophobicity of the conidium and cuticle. The presence and role of adhesives in the pathogenesis of nematophagous fungi (endoparasites, trapping fungi and egg parasites) has already been discussed (see How nematophagous fungi infect nematodes). Once germination is over, the germ tube secretes enzymes that facilitate adhesion. Germlings and ungerminated conidia of nematode egg parasites secrete several proteases (Lopez-Llorca *et al.*, 2002b). This has also been found for entomopathogenic fungi (Charnley, 2003). Lectins and other glycoproteins plus other proteins (including hydrophobins) also play a role in the early pathogenesis of FPI (Charnley, 1997;

Jansson & Lopez-Llorca, 2001). The FPI penetrate host barriers by means of appressoria (St. Leger *et al.*, 1989; Lopez-Llorca & Claugher, 1990), which concentrate both mechanical pressure and enzymatic degradation in a small area of the host. However, there are reports of direct penetration (without appressoria) for entomopathogenic fungi (Altre & Vandenberg, 2001) and nematode-trapping fungi form appressoria only when colonizing plant roots, but never when infecting nematodes (Bordallo *et al.*, 2002).

The fungal pathogenesis of insects and nematodes is facilitated by the secretion of similar extracellular enzymes, which degrade the main components of the host barriers. Entomopathogenic and nematophagous fungi first secrete proteases then chitinases (Bidochka *et al.*, 1999). This is highly adaptative, since the protein component in insect and nematode eggshell barriers is amorphous and shields chitin microfibrils (Andersen, 1979; St. Leger, 1993; Wharton, 1980; Bird & Bird, 1991). The proteases involved in the infection process are generally serine-proteases class II, and are extracellular and of subtilisin-type (Segers *et al.*, 1999). Characteristics of subtilisins may vary between species and isolates. Proteases of entomopathogenic fungi are synthesized in large amounts by virulent isolates. The synthesis is very fast and localized in the early infection stages (Charnley & St. Leger, 1991; Charnley, 1997). Regarding similarities between different species, basic proteases of several *B. bassiana* isolates were serologically identical, and the same was true for isolates of the close species *B. brogniartii* (Shimizu, Tsuchitani & Matsumoto, 1993). However, in spite of their similar substrate degradation capacities, *B. bassiana, P. fumosoroseus* and *M. anisopliae* proteases were not serologically identical, although common antigens could be found.

The species of *Pochonia* (formerly *Verticillium*) infecting nematode eggs produce chymoelastase-type proteases involved in the infection of nematode eggs (Jansson & Lopez-Llorca, 2001). *P. chlamydosporia*, *P. rubescens* and *M. anisopliae* proteases are serologically and functionally related (Segers *et al.*, 1995). Subtle differences in specificity of these and other enzymes may favour competence for diverse nutrient sources. This may also determine the potential host range. In this respect, the capacity to secrete chitin degrading enzymes distinguishes the invertebrate pathogenic species from the plant pathogenic species in the old *Verticillium* spp. genus (Bidochka *et al.*, 1999). Expression of chitinolytic enzymes is considered a key factor in infection processes of entomopathogenic, nematophagous and mycoparasitic fungi (Dupont, Segers & Coosemans, 2002). *Pochonia* spp. produce a major endochitinase (CHI43) that shares biochemical features with those entomopathogenic fungi (Tikhonov *et al.*, 2002). These key enzymes (proteases and chitinases) have been located in appressoria of both entomopathogenic and nematophagous fungi infecting their canonical hosts.

Lecanicillium lecanii case

We shall provide an overview of the fact that bona fide entomopathogenic and nematophagous fungi are found in nature infecting hosts other than their canonical. A striking example in the changing hosts scenario is the '*Lecanicillium lecanii* case', but it is by no means the only one. *L. lecanii* has been well documented as an entomopathogen (Inglis *et al.*, 2001) infecting mainly aphids, mites and whiteflies (Boucias & Pendland, 1998; Tanada & Kaya, 1993). However, *L. lecanii* has also been described as a mycopathogen (see Mycoparasitic NEF), as a nematophagous fungus (Meyer, Roberts & Wergin, 1998; Olivares-Bernabeu & Lopez-Llorca, 2002), and as a root endophyte (Meyer *et al.*, 1998; Benhamou & Brodeur, 2001).

Pochonia chlamydosporia is usually found infecting eggs and nematode females in agricultural soils of plant parasitic nematodes, where it is the main cause of natural suppressivity (Kerry & Jaffee, 1997). It has also been described as a rust parasite (Leinhos & Buchenauer, 1992) and as an entomopathogen (Bidochka *et al.*, 1999). We have found that it can experimentally colonize roots of both mono- and dicotyledonous plants as an endophyte (Bordallo *et al.*, 2002). *Pochonia* sp. was also retrieved from roots of several cereal crops grown in soils naturally suppressive to the cereal cyst nematode *Heterodera avenae* (Lopez-Llorca & Boag, 1993). However, the root endophytic behaviour of the fungus in nature remains to be clearly established.

Paecilomyces includes important entomopathogenic species such as *P. fumosoroseus* or *P. farinosus*. *Paecilomyces lilacinus* is nematophagous (Chen, Dickson & Mitchell, 1996; Olivares-Bernabeu & Lopez-Llorca, 2002), entomopathogen (Oborník, Klic & Zizka, 2000) and has even been described as an opportunistic human pathogen (Walsh & Groll, 1999). *B. bassiana*, a typical entomopathogen found worldwide, also parasitizes *Heterodera glycines* eggs (Chen *et al.*, 1996) and is an endophyte in corn (Wagner & Lewis, 2000).

These examples illustrate the fact that FPI not only are facultative parasites that can be grown in artificial media in the laboratory and degrade organic matter in soil. They also have a flexible parasitic/mutualistic life style where they can infect their canonical hosts and others, including plants.

We do not know whether, in nature, an isolate infecting a given host is able to infect another one. Such plastic behaviour can, however, be assessed under controlled laboratory conditions. We shall give two extra examples on the issue of 'changing hosts' in FPI discovered in our laboratory. *Drechmeria coniospora* is an endoparasitic nematophagous fungus that infects vermiform nematodes by means of adhesive conidia (Fig. 17.4A). The mode of infection of this fungus is well documented (Jansson, Hofsten & Meckenburg, 1984; Jansson, 1993; Dijksterhuis, Veenhuis & Harder, 1990). We have found that this fungus can – at least in vitro – also infect root-knot nematode eggs (Fig. 17.4B).

Beauveria bassiana, as has already been discussed, is an entomopathogen with a wide host range. It is frequently found infecting coleopterans (Fig. 17.4C) under natural conditions. We have experimentally infected cyst nematode eggs with this fungus (Fig. 17.4D). Just as for nematode egg parasites the entomopathogen formed appressoria on the nematode eggshell (Fig. 17.4D, arrows).

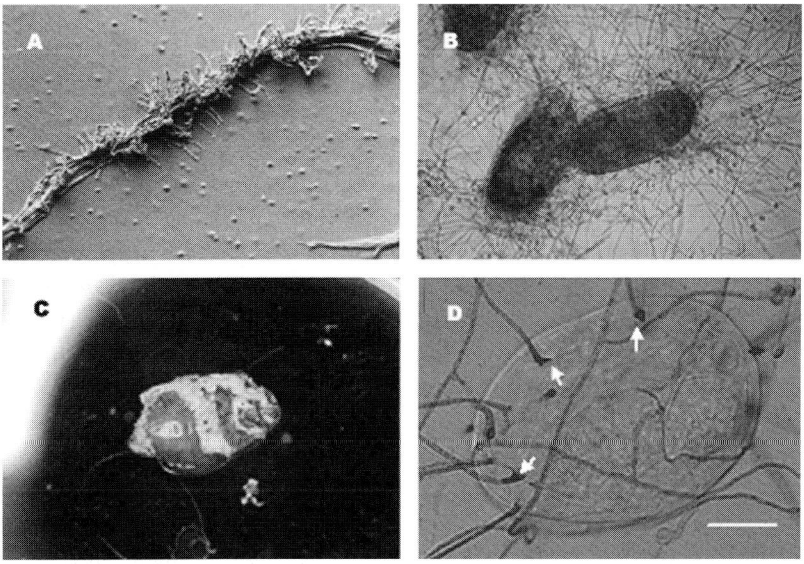

Figure 17.4 Canonical (A and C) and non-canonical mode of action of NEF. (A) *Drechmeria coniospora* infecting larva of the nematode *Panagrellus redivivus* (Jansson, Hofsten & Mecklenburg, 1984, courtesy of *Antonie van Leeuwenhoek*). (B) *D. coniospora* infecting egg of the nematode *Globodera rhosochiensis*. (C) *Beauveria bassiana* infecting adult insect coleopteran (Lopez-Llorca, 1992b, courtesy of *Boletín de la Sociedad Micológica de Madrid*). (D) *B. bassiana* infecting egg of the nematode *Globodera rhosochiensis*.

These examples illustrate the fact that due to the similarities between FPI, changing hosts is easy within the canonical (from nematode larvae to egg) and non-canonical (from insects to nematode eggs). There are, however, limitations. In our laboratory we have found that entomopathogenic fungi may easily infect nematode eggs, but nematode egg parasites are bad entomopathogens (Lopez-Serna, 2004). The reasons for this behaviour are unknown.

Multimodal fungal invertebrate pathogens, evolution and biocontrol – conclusions

The similarities in modes of action (enzymatic capacities and infection mechanisms) of entomopathogenic and nematophagous fungi may indicate that both groups are related phylogenetically. Some hypotheses point towards evolutionary divergence as the process evolved. In fact, these fungi have similar life styles and some of them, as has been discussed in this chapter, may infect hosts other than their own canonical hosts. As we have discussed above, FPI members of the 'old *Verticillium*' such as the entomopathogen *L. lecanii* or the nematophagous *P. chlamydosporia*, are clear examples of multimodal FPI, since these fungi infect (even in nature) non-canonical hosts. Most pathogenic fungi possess so-called wide spectrum enzymes (e.g. proteases, polysaccharidases), which can degrade many different substrates. Most of the proteases of FPI studied to date are wide spectrum enzymes, such as the well known Pr1 for *Metarhizium anisopliae* or *Beauveria bassiana* proteases. This does not rule out that the variability that may account for host preference can be found, as for *Pochonia* proteases. According to Bidochka *et al.* (1999), the old *Verticillium* genus may have suffered evolution in the production of enzymes, as an adaptation to the conditions imposed by new hosts. In this sense, entomopathogenic, nematophagous and mycoparasitic species differ from phytopathogenic fungi that they (the former) can produce chitin-degrading enzymes (chitinases and other enzymes). Proteases of nematophagous and entomopathogenic fungi are coded by several genes and can be polymorphic. FPI as facultative pathogens and saprotrophic fungi have adaptative polymorphism (Segers *et al.*, 1999). They display more genetic variability (some species such as *P. chlamydosporia* are often considered as species complexes) than obligate parasites or highly specialized pathogens, since they colonize more uniform environments. In this way, fungi may secrete enzymes different from the common ones, which may degrade other compounds, allowing the exploitation of diverse nutrient sources. This may in turn influence the virulence and potential host range (Segers *et al.*, 1995). According to St. Leger, Joshi & Roberts (1997), FPI displayed enzymatic adaptation to polymers present in the barriers of their hosts. As a result, plant pathogenic fungi display enzymes able to degrade polysaccharides, e.g. cellulose, xylan or cutin. In turn, these fungi do not show chitinolytic or proteolytic activity towards elastin or mucin. Entomopathogenic fungi are able to degrade a wide array of proteins, including elastin and mucin, and produce very few polysaccharidases, except chitinase, cellulase (*L. lecanii*) and few cutinases (*B. bassiana*). Saprotrophic fungi and opportunist pathogens produce the widest spectrum of enzymes capable of degrading proteins and polysaccharides, indicating their unspecialized nutritional status. Phylogeny studies indicate that *Paecilomyces* is polyphyletic, and that it is closely related (amongst others) to the old *Verticillium* and *Beauveria* genera (Obroník, Jirku & Dolezel, 2001). Phylogenetic studies of nematode-trapping fungi shows that there is a closer homology between fungi with the same type of trapping organ than traditional spore morphology (Ahrén, Ursing & Tunlid, 1998; Hagedorn & Scholler, 1999). The close relationships between different FPI (especially entomopathogenic fungi) could be due to the presence of a common ancestor, probably meiotic, able to form several mitotic stages. This may explain the capacity of the widespread multimodal behaviour of FPI.

Pleurotus forms basidiocarps and its standard means of living is saprophytic growth on decaying wood. The most common species, *P. ostreatus*, is an edible mushroom commercially cultivated. The fungus compensates for the lack of nitrogen in wood, its natural substrate, with its nematophagous habit (Thorn & Barron, 1984). The strong relation with plant tissues, as a natural wood decomposer, was also confirmed in roots when we found that, just like *N. robustus*, *P. djamor* penetrated early, and extensively colonized barley roots. The fungus was in fact very aggressive since some parts of the root appeared to be decorticated three weeks after inoculation (Lopez-Llorca *et al.*, 2006), as was *A. oligospora* in previous experiments (Bordallo *et al.*, 2002). *Arthrobotrys* spp. are wood colonizers (Persson, Olsson & Jansson, 2000) and their teleomorphs in *Orbilia* are also known as wood decomposers (Pfister, 1997).

In practical terms, the multimodal behaviour of NEF can widen their usage as biocontrol agents (i.e. broaden host range). Recent studies on the endophytic behaviour of NEF point to new biocontrol activities (e.g. modulation of plant defences) or even new (more targeted) means of delivery in agroecosystems. The latter may increase efficiency of the inoculum in adverse environments such as the rhizosphere. This multimodal behaviour may also shed light on the very nature of biocontrol and may help us to understand why a soil can be suppressive to, for instance, plant-parasitic nematodes.

Acknowledgements

We thank Dr Y. Persson, Dr M. L. Lopez-Serna and Ms S. Gomez for unpublished results supplied for this chapter. This work was partially financed by the AGL2004-05808/AGR project from the Spanish Ministry of Education and Science.

References

Åhman, J., Ek, B., Rask, L. & Tunlid, A. (1996). Sequence analysis and regulation of a gene encoding a cuticle-degrading serine protease from the nematophagous fungus *Arthrobotrys oligospora*. *Microbiology*, **142**, 1605–16.

Ahrén, D., Ursing, B. M. & Tunlid, A. (1998). Phylogeny of nematode trapping fungi based on 18S rDNA sequences. *FEMS Microbiology Letters*, **158**, 179–84.

Altre, J. A. & Vanderberg, J. D. (2001). Penetration of cuticle and proliferation in hemolymph by *Paecilomyces fumosoroseus* isolates that differ in virulence against lepidopteran larvae. *Journal of Invertebrate Pathology*, **78**, 81–6.

Andersen, S. O. (1979). Biochemistry of insect cuticle. *Annual Review of Entomology*, **24**, 29–61.

Asensio, L. (2004). Control biologic de plagues i malalties de palmeres per fongs entomopatógens. Ph.D. thesis, University of Alicante, Alicante, Spain.

Asensio, L., Lopez-Llorca, L. V. & Lopez-Jiménez, J. A. (2005). Use of light, scanning electron microscopy and bioassays to evaluate paratism by entomopathogenic fungi of the red scale insect of palms (*Phoenicococcus marlatti* Ckll., 1899). *Micron*, **36**, 169–75.

Azevedo, J. L., Maccheroni, W., Pereira, J. O. & Araujo, W. L. (2000). Endophytic microorganisms: a review on insect control and recent advances on tropical plants. *Electronic Journal of Biotechnology*, **3**, 40–65.

Barron, G. L. (1977). *The Nematode-Destroying Fungi. Topics in Mycobiology No. 1*. Guelph: Canadian Biological Publications Ltd.

Barron, G. L. (1992). Ligninolytic and cellulolytic fungi as predators and parasites. In *The Fungal Community: Its Organisation and Role in the Ecosystem*, eds. G. C. Carroll & T. D. Wicklow. New York: Marcel Dekker, pp. 311–26.

Benhamou, N. (2004). Potential of the mycoparasite, *Verticillium lecanii*, to protect citrus fruit against *Penicillium digitatum*, the causal agent of green mold: a comparison with the effect of chitosan. *Phytopathology*, **94**, 693–705.

Benhanou, N. & Brodeur, J. (2001) Pre-inoculation of Ri T-DNA transformed cucumber roots with the mycoparasite, *Verticillium lecanii*, induces host defence reactions against *Pythium ultimum* infection. *Physiological and Molecular Plant Pathology*, **58**, 133–46.

Bidochka, M. J., Kamp, A. M. & de Croos, A. (2000). Insect pathogenic fungi: from genes to populations. In *Fungal Pathology*, ed. J. W. Kronstad. Dordrecht: Kluwer, pp. 171–93.

Bidochka, M. J., St. Leger, R. J., Stuart, A. & Gowanlock, K. (1999). Nuclear rDNA phylogeny in the fungal genus *Verticillium* and its relationship to insect and plant virulence, extracellular proteases and carbohydrates. *Microbiology*, **145**, 955–63.

Bing, L. A. & Lewis, L. C. (1991). Suppression of *Ostrinia nubilalis* (Hübner) (Lepidoptera: Pyralidae) by endophytic *Beauveria bassiana* (Balsamo) Vuillemin. *Environmental Entomology*, **20**, 1207–11.

Bing, L. A. & Lewis, L. C. (1993). Occurrence of the entomopathogen *Beauveria bassiana* (Balsamo) Vuillemin in different tillage regimes and in *Zea mays* L. and virulence towards *Ostrinia nubilalis* (Hübner). *Agriculture, Ecosystems and Environment*, **45**, 147–56.

Bird, A. F. (1985). The nature of the adhesion of *Corynebacterium rathayi* to the cuticle of the infective larva of *Anguina agrostis*. *International Journal of Parasitology*, **15**, 301–8.

Bird, A. F. & Bird, J. (1991). *The Structure of Nematodes*. San Diego: Academic Press.

Bordallo, J. J., Lopez-Llorca, L. V., Jansson, H.-B., Salinas, J., Persmark, L., Asensio, L. (2002). Effects of egg-parasitic and nematode-trapping fungi on plant roots. *New Phytologist*, **154**, 491–9.

Boucias, D. G. & Pendland, J. C. (1998). Entomopathogenic fungi: fungi imperfecti. In *Principles of Insect Pathology*, eds. D. G. Boucias & J. C. Pendland. Dordrecht: Kluwer, pp. 321–64.

Burges, H. D. (1998). Formulation of mycoinsecticides. In *Formulation of Microbial Pesticides. Beneficial Microorganisms, Nematodes and Seed Treatment*, ed. H. D. Burges. Dordrecht: Kluwer, pp. 131–85.

Butt, T. M. (2002). Use of entomogenous fungi for the control of insect pests. In *The Mycota XI: Agricultural Applications*, ed. F. Kempken. Berlin: Springer-Verlag, pp. 111–34.

Charnley, A. K. (1997). Entomopathogenic fungi and their role in pest control. In *The Mycota IV: Environmental and Microbial Relationships*, eds. T. D. Wicklow & B. Söderström. Berlin: Springer-Verlag, pp. 185–201.

Charnley, A. K. (2003). Fungal pathogens of insects: cuticle degrading enzymes and toxins. *Advances in Botanical Research*, **40**, 241–321.

Charnley, A. K. & St. Leger, R. J. (1991). The role of cuticle degrading enzymes in fungal pathogenesis of insects. In *Fungal Spore and Disease Initiation in Plants and Animals*, eds. G. T. Cole & H. C. Hoch. London: Plenum Press, pp. 129–56.

Chen, S. Y., Dickson, D. W. & Mitchell, D. J. (1996). Pathogenicity of fungi to eggs of *Heterodera glycines*. *Journal of Nematology*, **28**, 148–58.

Clarke, A. J., Cox, P. M. & Shepherd, A. M. (1967). The chemical composition of the egg shells of the potato cyst-nematode, *Heterodera rostochiensis* Woll. *Biochemical Journal*, **104**, 1056–60.

Clarkson, J. M., Scree, S., Bailey, A., Cob, B. & Charnley, K. (1998). Fungal pathogenesis in insects. In *Molecular Variability of Fungal Pathogens*, eds. P. D. Bridge, Y. Couteaudier & J. M. Clarkson. Wallingford: CAB International, pp. 83–94.

De Nobel, H., Sietsma, J. H., van den Ende, H. & Klis, F. M. (2001). Molecular organization and construction of the fungal cell wall. In *The Mycota VIII: Biology of the Fungal Cell Wall*, eds. R. J. Howard & N. A. R. Gow. Berlin: Springer-Verlag, pp. 181–200.

Dijksterhuis, J., Veenhuis, M. & Harder, W. (1990). Ultrastructural study of adhesion and initial stages of infection of nematodes by conidia of *Drechmeria coniospora*. *Mycological Research*, **94**, 1–8.

Domsch, K. H., Gams, W. & Anderson, T-H. (1993). *Compendium of Soil Fungi*, Vol. 1. Eching, Germany: IHW-Verlag.

Dupont, A., Segers, R. & Coosemans, J. (2002). The effect of chitinase from *Verticillium chlamydosporium* on the egg of *Meloidogyne incognita*. *Mededelingen Faculteit Landbouwkundige en Toegepaste Biologische Wetenschappen Universiteit Gent*, **64**, 383–9.

Fry, S. C. (2001). Plant cell walls. *Encyclopedia of life sciences*. Nature Publishing Group/ www.els.net.

Gams, W. & Zare, R. (2001). A revision of *Verticillium* sect. *Prostrata*. III. Generic classification. *Nova Hedwigia*, **72**, 329–37.

Gómez-Vidal, S., Lopez-Llorca, L. V., Jansson, H.-B. & Salinas, J. (2006). Endophytic colonization of date palm (*Phoenix dactylifera* L.) leaves by entomopathogenic fungi. *Micron*, **37**, 624–32.

Hagedorn, G. & Scholler, M. (1999). A reevaluation of predatory orbiliaceous fungi. I. Phylogenetic analysis using rDNA sequence data. *Sydowia*, **51**, 27–48.

Hajek, A. E. & St. Leger, R. J. (1994). Interactions between fungal pathogens and insect hosts. *Annual Review of Entomology*, **39**, 293–322.

Hu, G. & St. Leger, R. J. (2002). Field studies using a recombinant mycoinsecticide (*Metarhizium anisopliae*) reveal that it is rhizosphere competent. *Applied and Environmental Microbiology*, **68**, 6383–7.

Inglis, G. D., Goettel, M. S., Butt, T. M. & Strasser, H. (2001). Use of hyphomycetous fungi for managing insect pests. In *Fungi as Biocontrol Agents: Progress, Problems and Potential*, eds. T. M. Butt, C. Jackson & N. Magan. Wallingford: CABI Publishing, pp. 23–69.

Inyang, E. N., Butt, T. M., Beckett, A. & Archer, S. (1999). The effect of crucifer epicuticular waxes and leaf extracts of the germination and virulence of *Metarhizium anisopliae* conidia. *Mycological Research*, **103**, 419–26.

Jansson, H.-B. (1993). Adhesion to nematodes of conidia from the nematophagous fungus *Drechmeria coniospora*. *Journal of General Microbiology*, **139**, 1899–906.

Jansson, H.-B., Hofsten, A. V. & Mecklenburg, C. V. (1984). Life cycle of the endoparasitic nematophagous fungus *Meria coniospora*: a light and electron microscopic study. *Antonie van Leeuwenhoek*, **50**, 321–7.

Jansson, H.-B. & Lopez-Llorca, L. V. (2001). Biology of nematophagous fungi. In *Trichomyces and Other Fungal Groups*, eds. J. K. Misra & B. W. Horn. Enfield: Science Publishers, pp. 145–73.

Jansson, H.-B. & Lopez-Llorca, L. V. (2004). Control of nematodes by fungi. In *Fungal Biotechnology in Agriculture, Food, and Environmental Applications*, ed. D. K. Arora. New York: Marcel Dekker, pp. 205–15.

Jansson, H.-B. & Nordbring-Hertz, B. (1979). Attraction of nematodes to living mycelium of nematophagous fungi. *Journal of General Microbiology*, **112**, 89–93.

Jansson, H.-B. & Nordbring-Hertz, B. (1988). Infection events in the fungus-nematode system. In *Diseases of Nematodes*, Volume 2, eds. G. O. Poinar & H.-B. Jansson. Boca Raton: CRC Press, pp. 59–72.

Jansson, H.-B., Tunlid, A., & Nordbring-Hertz, B. (1997). Nematodes. In *Fungal Biotechnology*, ed. T. Anke. Weinheim: Chapman & Hall, pp. 38–50.

Kavková, M. & Curn, V. (2005). *Paecilomyces fumosoroseus* (*Deuteromycotina: Hyphomycetes*) as a potential mycoparasite on *Sphaerotheca fuliginea* (*Ascomycotina: Erysiphales*). *Mycopathologia*, **159**, 53–63.

Kerry, B. R. & Jaffee, B. A. (1997). Fungi as biological control agents for plant parasitic nematodes. In *The Mycota*, eds. K. Esser & P. A. Lemke. Berlin: Springer-Verlag, pp. 204–18.

Leinhos, G. M. E. & Buchenauer, H. (1992). Hyperparasitism of selected fungi on rust fungi of cereal. *Journal of Plant Disease and Protection*, **99**, 482–98.

Lecuona, R. (1996). Control microbiano, utopia o realidad. In *Microorganismos Patógenos Empleados en el Control de Insectos Plaga*, ed. E. Leucona. Buenos Aires: Talleres Gráficos Mariano Mas, pp. 13–16.

Lodge, D. J., Fisher, P. J. & Sutton, B. C. (1996). Endophytic fungi of *Manilkara bidentata* leaves in Puerto Rico. *Mycologia*, **88**, 733–8.

Lopez-Llorca, L. V. (1990a). Purification and properties of extracellular proteases produced by the nematophagous fungus *Verticillium suchlasporium*. *Canadian Journal of Microbiology*, **36**, 530–7.

Lopez-Llorca, L. V. (1992b). Coleóptero radicícola infectado por el hongo entomopatógeno *Beauveria Bassiana*. *Boletín de la Sociedad Micológica de Madrid*, **17**, 13–17.

Lopez-Llorca, L. V. & Boag, B. (1993). A contribution to the ecology of fungi associated with female and cysts of the cereal cyst nematode *Heterodera avenae* in Eastern Scotland. *Nematologica Mediterranea*, **21**, 187–97.

Lopez-Llorca, L. V., Bordallo, J. J., Salinas, J., Monfort, E. & Lopez-Serna, M. L. (2002a). Use of light and scanning electron microscopy to examine colonisation of barley rhizosphere by the nematophagous fungus *Verticillium chlamydosporium*. *Micron*, **33**, 61–7.

Lopez-Llorca, L. V. & Carbonell, T. (1998). Use of almond mesocarp for production of the entomopathogenic fungus *Verticillium Lecanii*. *Canadian Journal of Microbiology*, **44**, 886–95.

Lopez-Llorca, L. V., Carbonell, T. & Gomez-Vidal, S. (2002). Degradation of insect cuticle by Proteases of *Paecilomyces farinosus*. *Mycological Progress*, **1**, 249–56.

Lopez-Llorca, L. V., Carbonell, T. & Salinas, J. (1999). Colonization of plant waste substrates by entomopathogenic and mycoparasitic fungi – a SEM study. *Micron*, **30**, 325–33.

Lopez-Llorca, L. V. & Claugher, D. (1990). Appressoria of the nematophagous fungus *Verticillium suchlasporium*. *Micron and Microscopica Acta*, **21**, 125–30.

Lopez-Llorca, L. V. & Fry, S. C. (1989). Dityrosine, trityrosine and tetratyrosine, potential cross-links in proteins of plant parasitic nematodes. *Nematologica*, **35**, 165–79.

Lopez-Llorca, L. V., Jansson, H.-B., Maciá Vicente, J. G. & Salinas, J. (2006). Nematophagous fungi as root endophytes. In *Soil Biology, Volume 9. Microbial Root Endophytes*, eds. B. Schulz, C. Boyle & T. N. Sieber. Berlin: Springer Verlag, pp. 191–206.

Lopez-Llorca, L. V., Olivares-Bernabeu, C., Salinas, J., Jansson, H.-B. & Kolattukudy, P. E. (2002b). Pre-penetration events in fungal parasitism of nematode eggs. *Mycological Research*, **106**, 499–506.

Lopez-Llorca, L. V. & Robertson, W. (1992a). Ultrastructure of infection of cyst nematode eggs by the nematophagous fungus *Verticillium suchlasporium*. *Nematologica*, **39**, 65–74.

Lopez-Llorca L. V. & Robertson W. M. (1992b). Immunocytochemical localization of a 32-kDa protease from the nematophagous fungus *Verticillium suchlasporium* in infected nematode eggs. *Experimental Mycology*, **16**, 261–7.

López Serna, M. L. (2004). Interrelaciones entre hongos patógenos de invertebrados: hongos nematófagos y hongos entomopathógenos. Ph.D. thesis, University of Alicante, Alicante, Spain.

Meyer, S. L. F., Roberts, D. P. & Wergin, W. P. (1998). Association of the plant-beneficial fungus *Verticillium lecanii* with soybean roots and rhizosphere. *Journal of Nematology*, **30**, 451–60.

Monfort, E. (2004). Interacciones tróficas entre hongos nematófagos, la rizosfera y sus patógenos fúngicos. Ph.D. thesis, University of Alicante, Alicante, Spain.

Monfort, E., Lopez-Llorca, L. V., Jansson, H.-B., Salinas, J., Park, J. O. & Sivasithamparam, K. (2005). Colonisation of seminal roots of wheat and barley by egg-parasitic nematophagous fungi and their effects on *Gaeumannomyces graminis* var. *tritici* and development of root-rot. *Soil Biology and Biochemistry*, **37**, 1229–35.

Morgan-Jones, G. & Rodrigues-Kábana, R. (1988). Fungi colonising cysts and eggs. In *Diseases of Nematodes*, Volume 2, eds. G. O. Poinar & H.-B. Jansson. Boca Raton: CRC Press, pp. 39–58.

Morton, C. O., Hirsch, P. R. & Kerry, B. R. (2004). Infection of plant-parasitic nematodes by nematophagous fungi – a review of the application of molecular biology to understand infection processes and improve biological control. *Nematology*, **6**, 161–70.

Nicholson, R. L. (1996). Adhesion of fungal propagules. In *Histology, Ultrastructure and Molecular Cytology of Plant–Microorganism Interactions*, eds. M. Nicole and V. Gianinazzi-Pearson. Amsterdam: Kluwer, pp. 117–34.

Nordbring-Hertz, B. & Mattiasson, B. (1979). Action of a nematode-trapping fungus shows lectin-mediated host-microorganism interaction. *Nature*, **281**, 477–9.

Nordbring-Hertz, B. & Stålhammar-Carlemalm, M. (1978). Capture of nematodes by *Arthrobotrys oligospora*, an electron microscopic study. *Canadian Journal of Botany*, **56**, 1297–307.

Obroník, M., Jirku, M. & Dolezel, D. (2001). Phylogeny of mitosporic entomopathogenic fungi: is there a genus *Paecilomyces* polyphletic? *Canadian Journal of Microbiology*, **47**, 813–19.

Obroník, M., Klic, M. & Zizka, L. (2000). Genetic variability and phylogeny inferred from random amplified polymorphic DNA data reflect life strategy of entomopathogenic fungi. *Canadian Journal of Botany*, **78**, 1150–5.

Olivares-Bernabeu, C. & Lopez-Llorca, L. V. (2002). Fungal egg-parasites of plant-parasitic nematodes from Spanish soils. *Revista Iberoamericana de Micologia*, **19**, 104–10.

Persson, C., Olsson, S. & Jansson, H.-B. (2000). Growth of *Arthrobotrys superba* from a birch wood food base into soil determined by radioactive tracing. *FEMS Microbiology Ecology*, **31**, 47–51.

Persson, Y., Veenhuis, M. & Nordbring-Hertz, B. (1985). Morphogenesis and significance of hyphal coiling by nematode-trapping fungi in mycoparasitic relationships. *FEMS Microbiology Ecology*, **31**, 283–91.

Pfister, D. H. (1997). Castor, Pollux and life histories of predacious hyphomycetes. *Mycologia*, **89**, 1–23.

Schulz, B., Boyle, C., Draeger, S., Römmert, A. & Krohn, K. (2002). Endophytic fungi: a source of novel biologically active secondary metabolites. *Mycological Research*, **106**, 996–1004.

Segers, R., Butt, T. M., Carder, J. H., Keen, J. N., Kerry, B. R. & Perderby, J. F. (1999). The subtilisins of fungal pathogens of inscects, nematodes and plants: distribution and variation. *Mycological Research*, **103**, 395–402.

Segers, R., Butt, T. M., Keen, J. N., Kerry, B. R. & Peberdy, J. F. (1995). The subtilisins of the invertebrate mycopathogens *Verticillium chlamydosporium* and *Metarhizium anisopliae* are serologically and functionally related. *FEMS Microbiology Letters*, **126**, 227–32.

Segers, R., Butt, T. M., Kerry, B. R. & Peberdy, J. F. (1994). The nematophagous fungus *Verticillium chlamydosporium* Goddard produces a chymoelastase-like protease which hydrolyses host nematode proteins in situ. *Microbiology*, **140**, 2715–23.

Shimizu, S., Tsuchitani, Y. & Matsumoto, T. (1993). Serology and substrate specificity of extracellular proteases from four species of entomopathogenic hyphomycetes. *Journal of Invertebrate Pathology*, **61**, 192–5.

Sneh, B., Humble, S. J. & Lockwood, J. L. (1977). Parasitism of oospores of *Phytophthora megasperma* var. sojae, *P. cactorum*, *Pythium*, and *Aphanomyces euteiches* by oomycetes, chytridomycetes, hyphomycetes, actinomycetes and bacteria. *Phytopathology*, **67**, 622–8.

St. Leger, R. J. (1993). Biology and mechanisms of insect-cuticle invasion by deuteromycete fungal pathogens. In *Parasites and Pathogens of Insects. Volume 2: Pathogens*, eds. M. E. Beckage, B. A. Federici & S. N. Thompson. New York: Academic Press, pp. 211–29.

St. Leger, R. J., Butt, T. M., Goettel, M. S., Staples, R. C. & Roberts, D. W. (1989). Production in vitro of appressoria by the entomopathogenic fungus *Metarhizium anisopliae*. *Experimental Mycology*, **13**, 273–88.

St. Leger, R. J., Joshi, L. & Roberts, D. W. (1997). Adaptation of proteases and carbohydrases of saprophytic, pathogenic and entomopathogenic fungi to the requirements of their ecological niches. *Microbiology*, **143**, 1983–92.

Tanada, Y. & Kaya, H. K. (1993). *Insect Pathology*. San Diego: Academic Press.

Thorn, R. G. & Barron, G. L. (1984). Carnivorous mushrooms. *Science*, **224**, 76–8

Tikhonov, V. E., Lopez-Llorca, L. V., Salinas, J. & Jansson, H.-B. (2002). Purification and characterization of chitinases from the nematophagous fungi *Verticillium chlamydosporium* and *V. suchlasporium*. *Fungal Genetics and Biology*, **35**, 67–78.

Tunlid, A., Jansson, H.-B. & Nordbring-Hertz, B. (1992). Fungal attachment to nematodes. *Mycological Research*, **96**, 401–12.

Tunlid, A. & Talbot, N. J. (2002). Genomics of parasitic and symbiotic fungi. *Current Opinion in Microbiology*, **5**, 513–19.

Varma, A., Verma, S., Sudha, Sahay, N., Bütehorn, B. & Franken, P. (1999). *Piriformospora indica*, a cultivable plant-growth-promoting root endophyte. *Applied and Environmental Microbiology*, **65**, 2741–4.

Veenhuis, M., Nordbring-Hertz, B. & Harder, W. (1985). An electron microscopical analysis of capture and initial stages of penetration of nematodes by *Arthrobotrys oligospora*. *Antonie van Leeuwenhoek*, **51**, 385–98.

Verdejo-Lucas, S., Ornat, C., Sorribas, F. J. & Stchiegel, A. (2002). Species of root-knot nematodes and fungal egg parasites recovered from vegetables in Almeria and Barcelona, Spain. *Journal of Nematology*, **34**, 405–8.

Verdejo-Lucas, S., Sorribas, F. J., Ornat, C. & Galcano, M. (2003). Evaluating *Pochonia chlamydosporia* in a double-cropping system of lettuce and tomato in plastic houses infested with *Meloidogyne javanica*. *Plant Pathology*, **52**, 521–8.

Verhaar, M. A., Hijwegen, T. & Zadoks, J. C. (1998). Selection of *Verticillium lecanii* isolates with high potential for biological control of cucumber powdery mildew by means of component analysis at different humidity regimes. *Biocontrol Science and Technology*, **8**, 465–77.

Wagner, B. L. & Lewis, L. C. (2000). Colonization of corn, *Zea mays*, by the entomopathogenic fungus *Beauveria bassiana*. *Applied and Environmental Microbiology*, **66**, 3468–73.

Walsh, T. J. & Groll, A. H. (1999). Emerging fungal pathogens: evolving challenges to immunocompromised patients for the tweny-first century. *Transplant Infectious Disease*, **1**, 247–61.

Wharton, D. A. (1980). Nematode egg-shells. *Parasitology*, **81**, 447–63.

Zare, R. & Gams, W. (2001). A revision of *Verticillium* section *Prostrata*. IV. The genera *Lecanicillium* and *Simplicillium* gen. nov. *Nova Hedwigia*, **73**, 1–50.

Index